Voice Over WLANs

Voice Over WLANs

The Complete Guide

Michael F. Finneran

AMSTERDAM • BOSTON • HEIDELBERG • LONDON
NEW YORK • OXFORD • PARIS • SAN DIEGO
SAN FRANCISCO • SINGAPORE • SYDNEY • TOKYO
Newnes is an imprint of Elsevier

Newnes is an imprint of Elsevier
30 Corporate Drive, Suite 400, Burlington, MA 01803, USA
Linacre House, Jordan Hill, Oxford OX2 8DP, UK

Recognizing the importance of preserving what has been written, Elsevier prints its books on acid-free paper whenever possible.

Library of Congress Cataloging-in-Publication Data
Finneran, Michael.
  Voice over WLANs : the complete guide / Michael Finneran.
      p.   cm.
  Includes bibliographical references and index.
  ISBN 978-0-7506-8299-2 (pbk. : alk. paper)   1. Wireless LANs.   2. Internet telephony.
I. Title.
  TK5105.78.F56 2007
  004.6'8—dc22

                                                            2007038489

British Library Cataloguing-in-Publication Data

A catalogue record for this book is available from the British Library.

ISBN: 978-0-7506-8299-2

For information on all Newnes publications visit our Web site at www.books.elsevier.com

07 08 09 10   9 8 7 6 5 4 3 2 1

Printed in the United States of America

Working together to grow
libraries in developing countries

www.elsevier.com | www.bookaid.org | www.sabre.org

ELSEVIER   BOOK AID
                International   Sabre Foundation

Contents

About the Author

Michael F. Finneran, President of dBrn Associates, Inc., is an independent telecommunications consultant, industry analyst, and writer with extensive experience in the specification, design, and installation of wired and wireless communication networks. He has worked with wireless technologies since the early-1970s, and currently teaches the MediaLive/BCR seminar programs on wireless networks, wireless LANs, and VoIP.

As an independent consultant, Mr. Finneran has provided assistance to a number of telecommunications carriers, equipment manufacturers, enterprise organizations, and government agencies. A frequent speaker at industry conferences including InterOp and VoiceCon, he has written numerous articles and White Papers on wireless technologies. For the past 25-years he has written the "Networking Intelligence" column for *Business Communications Review*, and he has contributed to *Computerworld*, *The Ticker*, and *The ACUTA Journal*.

In the training area, Michael has conducted over 2,000 networking seminars for vendors and users in the U.S., Latin America, Europe, Asia, and Africa. He has taught in the Graduate Telecommunications program at Pace University and conducted programs at the Center for the Study of Data Processing at Washington University in St. Louis.

Mr. Finneran is a member of the IEEE and holds a Masters Degree in Business from the J. L. Kellogg Graduate School of Management at Northwestern University.

Preface

Two of the major developments reshaping the telecommunications landscape are mobile wireless connectivity and the migration of voice telephone services to IP technology. Those two ideas come together in networks that carry voice services over a wireless LAN (VoWLAN). The purpose of this text is to provide network professionals with the technical background and practical guidance needed to deploy these networks successfully.

That is a tall order in that both wireless LANs and voice over IP (VoIP) are relatively new technologies and both are advancing at a breakneck pace. We can place the inception of the wireless LAN (WLAN) market at 1999 with the ratification of the 802.11b radio link protocol that boosted transmission rates to 11 Mbps. Since then, hundreds of millions of WLAN compatible devices have been sold around the world and the technology continues to evolve.

The cost of a Wi-Fi interface has dropped from several hundred dollars to essentially zero; a Wi-Fi card is now a standard component in virtually every laptop sold. The penetration of that technology is now expanding to cell phones, PDAs, digital cameras, and other types of consumer electronics. Data rates have grown to 54 Mbps, and the new 802.11n radio link could push that to hundreds of millions of bits per second. Along with that we have seen important developments in security and manageability that make WLANs a trusted technology element in enterprise environments.

VoIP has had a much longer and more varied adoption curve. The idea of using IP technology for voice transport was first proposed in the 1970s, though the routing technologies of that time had neither the capacity to support voice traffic nor the required quality of service capabilities. The concept resurfaced in the early 1990s with the idea of providing packet-based voice services over the nascent frame relay technology at companies like Stratacom.

The interest in wide area packet-based voice technology quickly shifted from the enterprise market to consumer VoIP services like Skype, where IP technology could provide cheap phone calls for those who were interested in price rather than quality. In the enterprise space, the focus shifted to local IP voice in the form of an IP PBX. The essential idea was to eliminate the separate, stand-alone PBX system that had existed for almost 100 years and move that voice switching function on to the LAN switch infrastructure. Now IP telephone handsets with Ethernet interfaces could establish voice connections through the LAN switch with the help of a telephony server, essentially eliminating the stand-alone PBX infrastructure. The idea of WLAN voice really grows out of that migration of local voice service to IP-based PBX solutions.

WLAN voice is essentially local VoIP service that is delivered to the user over a wireless LAN rather than a traditional wired Ethernet connection; indeed some have taken to calling it "wireless VoIP". Just as WLAN technology brought mobility to the local computing environment, WLAN voice holds the promise of providing mobility for voice telephone services. The potential productivity benefits of an integrated mobile voice capability far exceed the productivity enhancements of mobile data.

The first WLAN voice deployments appeared around 2003, but were restricted to a few specialized vertical markets. Health care represented the largest segment as hospitals and other health care facilities recognized the productivity benefits they could reap by equipping nurses and other members of their highly mobile workforce with voice handsets operating over their wireless LAN infrastructure. Patients could reach the nurse directly, make a request, and the service could be provided in one patient room visit with fewer wasted steps.

Materials handling was another natural vertical market for WLAN voice. Warehouse and production support was one of the first markets for data WLANs. Workers equipped with WLAN-capable bar code scanners and mobile computers were a critical part of systems to mechanize warehousing and to support just-in-time manufacturing systems. Adding a voice capability to that mobile data device allowed the worker to get more jobs done, more efficiently with that one mobile device.

The real explosion in WLAN voice will come when it moves past the production workers to management and knowledge workers. One of the biggest productivity impediments in office work is the all-too-frequent case where the person with the knowledge or the person with the approval authority cannot be reached at a critical time. What follows is a frenzied series of attempts to contact the party by cell phone, pager, email, and even room-to-room searches. Keeping people connected and easily accessible when they are away from their desks will be a major boon to productivity. Not only can the person be accessible, but with the presence capabilities of an IP PBX system, everyone can see immediately the communication options available (e.g., voice, video, text message, email, etc.).

The Challenges

If WLAN voice were cheap and easy, it would already be widely deployed. Network professionals are finding numerous challenges in implementing these networks. At the root of the difficulty is the fact that it involves the merger of two relatively new, and often poorly-understood, technologies. Millions of WLANs have been deployed but the vast majority of those networks use a single access point and provide service to a handful of users. A large-scale wireless LAN is a significant undertaking.

An enterprise wireless LAN uses a configuration similar to an indoor cellular network. Each access point or WLAN base station will cover a particular area and a large-scale deployment may have hundreds or even thousands of access points. The need to prioritize voice traffic to ensure good quality and provide enterprise-grade security means we will be deploying these large-scale networks with leading edge WLAN technologies. Further, we are only starting to come to grips with the testing, management, and support systems we will need to monitor and maintain service quality.

Similarly, VoIP is experiencing growing pains. Fortunately we now understand the major quality issues that must be addressed in a VoIP network, and more importantly, the performance levels we must maintain in each of those areas to meet a business user's expectations for quality and reliability. The leading edge users who have deployed IP PBXs over wired LAN have found there is a steep learning curve, and the first few installations take far longer than they had anticipated.

The major lesson those companies have learned is that their organization and skill sets also had to be overhauled, and not just their technology infrastructure.

Technical professionals could no longer be categorized as "voice people" and "data people". With regard to infrastructure, there are three critical areas to address in a WLAN voice deployment: voice, data, and radio. This book is not targeted specifically at any of those three areas, but rather looks to provide an overall description of the interaction among them. In essence, the goal is to inform voice people about data networking, data people about voice networking, and both groups about VoIP and WLAN technologies.

Overall Organization

This book is designed to provide both the technical background and practical considerations a network designer must take into account in planning a WLAN voice deployment. Providing enterprise-grade voice services in an IP environment is challenging by itself. When you add the inherent vagaries of radio transmission, particularly radio transmission over a contention-based WLAN network, the degree of difficulty goes up exponentially.

Voice Over WLANs: The Complete Guide is organized into four major parts.

First Part—Chapters 1 and 2: Introduction and Market Overview

This section is designed to provide a general overview of the WLAN and VoIP markets, their primary applications, and business justifications. The basic issues of bandwidth, interference, and capacity are described along with the particular challenges brought about by the use of a radio network.

Second Part—Chapters 3–6: WLAN Technology Background

Here, we will introduce the components and technologies that are used in wireless LANs. We will examine the equipment elements, the radio link technologies, and the details of the WLAN protocols. We will also look at the range of security challenges introduced and the various tools and techniques we can use to provide a secure wireless infrastructure.

Third Part—Chapters 7–9: Voice Network and VoIP Background

The other essential background area is VoIP. We will overview the TCP/IP concepts and protocols, with a particular focus on the elements that contribute to VoIP networks. We will also identify the major quality issues that must be addressed, the network parameters that affect them, and the performance levels we must achieve to meet user expectations. We will also provide an overall description of the typical equipment and service elements in an enterprise voice environment. Finally, we will describe the process of voice traffic engineering, or the mechanism by which we determine the volume of voice traffic a network can support.

Fourth Part—Chapters 10–14: Voice over WLAN Technology, Design, and Management

In the final section we will describe how all of these elements come together in a VoWLAN deployment. We will begin with a description of the components of a VoWLAN network and the requirements for WLAN voice handsets and other user equipment. Then we will describe the technical requirements for the WLAN infrastructure including such features as quality of service, security, and battery conservation. We will also review the major steps in the process of designing and implementing a voice-capable WLAN; this is where we will introduce the practical methods used to deal with the inherent challenges of indoor radio transmission. Finally we will describe the network management systems needed to provide reliable and cost-effective service on an ongoing basis. Along the way we will also look at the options for fixed-mobile convergence, or the idea of integrating Wi-Fi with cellular and possibly WiMAX wireless technology.

Using this Book

I have tried to lay out the topics in a logical fashion that will bring the reader along through a set of building blocks to understanding. I realize that many of our readers will be experienced network professionals who are used to "jumping ahead to the punch line." For those readers, I have tried to make each chapter a stand-alone element as far as possible. I have indicated where the preliminary information is located in case a jump-ahead reader should become lost in the more technical areas. I have also included a Glossary of Acronyms and a Glossary of Terms to ensure a basic understanding of the vocabulary used throughout.

Make no mistake about it, WLAN voice is still in its infancy, and we are effectively figuring this out as we go. While I had a range of texts to draw on for the general aspects of VoIP, radio technology, and wireless LANs, there were no texts specifically on the topic of WLAN voice. Having spent the last thirty-plus years figuring out how to do things that nobody knew how to do, I felt that I was perfectly qualified for a project such as this. Anyone who doesn't relish the challenge of delivering a functional solution in a largely uncharted environment should not sign up for a WLAN voice project. For those of us who enjoy the challenge that this type of project can bring, this book will be the best preparation for what lies ahead.

Voice Over WLANs: The Complete Guide will help the reader to understand the underlying technologies and critical challenges, and give the best practical perspective on what it will take to design, implement, and maintain a WLAN voice network. I hope this book helps readers to understand the technologies but, more importantly, I hope it helps network administrators build a WLAN voice network that really works.

Good luck.

Michael F. Finneran

Acknowledgments

I have drawn on the writings and experience of hundreds of network professionals in writing this book, but there are some who I would like to thank directly.

Thanks go to Matthew Gast, Principal Network Engineer at Trapeze Networks and author of the most authoritative text on wireless LAN technology, *802.11 Wireless LAN Networks*. Besides being a brilliant and dedicated engineer, Matthew is as gracious and supportive a person as ever you will meet. Thanks also to Craig Mathias, President of Farpoint Networks, one of the most insightful analysts in all aspects of the wireless market, and one of the few I look to for a meaningful assessment of developing technologies. My deep gratitude goes to Peter Thornycroft, Product Manager for WLAN Voice at Aruba Networks, who not only helped with information on Aruba's WLAN switch products, but offered insights on esoteric subjects ranging from 802.11e parameters to strategies for improving battery life—and we all like his accent. I am most appreciative of Kathy Small, former Marketing Manager of Mobility Solutions at Cisco Systems, whose help with the Cisco product line was always accurate, detailed, and objective.

I was greatly aided by Ben Guderian, VP of Marketing at Polycom/Spectralink, one of the real pioneers in the WLAN voice market and always a sober and experienced voice of reason. I learned a lot from Robert K. Morrow, President of Morrow Technical Services, whose technical understanding of radio systems is matched only by his down-to-earth ability to make these things understandable even to dullards like me. I am grateful to Vivek Khuller, President of DiVitas Networks, whose infectiously optimistic spirit is one of the sure signs this technology will be successful.

Thanks go to Lior Nir, Director, Product Marketing Voice Solutions for Nokia, who is always on top of his products and always willing to help. I also appreciate the help I got from Eric M. Ritter, Director of WLAN Solutions for

Research In Motion, who provides an unparalleled level of help and support on the Blackberry product line. I would also like to thank Luc Roy, Vice President of Product Planning for Siemens-Chantry Networks, and John DiGiovanni, Director of Marketing for Xirrus.

I must also thank my regular circle of advisors who are always so generous with their time and their understanding; with long time friends like these, even the simple can be made to sound intelligent. That list would certainly include:

- Gary Audin, President of Delphi Inc., who helped me into the consulting business many years ago and whose friendship and advice I have depended on ever since.

- Fred Knight, Publisher and General Manager of *Business Communications Review*, Chairman of VoiceCon, and someone I have been proud to call a friend for over two decades.

- Eric Krapf, Publisher of *Business Communications Review*, and my long time editor and sounding board.

- Buddy Shipley, President of Shipley Consulting, who amazes me with his breadth of practical networking experience and who can always point out what an idiot I am without coming right out and saying it.

- Dave Stein, Principal and Director of Operations for PlanNet Consulting, and a priceless source of practical guidance for those of us who might otherwise be lost in theoretical Never-Never Land.

- Ed Mier, CEO of Mier Consulting, one of the real pioneers in VoIP testing, and the editor of the first article I ever published.

- The thousands of attendees at my lectures and seminars over the years who have taught me so much and shared my love of this field.

- Harry Helms, Consulting Acquisitions Editor for Elsevier, who talked me into doing this in the first place and was a constant supporter throughout.

- Finally, Jeff Freeland, my Project Manager at Elservier who taught me what it takes to turn a pile of pages into a book and didn't shoot me along the way.

Most of all I have to thank my wife Jane and my son Derek who have put up with my insane work binges and resulting lapses in family responsibilities over these past several months. I couldn't throw myself into everything I do if I didn't have this wonderful family to come home to.

CHAPTER 1

The Convergence of Wireless LANs and VoIP

The growing popularity of both wireless LANs (WLANs) based on the IEEE 802.11 standards and voice solutions based on the Internet Protocol (i.e., VoIP) has made their convergence almost inevitable. Wireless LANs are becoming commonplace in enterprise organizations, home networks, and with broadband Internet services that use Wi-Fi access. At the same time, enterprise customers are now managing the transition of their traditional voice telephone systems to implementations that use IP, Ethernet, and LAN switching technologies. All the while, consumers are migrating from the traditional circuit switched telephones services offered by the traditional telephone companies to broadband VoIP services from cable TV companies and VoIP carriers like Vonage.

Both IP telephony and wireless LANs are relatively new technologies and each brings with it a number of issues regarding service quality, security, and reliability. IP was initially designed to support the requirements of data rather than voice traffic, and users have high expectations for sound quality and reliability when it comes to telephone service. Wireless LANs were also designed for data applications, and sending radio waves through free space in the face of interference and physical obstructions is still a challenging endeavor. So while there are cost and convenience benefits to be gained from carrying voice over a wireless LAN infrastructure, a successful outcome will require good design and careful planning.

The purpose of this text is to provide the essential background a network designer will need to build and manage a WLAN-based voice communication system successfully. That background will span a number of technical areas including: radio system design, IP telephony technologies, voice network planning and engineering, and network security.

In this first chapter we will provide an overview of several of the topic areas that will be critical to a Voice over Wireless LAN (VoWLAN) network design. First, we will provide an overview of wired and wireless LANs, and describe the major segments of the VoWLAN market. Then we will introduce the basic technology, configurations, and applications for 802.11 or Wi-Fi wireless LANs. We will also distinguish between licensed and unlicensed radio spectrum and describe some of the major challenges involved in building indoor radio networks. Finally we will investigate the basic technology of packet switched voice and the benefits to be gained from integrating voice and data services in both the local and wide area markets.

1.1 The WLAN Voice Market

Since their commercial debut in the early 1990s, wireless LANs have become a major element in consumer, enterprise, and carrier service markets. The major development in WLANs came with the adoption of the IEEE 802.11 standard in 1997. A standards-based solution allowed the development of a broad range of interoperable WLAN devices. The assurance of compatibility was provided by the Wi-Fi Alliance with their highly successful product certification programs.

The use of wireless LAN networks to carry voice traffic is a more recent development. Of the hundreds of millions of WLAN client devices in use today, less than one percent support voice.

The emerging WLAN voice market today can be divided into three major areas:

- Consumer Wi-Fi Phones

- Enterprise grade Wi-Fi Voice Systems

- Integrated WLAN/Cellular Networks

Consumer Wi-Fi phones are wireless telephone handsets that operate over an 802.11-compatible wireless LAN. These devices can be used in conjunction with a home-based wireless LAN or public Hot Spot services. Handset manufacturers are aligning themselves with VoIP service providers like Vonage and Skype who are marketing the Wi-Fi handsets as a way to increase the functionality of their services.

Enterprise grade Wi-Fi voice systems support multiple handsets and operate in conjunction with a controller or telephony server. The controller supports call setup, feature activation, and provides an interface to connect calls through to the wired voice PBX system. Polycom/SpectraLink, Cisco, Siemens, Symbol Technologies, and Vocera are the major manufacturers. Given the requirements of the enterprise market, these systems must operate over large-scale wireless LANs with hundreds of access points covering tens of thousands of square feet. Enterprise networks also require excellent voice quality, security, and the ability to quickly hand off calls between access points as a user moves through the coverage area. These enterprise systems will be the primary focus of this text.

Integrated Wi-Fi/cellular handsets have the capability to operate over either a cellular service or a private Wi-Fi network. The Wi-Fi capability may mimic a consumer Wi-Fi Cordless phone or it might be designed to operate in conjunction with an enterprise Wi-Fi voice system. The distinguishing characteristic of Wi-Fi/cellular phones is the degree of integration between the two elements. At the lowest level, you can have two separate wireless phones built into the same package, and the user selects which network the call will be placed over. Ideally, the capabilities would be integrated, and the phone would sense when a Wi-Fi network is available, and automatically hand off calls between Wi-Fi and the cellular networks. That latter capability is referred to as fixed-mobile convergence (FMC), and a truly integrated Wi-Fi/cellular service would require the cooperation of the cellular carriers. US-based cellular carriers are only now introducing these services, the first being a consumer offering from T-Mobile called HotSpot@Home.

1.2 Development of Wireless LANs

As computer networking transitioned from centralized mainframe systems to distributed client-server architectures, wired local area networks (LANs) based on the Ethernet or IEEE 802.3 standards have become a critical element in the enterprise data processing infrastructure. However, users were tethered to their desks with wired connections. The first phase in LAN mobility came about when users started bringing laptops to meetings, and organizations installed Ethernet ports in conference rooms to support them.

1.2.1 The Birth of Wireless LANs

The first true wireless LANs appeared in the early 1990s, though they were low-speed and geared toward specialized applications. Companies like NCR, Proxim, Photonics, and Spectix introduced wireless LAN products that operated at 1 or 2 Mbps and were used primarily in warehousing and materials handling applications. The WLAN terminal was often a special-purpose device like a barcode scanner. Many of the operating principles used in modern wireless LANs are derived directly from those early products.

1.2.2 Wireless LANs Go Mainstream

The first 802.11-series standards for wireless LANs were introduced in 1997, though the relatively slow 1- to 2-Mbps data rate was an obstacle to widespread acceptance. The market took off in 1999 with the introduction of the 11 Mbps 802.11b radio link. The first major application for wireless LANs was home networking where the wireless LAN base station or access point (AP) was built into a DSL/cable modem router manufactured by companies like Linksys (now a division of Cisco Systems), Netgear, and D-Link. See Figure 1-1. These routers allowed a number of wired and wireless devices to share one high-speed Internet connection, and the wireless connection provided mobility eliminating the need to install LAN cabling throughout the home. The initial 802.11 standards were weak in terms of security and were better suited for residential than for commercial installations.

1.2.3 Enterprise Wireless LANs

Wireless LANs have now moved into the mainstream for enterprise customers. Market research indicates that sales of Wi-Fi devices are doubling annually, and the newer products are clearly targeted toward the requirements of commercial installations. That acceptance was further accelerated by the prevalence of Wi-Fi interfaces in laptop computers. Apple Computer was an early adopter of Wi-Fi technology, and Intel's Centrino chip, which included a Wi-Fi interface, brought wireless LAN capability to the vast majority of Wintel laptops. Given their special requirements, hotels, hospitals, and universities have been in the forefront of this move.

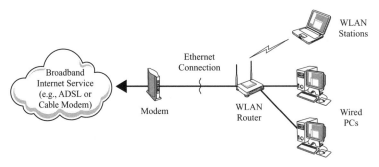

Figure 1-1: Home-based Wireless LAN

An enterprise wireless LAN will be a more complex installation than what someone would use at home. A number of access points would be distributed throughout the coverage area and connected through the enterprise's wired LAN switching system. Each access point would act like a base station in a cellular telephone network. An organization might provide coverage throughout the entire building or campus, or they might choose to provide service only in common areas like conference rooms and cafeterias. See Figure 1-2.

There are several potential benefits to be gained:

- **Convenience:** With Wi-Fi equipped laptops, users can access network resources from any location within the coverage area.

- **Mobility:** That same Wi-Fi interface can allow a laptop user to access public wireless networks or Hot Spots outside their normal work environment. Those public Hot Spot services may be free or subscription based, but they allow the user to have email, Web access, or remote access to the corporate LAN from coffee shops, hotels, or airports.

- **Productivity:** Users with wireless network access can remain in constant contact with their company's computer network resources as they move from

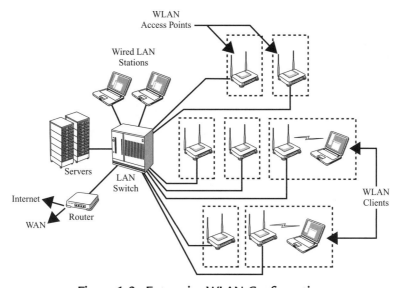

Figure 1-2: Enterprise WLAN Configuration

place to place. Employees can be more productive because they can remain in touch when moving around the office or on the road.

- **Speed of Deployment:** A WLAN can be deployed quickly whereas a wired network requires physical cables to be run to each work area.

- **Cost:** Wireless networks can be cheaper to deploy by avoiding the cost of installing wired cable runs to each work area (wired connections to each access point are still needed). Given the limited transmission capacity and ongoing reliability problems, most organizations are looking at WLANs as an adjunct to their wired infrastructure, not as a replacement.

- **WLAN Voice Service:** While wireless LANs were initially deployed to provide data access, they can also support voice calls. That would mean that a user could have continuous telephone system access and be reachable anywhere within the facility. Further, unlike costly cellular service where indoor coverage is problematic, a WLAN voice user could have calls dialed to their desk set forwarded directly to their WLAN phone.

1.3 Wireless LAN Applications

Enterprise users have taken a cautious approach regarding WLAN deployment in office environments due largely to concerns about the security of the information that would be carried on those networks. WLAN signals inevitably leak out of the facility, and so all transmissions must be encrypted to ensure their security. With the rather rudimentary encryption system used in the original WLAN standards, those transmissions were susceptible to eavesdropping, and unauthorized users could potentially get access to sensitive information on the LAN. Fortunately those security deficiencies have now been addressed, and with the new WLAN security standards we can implement wireless LANs that meet the security requirements of most enterprise organizations.

Office-based networks: Many enterprise organizations have implemented wireless LANs, but the requirement to support voice is changing the scope of those deployments. Most initial enterprise WLANs provided coverage in meeting rooms and public spaces like the cafeteria. That type of spot coverage will not be adequate for voice applications, as voice users are far more *mobile* than data users. To ensure that voice users can get access to the network anywhere within the facility, a voice WLAN will require pervasive deployment with adequate network capacity throughout the coverage area.

1.3.1 Distribution Center/Factory Floor Systems

One of the first applications for wireless LANs was in locations where it is difficult to install LAN cabling and mobility is important. Warehouse operators were among the first to use WLANs for data collection systems using handheld input devices like barcode readers. Some of those devices have now been augmented with voice capabilities, particularly push-to-talk voice.

1.3.2 Hospitals/Health Care

Given the mobility of the work force, wireless LANs have found great acceptance in the health care environment. Applications range from bedside data collection to physician order entry, and equipment location for stat carts. These organizations have also been the most progressive in deploying WLAN voice to improve patient care and to supplant intrusive paging systems.

1.3.3 Hospitality

Hotels, airports, and train stations are also major markets for wireless LANs, as they are places where business users tend to congregate. In hotels, WLANs are now being used to replace ADSL as the technology used to provide in-room broadband Internet access as well as providing wireless connectivity in meeting rooms and public areas. These networks have been configured to support data clients, however, with the advent of Wi-Fi voice handsets, these networks may wind up supporting voice whether they intend to or not.

1.3.4 Universities

Most colleges and universities have deployed WLANs in libraries, dorms, student union buildings, and outdoor areas to provide access to university computing facilities throughout the campus. Typically access to these networks is restricted to registered students. Universities are also big users of point-to-point systems to provide connectivity between buildings.

1.3.5 Home Networking

Home networking still represents about 50 percent of all WLAN equipment sold. The standard configuration uses an access point that is built into a DSL/cable modem router allowing several (normally up to 16) wired and wireless users to share one broadband

Internet connection. Wi-Fi cordless phones will allow these networks to support voice, most likely Wi-Fi access to VoIP telephone services like those from Vonage or Skype.

1.3.6 Hot Spots and Wireless Internet Service Providers (WISPs)

Closely related to the hospitality networks is the new class of broadband Internet providers called Wireless ISPs (WISPs), who provide broadband Internet access to Wi-Fi equipped clients. These carriers can be divided between free and subscription services. While the subscription-based services like T-Mobile and Wayport were introduced with great fanfare, they have not been a resounding business success. The providers do not report profit and loss information, but it appears they have not yet reached profitability. This business model is now being put under greater pressure, as the WISPs will be competing with a growing number of free Hot Spot services.

1.3.7 Citywide Municipal Wi-Fi Mesh Networks

Given the success of Wi-Fi in the local area, a number of municipalities are pursuing the idea of deploying Wi-Fi Mesh networks that would cover entire metropolitan areas. These mesh networks would support government agencies like police and fire departments, but could also provide a wireless alternative to DSL or cable modem broadband Internet service. While they were introduced with great promise, it does not appear that these networks will become serious competitors to wide area cellular or WiMAX services.

1.4 Health Issues with Wireless Devices

One major unresolved issue hanging over the wireless market is the potential of health problems created by exposure to radio frequency radiation. Much of the "proof" comes in the form of anecdotal evidence that has linked radio frequency exposure to maladies ranging from increased incidence of certain forms of cancer to lower-level effects like headaches, irritability, and loss of concentration.

While no health concern should be minimized, there is simply not enough data at this point to make a definitive statement regarding the health effects of radio exposure. Some of these reports identify ailments like irritability or loss of concentration that would be difficult to substantiate in controlled conditions. Further, a comprehensive scientific study would require the ability to isolate other factors while examining the impact of exposure to different radio frequencies, at different power levels, for different durations, and at different stages of life. There are also reports that appear to be little more than exercises in public relations. In one case, a doctor claimed that using laptops could impact male fertility given the heat generated by the laptop's battery if the user actually used the device in his lap. That announcement was carried by all of the major network news agencies!

In short, all we can say with regard to safety at this point is that it is an area that should be monitored. The scientific community will have to provide us with more useful information regarding the nature and severity of the risks before we can take meaningful steps to mitigate those concerns. Today, users who harbor these concerns can use headsets, hold the antenna at a greater distance from the body, or forego the use of radio devices entirely. For the time being, most users are unwilling to deny themselves the convenience that radio communications brings in light of the rather tenuous evidence. Perhaps that position will be found to be foolhardy in the future, but there is simply not enough definitive proof to make such a judgment today.

1.4.1 Government Information: FCC and FDA

If you have concerns in the area of cell phones and health, the FDA and the FCC have created a joint Web site titled, "Cell Phone Facts—Consumer Information on Wireless Phones," which states:

> The available scientific evidence does not show that any health problems are associated with using wireless phones. There is no proof, however, that wireless phones are absolutely safe. Wireless phones emit low levels of radiofrequency energy (RF) in the microwave range while being used. They also emit very low levels of RF when in the stand-by mode. Whereas high levels of RF can produce health effects (by heating tissue), exposure to low-level RF that does not produce heating effects causes no known adverse health effects. Many studies of low-level RF exposures have not found any biological effects. Some studies have suggested that some biological effects may occur, but such findings have not been confirmed by additional research. In some cases, other researchers have had difficulty in reproducing those studies, or in determining the reasons for inconsistent results.

To track discoveries in this area, you can access the joint FDA/FCC Web site at http://www.fda.gov/cellphones.

1.5 Wireless LAN Organizations

Two primary organizations have been critical to the development and maintenance of WLAN standards, the IEEE and the Wi-Fi Alliance.

1.5.1 Institute of Electrical and Electronics Engineers (IEEE)

The Institute of Electrical and Electronics Engineers (IEEE) is the professional organization responsible for the development of local area network standards worldwide. That work is done through the IEEE's 802 committees. There are a number of committees within IEEE 802, and responsibility for wireless LANs fall under the IEEE 802.11 committee. Within 802.11 there are numerous subcommittees that address particular modifications to the basic standard; those additions to the standard and the groups that develop them are identified by letter suffixes (e.g., 802.11b, 802.11i, 802.11e, etc.). We will refer to many of those standards throughout the text, and there is a list of the major 802.11 standards in the Appendix. Those developments are ongoing.

1.5.2 The Wi-Fi Alliance

The other major entity is a nonprofit vendor group called the Wi-Fi Alliance. The Wi-Fi Alliance was originally called the Wireless Ethernet Compatibility Alliance (WECA), though the name was subsequently changed to *Wi-Fi*, short for *Wireless Fidelity*.

The primary task of the Wi-Fi Alliance is to foster the growth of the WLAN market by ensuring interoperability between products from different vendors. The IEEE standards typically define a wide range of capabilities within a standard and specify that different capabilities are either Mandatory or Optional. As a result, countless incompatible implementations can be designed, all of which are compatible with the IEEE standard. The technical committees within the Wi-Fi Alliance review the IEEE-defined standards, and select the options to be included in the standard implementation.

As the number of devices supporting wireless LAN interfaces increases, it becomes virtually impossible to test each new product against every existing product. To address this, the Wi-Fi Alliance develops a comprehensive test suite for each IEEE-defined capability. Devices are tested against this regimen, and if the device passes, we can say with a high degree of assurance that it will interoperate with every other Wi-Fi Certified device. Enterprise buyers should always look for that Wi-Fi Certification on any WLAN device they buy.

The high degree of interoperability is one of the elements that has made Wi-Fi so successful. There is always tension in any vendor alliance, as different vendors will seek to develop a competitive advantage pushing their proprietary solution. Further, vendors may develop non-compliant devices that will not only be incompatible, they may create interference problems for compliant devices.

In approving new implementations, the Wi-Fi Alliance rigorously adheres to the requirement for *backwards compatibility*, which ensures that any new device or function will not disrupt the performance of existing devices. In some cases the Wi-Fi Alliance has developed an interim standard where there has been a pressing market need, and the lack of an IEEE standard is holding up the process; Wi-Fi Protected Access and Wi-Fi Multimedia are examples of such interim standards. However, even in those special cases the Alliance still insists on the basic requirement for backwards compatibility with the final standard.

1.6 WLAN Configurations

Wi-Fi or IEEE 802.11 standard wireless LAN can be built in four primary configurations, though not all of these are fully defined in the standards:

1. Basic Service Set/Extended Service Set (ESS)

2. Ad Hoc/Peer-to-Peer

3. Point-to-Point

4. Mesh Architecture

1.6.1 Basic Service Set (BSS)/Extended Service Set (ESS)

Basic Service Set (also called Infrastructure Mode) is the WLAN configuration where a number of client devices in a service area or *cell* communicate with an Access Point (AP) that is in turn connected to the wired LAN. All transmissions flow client-to-access point or access point-to-client; the clients do not communicate directly with one another. In consumer applications, the access point is typically built into a DSL/cable modem router so that multiple wired and wireless devices can share one broadband Internet connection. In enterprise WLANs, a number of access points are used to cover different parts of the area. Those access points are interconnected through the wired LAN that the WLAN standards refer to as a *distribution system* (DS). A configuration

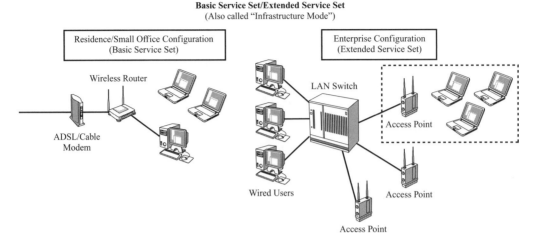

Figure 1-3: WLAN Configurations

using multiple interconnected access points all with the same network name or System Services Identification (SSID) is called an Extended Service Set. BSS and ESS configurations are used for WLAN voice, and we will focus on these arrangements primarily. See Figure 1-3.

1.6.2 Ad Hoc Network/Independent Basic Service Set (IBSS)

Two client devices equipped with wireless LAN cards can also communicate directly with one another in what is called an Ad Hoc or peer-to-peer network; the standards refer to this configuration as an Independent Basic Service Set (IBSS). All that is required for this configuration is two WLAN equipped devices that are optioned to support Ad Hoc networking. As peer-to-peer connections are unencrypted, commercial organizations typically have policies banning their use. In any event, users should be cautioned against connecting to any unknown device on a peer-to-peer basis, and the IT department should disable the peer-to-peer capability in any company-provided laptop.

1.6.3 Point-to-Point 802.11

The 802.11 radios can also be configured with directional antennas to provide the equivalent of a point-to-point radio link. Point-to-point 802.11 networks are typically used as bridged connections to link LANs in different buildings. With directional antennas at each end, the 802.11 radio link can be extended for 10 to 20 miles depending on the antennas used and the quality of the radio path.

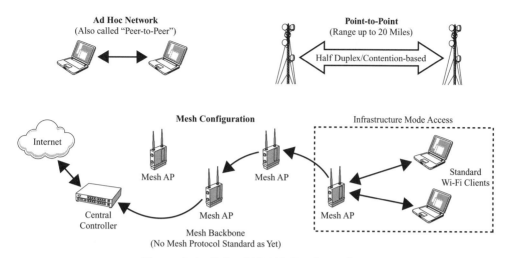

Figure 1-4: Other WLAN Configurations

1.6.4 Wi-Fi Mesh

A mesh is a specialized WLAN configuration that is not covered by the standards as yet. In a mesh configuration, client devices associate with a special access point much like a traditional BSS network. However, only one of the mesh access points needs a connection to the wired network. The client messages are forwarded access point-to-access point to the location with the wired connection (while client devices could be involved in the relay, that is rarely done today). This configuration provides a cost-effective way to extend WLAN coverage to a campus or even an entire metropolitan area. Mesh configurations may not be appropriate for voice, as the transit delays will increase with the number of access points involved. Further, there can be a funnel effect where contention delays increase as the frames get closer to the wired connection. All current mesh implementations use a proprietary backhaul protocol, but there is a mesh standard in development designated 802.11s. See Figure 1-4.

1.7 Wireless LAN Design Issues

Each wireless LAN will require a radio channel on which to operate. WLANs operate in unlicensed frequency bands, where there is no protection from interference caused by other users. A residential installation will typically require one access point, one channel, and the only concern is interference from wireless LANs in adjacent apartments or homes. A large commercial installation using the Extended Service Set will require a number of access points distributed throughout the coverage area, each with its own channel, and interconnected through a wired LAN. Wireless LAN devices will scan all of the channels and attempt to join the network whose access points provide the strongest signal and the least interference (i.e., the best signal-to-noise ratio).

1.7.1 Non-Interfering Channels

One of the critical variables in a wireless LAN is the number of channels that are available. The 802.11 standards define radio link interfaces that operate in either the 2.4 GHz Industrial, Scientific, Medical (ISM) or the 5 GHz Unlicensed Network Information Infrastructure (U-NII) unlicensed radio bands. The number of channels available will be based on the bandwidth for each channel (i.e., the range of frequencies the channel occupies) and the amount of radio spectrum allocated in the band. One basic rule of channel layout is that we attempt to avoid reusing a channel in an adjacent cell; so having more channels makes it easier to lay out the cells without creating interference.

The 802.11b and g standards specify channel bandwidths of 22 MHz and 20 MHz, supporting maximum data rates of 11 Mbps and 54 Mbps, respectively. However, standards 802.11b and g operate in the relatively narrow 2.4 GHz band. In most countries (including the US), the 2.4 GHz unlicensed band has only 83.5 MHz of spectrum available, so it can accommodate only 3 non-interfering channels. The 802.11a radio link provides a 54 Mbps data rate in a 20 MHz channel, but it operates in the more spacious 5 GHz band. In the US, the 5 GHz band has 555 MHz available, and can support 23 non-interfering channels. The radio links are summarized in Table 1-1.

1.8 The Packet Telephony Revolution

The underlying technology for voice on Wi-Fi is packet switching, in particular IP-based packet switching. The original switching technology used in the public telephone

Table 1-1: IEEE 802.11 Radio Link Interface

Radio Link Standard	Maximum Transmission Rate (Mbps)	Individual Channel Bandwidth (MHz)	Frequency Band (GHz)	Available Non-Interfering Channels
802.11b	11	22	2.4	3
802.11g	54	20	2.4	3
802.11a	54	20	5	23

network is called circuit switching, where a dedicated transmission path is set up and maintained between the two parties for the duration of a call. That design decision was made in the early days of telephony, when the switching mechanism might be as rudimentary as an operator connecting parties manually through a cord board.

Packet communications was developed for data communication systems in the 1960s, and it was recognized that it could also provide a more efficient mechanism for carrying phone calls. The basic idea of a packet switching network is to take chunks of digital information from different users, append an address to each, and intersperse them over a single shared communication channel. Those addressed chunks of information are called *packets*. The machines that route those addressed packets over the transmission links to their destinations are called *packet switching nodes* or *routers*.

There are several different technologies or services that fall under the general heading of packet switching, and those would include: X.25, frame relay, asynchronous transfer mode (ATM), and of course IP-based systems. Local area networks like Ethernet and Wi-Fi also operate on those same principles, though in LANs those addressed units of transmission are called *frames* rather than packets.

The technical and economic advantages of packet switching come from its ability to allocate the available transmission capacity dynamically, thereby allowing us to support a much larger volume of traffic while building or renting a much smaller amount of real transmission capacity. As we allocate that transmission capacity dynamically, each user gets the capacity that they need at the moment they need it.

There are two basic ideas that underlie all packet switching systems.

1. **Dynamic Allocation:** (also called "dynamic bandwidth allocation" or "statistical multiplexing") This means that a user is assigned transmission capacity only when the user actually has information to send.

2. **Over-Subscription:** Over-subscription means that the virtual capacity of the network (i.e., the total amount of transmission that could occur at any instant) will exceed the real capacity. Essentially, the network design is based on the statistical bet that not everyone will need to send at the same time.

1.8.1 Dynamic Allocation and Voice Activity Detection

The initial appeal of packet switching for voice came from the fact that there is an inherent inefficiency in using a circuit switched path for a telephone call: the two parties do not speak continuously. In a telephone conversation, only one party speaks at a time. Further, there are points in the conversation where neither party is speaking. If you were to analyze the pattern of activity on a typical telephone call, you would find that the communication channel is active roughly 40 percent of the time in each direction; typically the voice activity occurs in speech bursts originating from one end or the other. The idea of using packet switching for voice involves recognizing that activity pattern, and using a switching technology that can allow us to carry more conversations by taking advantage of the idle periods. In packet voice systems, that capability is called voice activity detection (VAD) or silence suppression as no packets would be sent during pauses in the conversation.

While the idea of dynamically interspersing voice and data packets over a shared facility generated the initial interest in packet voice, in practice it is not often used. Typically, WLAN voice is used in conjunction with an IP PBX, which is essentially a wired LAN switching system that also carries telephone calls. In a wired LAN, each user can be provided with an enormous amount of transmission capacity; typically each user is connected with a dedicated, full duplex 100 Mbps Ethernet connection. Voice consumes less than 0.1 percent of that capacity! As silence suppression adds complexity, users typically do not activate the feature in an IP PBX. As a result, when the voice calls are extended over the WLAN, the capacity is dedicated for the duration of the call regardless of whether the user is speaking.

The potential to use voice activity detection is an ongoing issue in wireless LANs. Unlike a wired LAN, WLANs provide far less transmission capacity. Further, a single WLAN channel will be shared by all users in that area. As a result, VAD and other mechanisms to utilize the available capacity efficiently may have far greater value in WLAN voice.

Time Assigned Speech Interpolation—TASI

The idea of using dynamic allocation to improve the efficiency and reduce the cost of telephone systems is by no means new. In the 1960s, Bell Laboratories developed an analog dynamic allocation voice system for submarine cables called Time Assigned Speech Interpolation (TASI). TASI involved installing a device at each end of the cable system that sensed voice activity and quickly switched those bursts of voice activity onto a smaller number of real channels. When it sensed

> ## Time Assigned Speech Interpolation—TASI (*Continued*)
>
> activity on an input channel, the TASI device inserted an address in front of the speech burst and quickly switched it to one of the "real" channels. The TASI device at the far end read the address, and quickly switched the speech burst onto the correct output. The system increased the number of calls that could be support by roughly 50 percent, though it could clip off part of the speech burst if too many people happened to be speaking in the same direction at the same time.

So while we do not get the efficiency gain of dynamic allocation, WLAN voice is subject to many of the drawbacks of packet voice. In particular, those issues would include:

- **Transit Delay:** Recognizing the fact that too many users may try to send at the same time, WLAN access points, LAN switches, and routers include buffers (i.e., temporary data storage) to accommodate those bursts of activity. Packet forwarding devices can only process one packet at a time, and when a packet arrives, there may be a number of packets already in the router's buffer waiting to be serviced. Wireless LANs also introduce unpredictable delays as users vie for access to the shared radio channel. Excessive transit delay is annoying to voice users as it impacts the flow of a voice conversation.

- **Jitter:** The other timing issue introduced by packet buffering is variable delay. The variation in packet-to-packet transit delay is called jitter. Left untreated, jitter will destroy the quality of the speech signal. In IP voice we use a special protocol called the Real time Transport Protocol (RTP) to mask that jitter effect.

- **Packet Loss:** While buffers are a necessary element in a packet network, there is a finite amount of buffering capacity in each router and in the end device. If the buffer capacity in any device is exceeded, it will drop some of the packets. When too many packets are dropped, the voice signal will degrade to the point where it is difficult or impossible to understand the speaker.

1.8.2 Quality of Service

Quality of Service (QoS) is one of the critical features that we use to improve the performance of voice services in a packet switching network. Data transactions are far more tolerant of delay, jitter, and loss than voice transmissions; if a data message is lost, a protocol like TCP will retransmit it automatically. With voice services, on the other hand, the idea of retransmitting lost packets is impractical. The retransmitted voice packets would arrive too late to have any relevance to the conversation.

If voice and data services are sharing the same network facilities, we would like to provide preferred handling for the voice packets over the data packets—that is what we mean by QoS. In a traditional WLAN or IP service, all packets are treated equally, and all packets would experience loss, delay, and jitter to the same degree. The term QoS refers to any mechanism that can be used to provide priority handling for one class of traffic over another. The goal of priority handling would be to deliver a higher percentage of the preferred packets (i.e., the voice packets) while minimizing the overall delay and jitter.

The actual means of providing quality of service will differ based on the technology we are dealing with. For example, routers typically maintain separate queues for the different traffic classes. In a Wi-Fi network, we have shorter pre-transmission waiting intervals for voice than for data frames. Regardless of the mechanics of the solution, if the network were to provide quality of service, the packets would be sorted based on a priority scheme (e.g., voice, video, and data), and the switching process would be modified to give preferred service to the higher priority traffic.

1.9 Local Area IP Telephony: IP PBX

The term Voice over IP or VoIP is used generically today, however it actually describes several distinct applications that address both local area and wide area services. The configuration and incentive for pursuing each of these strategies is different.

1.9.1 Local VoIP: The IP PBX

In the local area, enterprises are looking to reduce the cost of the communication infrastructure within their facilities by eliminating the traditional voice PBX system and using their LAN switch-based data network to carry voice calls. Enterprise organizations have traditionally operated two physically separate infrastructures within their facilities, one for voice and another for data. The voice system revolves around a stand-alone voice PBX with a cabling infrastructure and handsets that are dedicated to providing voice communication services and connecting users to the public telephone network. Data services are provided over a completely separate cable plant using LAN switches that connect PCs to shared printer, servers, and to the Internet.

Enterprise voice is now migrating to IP PBXs, where users are equipped with telephone sets that support an IP/Ethernet interface. Those IP telephones encode voice in a digital format, generate IP packets, and connect to the LAN switch over an Ethernet interface. The user's PC connects to their IP/Ethernet phone that in turn connects to the Ethernet interface, so the voice and data devices are sharing the cabling and the port on the LAN switch. To provide voice service, the LAN switch will require a telephony server (e.g., Cisco's Call Manager) to manage connections and support the telephony features. The major benefit users have found in this configuration is reducing their cabling costs while allowing new integrated voice/data applications to be developed.

1.9.2 Hybrid IP PBX

While LAN switch vendors like Cisco have hyped the IP PBX solution, the traditional PBX vendors (e.g., Avaya, Nortel, Siemens, etc.) have developed a configuration to migrate their existing users to IP. All of their traditional or TDM PBXs now support modules that will allow those legacy systems to support IP telephones. Rather than scrapping the entire PBX system, the user can invest in a card set that allows them to support the new IP telephone sets on their existing PBX.

Proponents point out that there is no reason to scrap a voice system that still has years of useful life ahead of it. Further, when we look at how costs are allocated in a PBX, two-thirds of the cost could go toward the handsets while one-third is tied up in the central PBX. With the hybrid strategy, the customer can cap their investment in digital handsets that will only work with the traditional PBX, and begin acquiring IP-based devices. Implicit in this approach is the assumption that all PBX systems will eventually be IP-based, and that legacy vendors do have full IP PBX solutions to offer as well.

The idea of enterprise WLAN voice grows out of this migration to IP PBXs. Just as a WLAN can provide mobility to data devices in the local area, WLAN voice handsets can provide a mobile voice capability. One of the major financial incentives of WLAN voice will be to reduce cell phone usage while in company facilities. Special care must be taken in the design of the wireless LAN to ensure that it can support the volume of voice traffic that is produced while maintaining adequate performance for delay, jitter, and packet loss. See Figure 1-5.

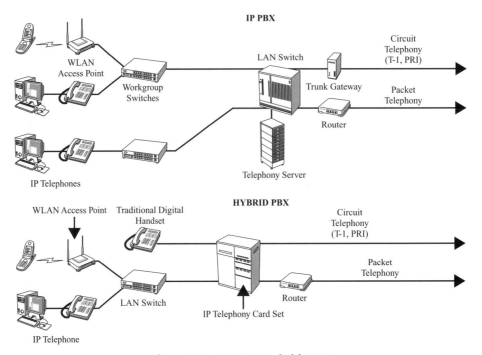

Figure 1-5: IP PBX/Hybrid PBX

1.10 Wide Area VoIP

Completely independent of the local network architecture is the challenge of integrating voice and data services in the wide area. The primary focus of wide area IP telephony is cost savings on long distance telephone calls or *toll replacement*; that means we would route phone calls over a wide area IP network rather than subscribing to traditional long distance telephone services. There are two major segments in the wide area VoIP market:

- Enterprise Services
  - Private Router Networks
  - IP Trunking

- Consumer/Small Business Services
  - Packet Telephony Services
  - Cable Telephony
  - Telco "Triple Play"

1.11 Enterprise VoIP Networks

One potential VoIP implementation in enterprise environments is to use the existing wide area router networks to carry voice traffic between sites. That voice traffic would use the same routers and flow over the same private line, frame relay, or IP network services that currently carry data traffic. Given the relatively low capacity of wide area communication links, QoS mechanisms are essential to minimize delay, jitter, and loss for voice packets. Even with QoS, the backbone links may still need to be upgraded to maintain adequate performance for both voice and data applications.

Despite the potential cost savings, this strategy has not yet been widely pursued by enterprise customers. The lack of acceptance is based equally on technical and business criteria. From a technical standpoint, it was more difficult to deliver the service quality that voice users were accustomed to over those wide area networks. More importantly, it was difficult to demonstrate a cost justification. The price of traditional circuit switched services declined tremendously during the 1980s and 1990s, with the result that large users are now paying about 1-1/2 cents per minute for long distance calls; with those prices, it is difficult to make a strong financial case for enterprise wide area VoIP. See Figure 1-6.

1.11.1 IP Trunking Service

While little voice traffic is finding its way onto the enterprise wide area router network, the carriers are now offering long distance telephone services that utilize IP technology. The telephone companies are already making extensive use of VoIP services to carry calls between their central offices. The idea of an IP Trunking Service is to connect enterprise customers directly to those VoIP-based backbone services by a dedicated 1.5 or 44.7 Mbps link. Rather than connecting to the carrier with a traditional circuit-based telephony interface like an ISDN Primary Rate Interface (PRI), voice calls are forwarded to the carrier in an IP-based packet stream. Those calls can still be delivered to subscribers with traditional circuit-based telephone connections. The appeal of these services is that the carriers will typically offer a far lower cost per minute than with traditional long distance services. See Figure 1-7.

1.12 Consumer Packet Telephony Services

While the idea of enterprise wide area IP voice is just catching on, consumer-oriented IP telephone services have found far greater acceptance. Despite the fact that IP

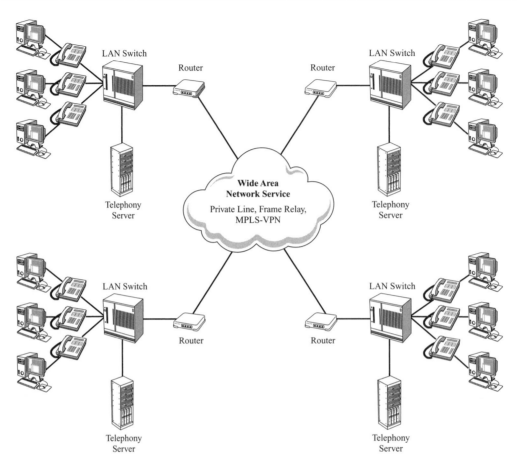

Figure 1-6: VoIP on Enterprise Wide Area Backbones

Telephony providers typically route voice packets over basic Internet services, the voice quality is generally indistinguishable from the traditional circuit switched telephone services. Some customers have reported persistent service problems with IP telephone services, so you may want to approach the decision with caution. The carriers operate a number of gateway facilities where they interconnect to the public telephone network so they can deliver calls to traditional telephone network subscribers.

Carriers like Vonage offer unlimited local and long distance calling for a flat rate, $24.99 per month. Skype began as a free service for PC-to-PC Internet calls using a softphone client the user would download from Skype's Web site. Skype was

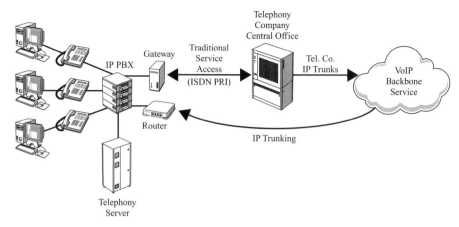

Figure 1-7: IP Trunking Service

subsequently acquired by eBay, and now has the ability to deliver calls to traditional telephone subscribers through a service called Skype-out. With SkypeOut, users get unlimited calls to the US and Canada for $29.95 per year.

To use either of these services, the customer must have a broadband Internet connection like cable modem or ADSL. With Vonage, a special interface box called a *Phone Adapter* is provided. That interface box digitizes the voice and generates the IP packets, which are interspersed with data packets on the broadband access connection. Vonage also offers WLAN handsets that operate over the user's home-based WLANs or public Hot Spots. While Skype has traditionally focused on using a free softphone client running the customer's PC, many users balked at using a PC and a headset to make phone calls. To address this, Skype now offers a range of phone adapters, handsets, and Wi-Fi phones. See Figure 1-8.

1.12.1 Cable Telephony

Cable TV companies have also recognized the revenue potential of offering telephone service over their broadband cable modem access networks. As of early 2006, over six million traditional lines have migrated from the incumbent telephone companies to cable telephony. When a customer subscribes to cable voice service, the cable company provides a new cable modem that includes the voice interface (i.e., the same functionality as the phone adapter provided by carriers like Vonage and Skype is built into that special cable modem).

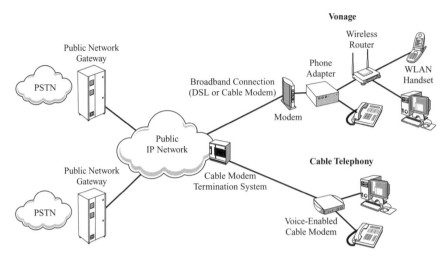

Figure 1-8: Consumer Packet Telephony Services

That voice/data cable modem digitizes the voice signal, generates IP packets, and intersperses them with data packets over the broadband Internet connection. As modern cable modem systems using version 1.1 or 2.0 of the Data Over Cable Service Interface Specification (DOCSIS) can support QoS, the service quality is typically indistinguishable from traditional telephone services. In most cases the cable company interconnects calls to the traditional telephone network at their head end facility so the calls are not transported over Internet facilities. The cable companies typically discount the voice service if the customer already has broadband Internet and digital cable service.

1.12.2 Telco Triple Play—FTTN/FTTP

Not wishing to be left behind in the migration to IP telephony, the local telephone companies are now deploying networks that will support integrated broadband Internet, cable TV, and IP voice services. The telco networks use either fiber to the neighborhood (FTTN) or fiber to the premises (FTTP) configurations. In FTTN, the services are delivered over a fiber link to a neighborhood fiber-based hub; the connection to the customer runs over copper wire using a next-generation DSL technology called Very high-speed DSL (VDSL). The plan is to support a data rate of 25 Mbps over that VDSL link for a range of 4,000 feet allowing IP voice, broadband Internet, and cable television services to be delivered. A customer terminal is provided that includes the VDSL modem and interfaces for voice, Internet, and cable TV services. See Figure 1-9.

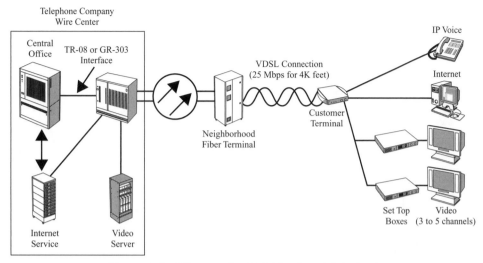

Figure 1-9: Fiber to the Neighborhood (FTTN)

In fiber to the premises (FTTP) systems, the fiber link extends all the way to the customer's home. The customer terminal terminates the fiber and provides interfaces for voice, Internet, and cable TV services. The fiber connection uses a multi-drop optical technology called a Passive Optical Network (PON) where up to 32 homes can be connected on a single link. The video and upstream/downstream PON channels are carried using three different optical wavelengths (i.e., three different "colors" of light). In either configuration, the voice service is provided using IP technology. See Figure 1-10.

The basic message is that the migration from traditional circuit switched telephone services to IP-based telephony has clearly begun. However, it is important to distinguish which application of IP telephony we are referring to as the transition is occurring at different rates and for different reasons in the local versus the wide area environment.

1.13 Conclusion

Two major trends in networking are converging to provide voice over wireless LANs. On one hand we have customers investing in wireless networks to provide mobility for data devices, and that same infrastructure can now be shared with voice handsets. On the other hand, voice networks are in the midst of a major technology shift from traditional circuit switching to IP-based packet switching. The result is that a customer

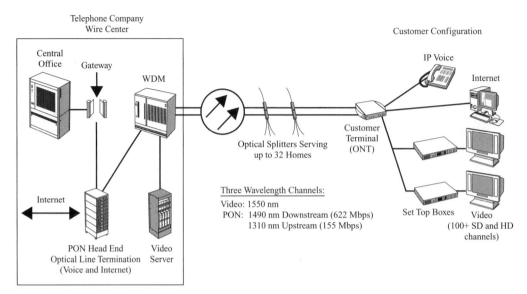

Figure 1-10: Fiber to the Premises (FTTP)

can now have one cost effective IP-based wired and wireless network that can support both voice and data applications.

There will be challenges in delivering these capabilities to be sure. Packet networks and wireless LANs were designed for the requirements of data not voice devices, and delivering the service quality users expect will require QoS mechanisms to minimize delay, jitter, and loss for voice packets. While QoS will improve the performance of voice services, it will also degrade the service offered to data users. The wireless LAN infrastructure will have to be expanded from localized or spot coverage to pervasive networks that are accessible throughout the entire facility. Also, we have to ensure that we incorporate adequate security measures to curtail theft of service or eavesdropping on WLAN voice calls.

All of these challenges can be met, however the network designer will require a firm understanding of the concepts and a detailed knowledge of the capabilities and limitations of the various solutions. Now it is time to look at some of the technology that makes all of this possible.

CHAPTER 2

Radio Transmission Fundamentals

To deal intelligently in wireless LANs you will need some fundamental background in transmission systems, particularly those that depend on radio. In this chapter we will focus on the factors that impact radio transmission systems, and the physical limits we face in sending digital information or "bits" over a radio channel. The basic relationships are spelled out in a formula called Shannon's Law, which identifies the three critical factors that will determine the maximum transmission capacity of any transmission link: bandwidth, noise, and received signal power. In particular we will look at how those elements interact, and their impact on radio networks.

We will also investigate the general difficulties involved in radio communications, including the range of impairments that will affect radio signals, especially radio systems operating in indoor environments. We will conclude by introducing the major elements in a wireless LAN including access points, antennas, and NICs, and wireless LAN Switches, which are becoming the key element in enterprise WLANs.

2.1 Defining Transmission Capacity and Throughput

Capacity is the most basic factor we deal with in any transmission system. Wireless LANs are relatively low-capacity communication systems so it is particularly important that we recognize the factors that impact their capacity. Network specialists often use the term "speed" to identify transmission capacity, however the correct term is actually "rate." The transmission rate of a digital transmission system is measured in bits per second. In determining the performance of a communications network, the *speed* is not nearly as important a factor as the *rate*. Technically, radio signals travel at the speed of light. However, to deliver a Web page that contains 100,000 bytes (i.e., 800,000 bits of information) to a PC, what counts is the number of bits per second sent, not how fast those bits are flying through the air! See Figure 2-1.

Unfortunately, it is not possible to compute the time needed to deliver a message over a communications network by taking the number of bits that have to be sent and

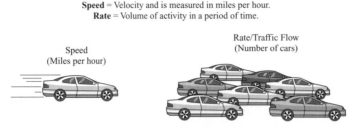

Speed = Velocity and is measured in miles per hour.
Rate = Volume of activity in a period of time.

Speed
(Miles per hour)

Rate/Traffic Flow
(Number of cars)

Figure 2-1: Speed Versus Rate

dividing by the transmission rate (e.g., 800,000 divided by 1,000,000 bits per second = 0.8 seconds). A transmission protocol defines the rules for transmitting that information over a communications link. In 802.11 wireless LANs, the protocol specifies that the bits are first divided into chunks (normally up to 12,000 bits per frame), and then additional header and trailer fields are added. The protocol also specifies that the station must pause for a defined interval before it sends and that the receiver must return an acknowledgment (referred to as the ACK) for each good frame it receives. If two transmitters collide on the radio channel, they back off a random amount and retransmit.

As you can see there is a lot more happening than the transmitter simply pounding bits out onto the channel. That is where the concept of *throughput* comes into play. Throughput defines the amount of data that can be sent over a communications channel in a given period of time (normally a second) including all of the overhead and delay elements introduced by the protocol. In a wireless LAN, the maximum throughput is generally about half of the raw transmission rate.

While the throughput will impact the performance that a user sees on the network, it all begins with the raw transmission rate, or the number of bits per second a station can send.

2.2 Bandwidth, Radios, and Shannon's Law

To understand the capabilities and constraints of a radio transmission system, it is necessary to start at the fundamentals. In our case, that begins with the real meaning of the term *bandwidth*. Bandwidth is the basic measure used to define the information carrying capacity of a channel. *Bandwidth defines the range of frequencies that can be carried on a channel.* When we are talking about radio systems, bandwidth refers to the swath of radio frequency the signal will occupy. The first thing to recognize is that bandwidth is a measure of analog capacity, not of digital capacity.

2.2.1 Bandwidth Is an Analog Measure!

Frequency is a characteristic of an analog signal, and it refers to the number of complete cycles of the signal that occur within a unit of time; by convention, the unit of time we use is a second. So a signal's frequency is measured in cycles per second. That phrase has too many syllables in it so in the early Twentieth Century, the International System of Units defined the term Hertz (abbreviated as the unit Hz) as the standard measure of frequency. The number of Hertz is equal to the number of cycles per second so the terms can be used interchangeably. The use of the term Hertz is in honor of Heinrich Rudolf Hertz (1857–1894), who is the man credited with the discovery of radio waves.

2.2.2 Bandwidth of Digital Facilities?

Before we go too far, it is important to clear up some sloppy terminology. Bandwidth is regularly used (or misused) to refer to digital transmission capacity. That is a rather egregious misuse of the term. The transmission rate of a digital is measured by the number of bits that can be sent in a period of time (i.e., *bits per second*). There is a relationship between bandwidth and bits per second, but we will get to that in a moment. In the meantime, we will be using the term *bandwidth* in its original analog meaning, and digital transmission capacities will always be referenced in bits per second.

As it defines the *range of frequencies* that can be sent over a channel, bandwidth is simply the difference between the highest frequency and the lowest frequency carried. A single 802.11b transmission channel that supports a digital transmission rate up to 11 Mbps requires a bandwidth of around 22 MHz (i.e., 22 million cycles per second). An 802.11g transmission channel can carry a data rate of 54 Mbps, but it requires only a 20 MHz channel. That is not a mistake; even though the 802.11g channel is about 10 percent smaller (i.e., 20 MHz versus 22 MHz), it can carry a data rate almost five times as great. The difference is that 802.11g devices use a much more efficient signal encoding system.

2.2.3 Shannon's Law: Bandwidth and Noise

A fellow named Claude Shannon spelled out the basic limits of transmission mathematically in the 1940s in what is called Shannon's Law. Shannon, who worked at Bell Laboratories, developed a formula that allows us to compute the maximum transmission capacity of any transmission channel; that capacity is measured in bits per second. The formula describes the relationship of the three factors that define the maximum transmission capacity of a communications channel: bandwidth, received signal power, and noise.

Shannon's formula is written:

$$C = W \log_2(1 + S/N)$$

where

C is capacity measured in bits per second,

W is the bandwidth measured in Hz

S is the received signal power

N is the power of the noise or interference seen at the receiver

The combined term "S/N" is the signal-to-noise ratio (measured in dB)

If you would like to download a copy of Shannon's original 1948 paper, go to: http://cm.bell-labs.com/cm/ms/what/shannonday/paper.html.

2.2.4 Relationships

Without getting into the math, Shannon's Law defines a few very important relationships (see Figure 2-2):

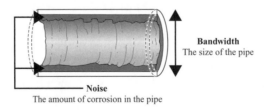

Figure 2-2: Shannon's Law

1. **Bandwidth (W):** The greater the bandwidth of the channel, the more bits per second (C) it will be able to carry.

2. **Signal-to-Noise Ratio (S/N):** The more noise that is present (relative to the receive signal power), the fewer bits per second a given size channel will be able to carry without creating errors.

3. **Maximum Transmission Capacity:** There is a maximum transmission capacity for any channel, and that capacity can be computed mathematically if we know:
   a. The bandwidth of the channel,
   b. The power of the received signal, and
   c. The level of noise that will be present.

2.3 Bandwidth Efficiency

Now that we have our terminology straight, we can put it to some practical use. What defines the digital transmission capacity of a channel (radio or otherwise) is:

- The bandwidth of the channel (i.e., the range of frequencies it will carry)

- The bandwidth efficiency or the number of bits we can carry on each cycle of available bandwidth

Bandwidth efficiency is the number of bits that can be carried on one cycle of bandwidth; bandwidth efficiency is measured in *bits per second per Hertz* (bps/Hz). For a channel with a bandwidth of 10 MHz and an encoding system with a bandwidth efficiency of 2 bits per second per Hz, the transmission rate will be 20 Mbps (i.e., 10 MHz × 2 bps/Hz). For example, an 802.11b interface uses a 22 MHz channel and carries a maximum data rate of 11 Mbps. That works out to a bandwidth efficiency of 0.5 bps/Hz (i.e., 11 Mbps ÷ 22 MHz). An 802.11g interface that uses a 20 MHz channel and carries a maximum data rate of 54 Mbps provides a bandwidth efficiency of 2.7 bps/Hz (i.e., 54 Mbps ÷ 20 MHz).

The challenge to improving factor to bandwidth efficiency is the amount of noise or interference relative to the received signal power. A stronger receive signal can be interpreted even in the presence of significant levels of noise. However, if the receive signal power decreases and the background noise remains constant, it will be more difficult to read the receive signal accurately enough to decode it. Hence, the measure of noise we use is the Signal-to-Noise (S/N) ratio.

The important thing to know about S/N ratio is that bigger numbers are better! A ratio is a fraction, and you want more "good" signal on the top and less "bad" noise on the bottom—either one of those factors affects the result. You really can't talk about signal power without noise and vice versa. For example, a WLAN client does not attempt to associate with the access point whose signal is strongest, but rather the one with the best S/N ratio.

A higher S/N ratio is how to describe a better channel, and a better channel can provide greater bandwidth efficiency. In order to encode information efficiently (i.e., more bps/Hz), the transmitter will have to generate a very complex analog signal. Inevitably, that complex waveform will be less robust, that is, less capable of operating error-free on a noisy channel. Carrying more bits on one cycle of bandwidth means a requirement for an encoding system that includes more and subtler distinctions the receiver will have to discern. Even minor distortions introduced in the channel will cause the receiver to make mistakes. The designers can make the signal more robust, but that means the signal will carry less information (i.e., fewer bits per second).

You can get an understanding of the principles by thinking of the problem of speaking to someone in a noisy room. Even if you are standing next to someone, you have to raise your voice in order to be heard (i.e., increased signal power). If you are standing on the other side of the room, you will have to shout in order to be heard (i.e., a poorer signal-to-noise ratio). When you are shouting from the other side of the room, you will also have to slow down your rate of speech in order for the other party to understand what you are saying (i.e., poorer bandwidth efficiency). See Figure 2-3.

Figure 2-3: Efficiency Versus Robustness

Not all encoding systems are created equal in terms of efficiency and robustness, and engineers are constantly working to develop more efficient coding systems that will continue to work without error in the presence of higher levels of noise. The reason that an 802.11g radio link is more efficient than an 802.11b radio link is that 802.11g uses a coding system called Orthogonal Frequency Division Multiplexing (OFDM). WLAN designers found that 802.11b was particularly susceptible to an impairment called *multipath*, which is very pronounced in indoor environments. They designed OFDM to mitigate that multipath effect.

2.3.1 Adaptive Modulation

Radio designers recognize that the radio link is not a perfect transmission path, and that received signal power and noise can fluctuate. To address the changing nature of the radio path, WLANs use a technique called *adaptive modulation*. Adaptive modulation means that the transmitter will reduce its transmission rate when it encounters a degraded channel. In poorer channel conditions, the bandwidth of the signal remains the same, but the transmitter shifts to a more robust, though lower bit rate, transmission technique.

2.3.2 Efficiency Tradeoff

The bottom line is that you can't have it both ways. A transmission system will be either *efficient* (i.e., carry more bits per second per Hertz) or *robust* (i.e., be able to operate error-free over an impaired channel). Adaptive modulation allows a device to make the best use of the channel that is available at that moment. The bandwidth efficiency is a function of noise, and engineers are continuously improving the robustness of transmitters operating over noisy channels.

2.4 Forward Error Correction (FEC)

The other factor that can affect the efficiency continuum is forward error correction (FEC). The idea of an FEC system is to combine redundant information with the transmission, and then allow the receiver to detect and correct a certain percentage of the errors using a probability technique. The addition of FEC coding increases the number of bits that must be transmitted over the channel, which in turn requires a more efficient (and hence "less reliable") transmission system. However, that is more than offset by the error performance gain provided by the FEC. That performance gain is typically expressed as the equivalent improvement in the signal-to-noise ratio that results. Both 802.11a and 802.11g radio link interfaces use FEC coding; the original 802.11 and 802.11b radio links do not. See Figure 2-4.

2.4.1 802.11a and 802.11g Convolutional Coding

There are a number of different FEC techniques that can be employed, and the 802.11a and 802.11g wireless LANs use a system called *convolutional coding*. Convolutional

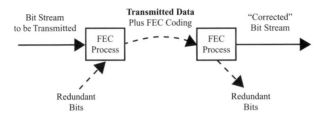

Figure 2-4: Forward Error Correction

coding works by taking the transmitted bit sequence, selecting a group of the most recently occurring bits, distributing them into two or more sets, and performing a mathematical operation (convolution) to generate one coded bit from each set as the output. The decoder in the receiver runs a comparison of the coded bit sequence and retains the best matches. With that information, the receive FEC process can make a maximum likelihood estimation of the correct decoded bit sequence.

2.4.2 FEC Overhead: 3/4 or 1/2 Coding

The coding rate of an FEC system is expressed as a fraction that represents the number of uncoded bits input at the FEC encoder and the number of coded bits output to the transmitter. The 802.11a and 802.11g interfaces use either 3/4 (where each 3 uncoded bits are sent as 4 coded bits—33 percent overhead) or 1/2 (where 1 uncoded bit is sent as 2 coded bits—100 percent overhead).

2.4.3 Other FEC Techniques

There are also more complicated FEC techniques like Trellis, Reed-Solomon, Viterbi, and Turbo Coding, however, none of these are specified in the wireless LAN standards. In each case, the basic issues are the amount of additional information that must be sent, and the resulting ability to correct errors. The result of FEC coding is that systems are now approaching their theoretical maximum capacities as defined by Shannon's Law.

2.5 Radio Regulation

The first design issue to recognize with regard to wireless LANs is that they operate over radio channels, and there are regulatory issues that govern the use of the radio spectrum. When radio transmitters first produced in the early Twentieth Century, many of the people experimenting with radio were amateurs (a harbinger of Ham radio). Often those amateur radio transmissions interfered with government and maritime users. The government recognized the need to protect important radio transmissions like ship-to-shore and distress calls, and the first regulations for the use of the US radio spectrum was drafted in "An Act to Regulate Radio Communication" adopted August 13, 1912. It limited amateur users to a wavelength of 200 m (i.e., 1.5 MHz).

2.5.1 Federal Communications Commission (1934)

In the early years, the major applications for radio were maritime communications, and so, responsibility for enforcement of United States radio laws was assigned to the Commerce Department's Bureau of Navigation. When the Federal Communications Commission (FCC) was created with the Communications Act of 1934, that responsibility was transferred to the FCC. Among the responsibilities of the FCC are the allocation and management of the radio spectrum in the US, the granting of licenses for radio transmitters, and coordination with international regulatory bodies.

2.5.2 International Telecommunications Union—Radio (ITU-R)

Following the lead of the US, most governments around the world established regulatory authorities to address radio technology. In 1927, the Consultative Committee on International Radio (CCIR) was established at a conference in Washington, DC to coordinate the activities of those regulatory bodies. The CCIR was given responsibility for coordinating the technical studies, tests, and measurements and for drawing up international radio standards. The CCIR was organized under the International Telecommunications Union, a branch of the United Nations. In 1992, the CCIR's name was changed to The International Telecommunications Union—Radio (ITU-R). The ITU organizes a World Administrative Radio Conference (WARC) periodically to address major issues on an international basis.

2.6 Licensed Versus Unlicensed Radio Spectrum

Allocation of the radio spectrum is done on a country-by-country basis, and hence the spectrum available for wireless LANs will vary based on the jurisdiction in which they are installed. The process of allocating the radio spectrum involves the balance of competing claims. The government often lays claim to significant swaths of spectrum for various military and public safety applications. Radio and television broadcasters are allocated spectrum, but in return must adhere to decency requirements and provide other services like emergency broadcasting capabilities. Cellular telephone companies must buy licenses for the radio channels they use in the various markets where they provide service. There are also frequency bands that are unlicensed (also called "license exempt") that are available to all users so long as they adhere to certain operating requirements.

To ease the development of radio products that can be sold around the world, the ITU-R attempts to coordinate spectrum allocation and regulation. However, the ITU-R cannot order countries to comply with its recommendations; it can only work to coordinate the activities of the various regulatory agencies. Ideally manufacturers would like to have one set of allocations and operating regulations throughout the world rather than having to tailor their products for each regulatory jurisdiction.

2.6.1 Major Licensed Frequency Bands

At the most basic level, regulatory bodies allocate radio spectrum into licensed and unlicensed bands. In licensed bands, one organization is given exclusive use of a specific range of radio spectrum in a defined geographic region and the use of that spectrum is governed by regulations designed to limit the interference between different license holders. Typically, those licenses are sold at auctions (in the US, the FCC conducts those auctions), and radio licenses in major markets sell for tens of millions of dollars. The major US licensed frequency bands are listed in Table 2-1.

Table 2-1: Examples of US Licensed Radio Bands

Frequency Band	Service
535 KHz to 1.705 MHz	AM Broadcast Radio
88–108 MHz	FM Broadcast Radio
54–216 MHz (Non-contiguous)	VHF Broadcast TV (Channels 2 to 13)
824–849 MHz and 869–894 MHz	AMPS Cellular Telephone
1.850–1.910 GHz and 1.930–1.990 GHz	PCS Cellular Telephone
1.710–1.755 GHz and 2.110–2.155 GHz	Cellular Advanced Wireless Service (AWS)
2.31–2.36 GHz	Satellite Digital Audio Radio Service
2.495–2.690 GHz	Broadband Radio Service (Licensed WiMAX)
76–77 GHz	Vehicular Radar

2.6.2 Unlicensed Spectrum in the US—The ISM and U-NII Bands

The regulatory bodies also leave swaths of spectrum unlicensed and allow anyone to build products that generate radio signals in those bands. Unlicensed spectrum is available for free to all users, so long as those devices limit their transmission power to specific limits and adhere to other published requirements. As a result, transmissions in the unlicensed bands are subject to interference from other users. In the US, the FCC has allocated three unlicensed bands. Initially they allocated 26 MHz in the 900 MHz band that was used for applications ranging from cordless phones to garage door openers. In 1985, they allocated an additional 83.5 MHz of unlicensed radio spectrum in the 2.4 GHz band designated Industrial, Scientific, and Medical (ISM) band. In 1997, they opened another 300 MHz of unlicensed spectrum in the 5 GHz band designated the Unlicensed National Information Infrastructure (U-NII) band. The 5 GHz band was allocated an additional 255 MHz in November 2003, bringing the total to 555 MHz. See Table 2-2.

2.6.3 Wireless LAN Spectrum

IEEE 802.11 wireless LANs operate in the ISM and U-NII bands generally referred to as the 2.4 GHz and 5 GHz bands. As these radio bands are unlicensed, there is no protection from interference among different products operating at those frequencies. As a result, WLANs could encounter interference from cordless phones, Bluetooth devices, baby monitors, and even microwave ovens. In practice, what we have found is that the greatest source of interference is other wireless LANs. For the moment, the 5 GHz band suffers less interference as far more products that have been developed for the 2.4 GHz band; however, nothing says that advantage will continue over time.

Table 2-2: US Unlicensed Radio Bands

Frequency Range	Name	Bandwidth (MHz)	Non-Interfering WLAN Channels
902–928 MHz	Industrial, Scientific Medical (ISM)	26	N/A
2,400–2,483.5 MHz	Industrial, Scientific Medical (ISM)	83.5	3
5150–5850 MHz (Non-continuous)	Unlicensed National Information Infrastructure (U-NII)	555	23

**Table 2-3: Comparison of US and International
Allocation in 2.4 GHz Band**

Country	Band Allocated (MHz)	Non-Interfering Channels
US and Canada	2400.0–2483.5	3
Japan	2400.0–2497.0	4
France	2446.5–2483.5	2
Spain	2445.0–2483.5	2
Rest of Europe	2400.0–2483.5	3

2.7 Unlicensed Spectrum in the Rest of the World

Not all countries have allocated the same bands for unlicensed operation, which makes it difficult for manufacturers to build wireless LAN products that can be sold worldwide. WLAN products sold in each country must restrict their operation to the bands that are allocated, and there may be additional regulations regarding their technical specifications. Depending on the range of frequencies that have been allocated by the particular regulatory authority, there may be more or fewer channels available. See Table 2-3.

2.7.1 European Union 5 GHz Spectrum

Allocation and operating rules for the 2.4 GHz band are fairly standard around the world, but the same cannot be said of the 5 GHz spectrum. Due to concerns regarding interference with military radars, regulators in the European Union have imposed additional requirements for transmissions in the 5 GHz band; those requirements are called Transmit Power Control (TPC) and Dynamic Frequency Selection (DFS). The European Telecommunications Standards Institute (ETSI) requires that 802.11a LAN products built for use in Europe conform to these specifications; those requirements are addressed in an extension to the 802.11 standard called 802.11h.

2.7.2 International 5 GHz Spectrum

As it was allocated more recently, there is also more variety in the allocations at 5 GHz. The EU regulatory body agreed to a standard 5 GHz implementation for all member states using the bands 5.159–5.735 GHz and 5.470–5.725 GHz. At the moment, some countries do not allow any Wi-Fi networking in the 5 GHz band. That list includes

Table 2-4: International 5 GHz Assignment

Frequency Band (GHz)	US	Europe	Japan
4.920–4.980			√
5.040–5.080			√
5.150–5.250	√		√
5.250–5.350	√		
5.470–5.725	√	√	
5.725–5.825	√		

Israel, Kuwait, Lebanon, Morocco, Thailand, Romania, Russia, and the United Arab Emirates (UAE). Ecuador, Peru, Uruguay, and Venezuela allow 5 GHz networking only in the upper band (i.e., 5.725 GHz to 5.825/5.850 GHz), providing four or five 802.11a channels. The World Radio Conference is moving to increase and standardize the 5 GHz band on a global basis. See Table 2-4.

2.8 General Difficulties in Wireless

The biggest issues in a wireless WAN are the quality and capacity of the transmission channel, as all users in that area will be operating on the same channel. Radio signals lose power more rapidly and are subject to more interference than anything we typically deal with in communications over physical media (e.g., copper wires, coaxial, or fiber optic cables). As we noted earlier, the key parameter that determines the transmission rate on a wireless LAN is the signal-to-noise (S/N) ratio. Distance, obstructions, and multipath effects reduce or attenuate the power of the radio signal while interference from other systems operating on the same channel increase the noise level. Further, higher frequency signals lose power at a faster rate than lower frequency signals.

The basic formula for computing radio path loss is:

$$P_r = P_t G_t G_r \left(\frac{\lambda}{4\pi}\right)^2 \frac{1}{d}^n$$

where:

P_r = receive signal power

P_t = transmit signal power

G_t and G_r = transmit and receive antenna gain

λ = wavelength of the signal (12.5 cm at 2.4 GHz, 5.5 cm at 5 GHz)

d = the distance in meters

n = the path loss exponent

The key variable in the path loss formula is the path loss exponent (n). This is where we account for variations in the environment that will affect signal loss. Free space signal loss is fairly predictable and is represented by a path loss exponent of 2. Radio propagation in indoor environments is far less predictable and varies widely based on the composition and density of walls, furniture, and other contents. The path loss exponent in open office environments may be close to free space, but could be as high as 6.

Among the major factors to be considered are distance and path loss, signal frequency, environmental obstacles, propagation between floors, multipath and inter-symbol interference, co-channel interference, the Doppler effect, and other environmental factors.

2.8.1 Distance and Path Loss

In free space, radio waves obey the inverse-square law, which states that the power density of the wave is proportional to the inverse of the square of r (where r is the distance, or radius, from the source). That means that when you double the distance from a transmitter, the power density of the received signal is reduced to one-fourth of its previous value.

2.8.2 Signal Frequency

Signal frequency is also a factor, and the loss increases as the frequency of the signal increases (i.e., a 5 GHz transmission will experience roughly twice the signal loss of a 2.4 GHz transmission traveling over the same distance). The maximum transmit power level of a radio operating in an unlicensed band is defined by regulation.

2.8.3 Environmental Obstacles

A radio signal traveling through free space will lose power based on properties that are well understood. However, you cannot accurately estimate signal loss in indoor environments by simply measuring distances. Radio signals lose differing amounts of power as they pass through different materials; those losses are greater at 5 GHz than at 2.4 GHz. Metal attenuates a radio signal far more than wood, glass, or sheetrock; metal objects in close proximity to the transmitting antenna (i.e., within one wavelength) create a major impediment. In the basic path loss formula above, the impact of environmental obstacles is taken into account with the path loss exponent. A signal traveling through free space will have a path loss exponent of 2, but indoors the path loss exponent can vary from 2 to 6! So even though we have a formula for path loss, the path loss exponent introduces a major swag factor. The basic message is that while you can predict radio path loss in free space fairly accurately, you cannot count on a mathematically computed loss for indoor environments.

2.8.4 Propagation Between Floors

Many enterprise wireless LANs will have to be installed in multi-story commercial buildings, where floor-to-floor propagation becomes an issue. To a large extent, the level of signal attenuation between floors will depend on the building materials used. The steel plank construction used in older buildings creates far more attenuation than the reinforced concrete construction used today. Interestingly, signal attenuation does not increase in a linear fashion with the number of floors. (See Table 2-5.) The greatest attenuation occurs with the first floor and then diminishes. This phenomenon is caused by the radio signal's diffracting and scattering along the sides of the building.

Table 2-5: In-Building Floor-to-Floor Attenuation

Separation	Additional Attenuation
One Floor	15 dB
Two to Four Floors	6 to 10 dB per Floor
Five or More Floors	2 to 3 dB per Floor

2.8.5 Multipath and Inter-Symbol Interference

Radio signals do not travel in a direct path from the transmitter to the receiver. Rather, they reflect, diffract, and scatter when they encounter material obstructions. On the positive side that means that with radio signals below 6 GHz you do not require a line of sight between the transmitter and the receiver in order to communicate. At higher frequencies, radio signals tend to travel in a straight line so the antennas must be aligned with a line of sight path between them. On the negative side, the echoes of the signal created by the reflection and scattering creates two major forms of interference, multipath and inter-symbol interference (ISI). Multipath is created as the radio waves bounce off flat, solid objects in the environment with the result that the receiver may detect the original signal as well as echoes of the signal that arrive microseconds later. As those echoes will be at the same frequency but out of phase with the original, the signals and echoes can effectively cancel each other out. This phenomenon can even affect devices located in close proximity to the base station. You will find that you can sometimes improve the received signal strength significantly by moving a device only a few inches one way or the other thereby taking it out of that phase cancellation zone. See Figure 2-5.

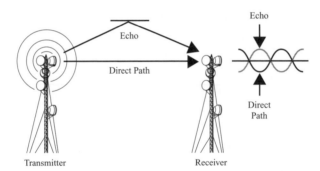

Figure 2-5: Multipath Cancellation

Transmission echoes are delayed for different periods based on the distance the various signals must travel from the transmitter to the receiver. The difference between the fastest and slowest signal image received is called the *delay spread*. As that delay increases it becomes more problematic because the echoes can interfere with bits that were sent on subsequent transmission symbols; that impairment is called inter-symbol interference. WLAN devices have different tolerances for ISI, as designers can include equalizers in the receiver to compensate for it.

2.8.6 Co-Channel Interference

In any cellular network, the same frequencies must be reused in different parts of the coverage area. As a general rule, frequency channels should not be reused in adjacent cells, as the transmission on that channel in Cell A will interfere with the party trying to use that same channel in adjacent Cell B. This is particularly problematic if you have only three channels to work with as we do in the 2.4 GHz band. As we will inevitably be reusing channels in different parts of the coverage area, if a clear path exists, then a transmission on the same channel in a non-adjacent cell might still reach a user in another part of the service area; that phenomenon is called *co-channel interference*. At a minimum, co-channel interference will degrade the S/N ratio causing affected stations to reduce their transmission rates. As cells are added to the network, we will have to reassign channels and adjust the transmit power in access points using the same channel to limit that co-channel interference. See Figure 2-6.

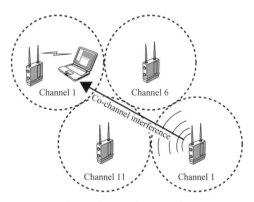

Figure 2-6: Co-Channel Interference (2.4 GHz Network)

2.8.7 The Doppler Effect

The Doppler effect can cause frequency shifts in the received signal if the mobile station transmitter is moving toward or away from the base station. The combined effects of multipath and Doppler can cause fading at particular frequencies. Fortunately, most wireless LAN users will be stationary or moving at relatively low speeds (i.e., ≤3 mph) so the Doppler effect will be minimal.

2.8.8 Other Environmental Factors

Radio signals cannot penetrate masses of dirt and rock, so if we are planning to provide outdoor coverage, buildings, hills, and other environmental obstacles will affect coverage. Water will also attenuate a radio signal. In providing outdoor coverage, water in leaves must be taken into account. If there are deciduous trees in the path, the amount of signal loss encountered will vary at different times of the year; that will affect both the coverage and the level of co-channel interference. The human body is approximately 60 percent water, and we will likely have a number of humans moving about in the coverage area.

2.9 Basic Characteristics of 802.11 Wireless LANs

Now that we have looked at the fundamental building blocks, we can make some general observations regarding the capabilities and limitations of wireless LANs. There are a few important characteristics shared by all of the 802.11 wireless LANs that distinguish them from typical wired implementations. Those characteristics have an impact on their ability to carry voice calls and on the number of simultaneous calls they will be able to support.

- **Half Duplex:** There is only one radio channel available in each wireless LAN, so only one device, the access point or a client station, can be sent at a time.

- **Shared Media:** Where each wired user is typically connected to his or her own port on the LAN Switch, a number of WLAN clients will typically associate with a single access point and share one radio channel. All stations associated with that access point will vie for access to the channel, so the time required to access the channel will be a random function. For voice services, that channel contention will affect transit delay and jitter.

- **Capacity/Range Relationship:** The maximum capacity of the WLAN is based on which radio link interface is used (i.e., 802.11a, b, g, n). However, not all stations will be able to operate at the defined maximum data rate, and each radio link interface defines a number of fallback rates. The data rate for each device is based on its distance from the access point, impairments in the radio path, and the level of interference on the radio channel. Lower transmission capacity reduces the number of simultaneous calls an access point can handle.

Table 2-6: IEEE 802.11 Radio Link Interfaces

Standard	Maximum Bit Rate (Mbps)	Fallback Rates	Channel Bandwidth	Non-Interfering Channels	Transmission Band	Licensed
802.11b	11	5.5, 2, or 1 Mbps	22 MHz	3	2.4 GHz	No
802.11g	54	48, 36, 24, 18, 12, 11, 9, 6, 5.5, 2, or 1 Mbps	20 MHz	3	2.4 GHz	No
802.11a	54	48, 36, 24, 18, 12, 9, or 6 Mbps	20 MHz	23	5 GHz	No
802.11n (Draft)	≤289 (at 20 MHz) ≤600 (at 40 MHz)	Down to 6.5 Mbps (at 20 MHz)	20 or 40 MHz	2 in 2.4 GHz 11 in 5 GHz (40 MHz)	2.4 or 5 GHz	No

- **Different Transmission Rates, Same Channel:** In a wireless LAN, different users can operate at different bit rates on the same, shared channel. Those lower bit rate users affect the performance of higher bit rate users, as the low bit rate users will keep the channel busy longer as they send a message.

- **Transmission Acknowledgments:** As WLAN stations cannot "hear" while they are sending, the recipient must return an acknowledgment for every correctly received frame. That acknowledgment process reduces the effective transmission capacity or *throughput* of the WLAN channel.

- **Protocol Overhead:** While the transmission rate over the wireless channel may be 11 or 54 Mbps, the requirement to send acknowledgments and other overhead features in the protocol reduce the throughput to about 50 percent of that rate. See Table 2-6.

2.10 Conclusion

We have now taken a first serious look at the technology that drives WLANs. Shannon's Law and the basic concepts of bandwidth, signal power, and noise provide the foundations. However, as WLANs use unlicensed radio channels, there is no protection from interference. As time goes on there is a concern that the unlicensed radio bands will become so overcrowded that the spectrum will become essentially useless. Before we reach that point, radio technology should advance to the point that the radio devices will attempt to avoid one another, reduce their transmit power, or take other steps to collaboratively share the available spectrum. In the meantime, with each new generation of WLAN technology, the engineers are incorporating techniques that will allow radio devices to operate at higher bit rates and greater reliability over these unlicensed radio channels.

The basic transmission channel is only the first step in understanding wireless LANs however, as we must also look at the range of impairments that will affect all radio transmissions. Further, the design of the 802.11 protocols defines the overall operating environment, which will affect the effective throughput and other parameters of the network performance.

We also introduced the general operating principles and characteristics for WLANs. In the next few chapters, we will look at the different parts of the wireless LAN standards in greater detail, and introduce the basic mechanisms used to plan, secure, and maintain that infrastructure.

CHAPTER 3

Wireless LAN Components/WLAN Switches

The task of building a WLAN for voice or data applications involves specifying requirements, developing an initial design, purchasing and installing components, tuning the installation, and providing the ongoing systems for management and maintenance. In this chapter we are going to look at the major elements needed to build that WLAN infrastructure.

A wireless LAN for home or small business use is a relatively simple configuration. The access point is built into the DSL/cable modem router, wireless network interface cards (NICs) are needed for any devices that will connect wirelessly, and there are some relatively simple setup procedures. Those procedures are "simple" for people with a networking background, but most non-technical users can't seem to figure out how to turn on the security or encryption features!

Commercial installations present a far greater challenge. Rather than one access point, we might need hundreds to cover a large building or campus. Security and privacy are mandatory requirements, and those systems must interoperate with the organization's overall security architecture. Further, we will need tools and systems that allow us to maintain and troubleshoot this complex infrastructure.

The requirement to support voice is a critical differentiator in WLANs. Where it might be adequate to provide data services in selected areas (e.g., conference rooms and public spaces) voice support requires pervasive coverage throughout the facility. If there is not sufficient capacity in a data network, users will get poor performance, but they will still be able to connect. Poor coverage in a voice network means degraded voice quality, disconnects, or the inability to make or receive calls. Finally, data devices are typically stationary. Based on their experience with cell phones, users expect mobility with a voice service, so the network will have to be able to hand off calls from access point to access point as a user moves through the coverage area. We also have to be sure there is adequate capacity in the cell the user is moving to or the call will disconnect.

In this section we will describe the major elements that are used to build a wireless LAN. Those will include network interface cards, access points, and possible external antennas. The major development in enterprise wireless LANs are networks of centrally controlled access points called WLAN switches. These systems provide the configuration, management, and support systems required to build a large-scale enterprise grade WLAN infrastructure. It is possible to build large-scale networks without one of these centrally controlled systems, but it will be far more complex and will probably not provide the type of reliable, pervasive coverage and capacity needed to support high quality voice services.

3.1 Elements in a Wireless LAN

The WLANs we will be using for voice use the Basic Service Set (BSS) or Extended Services Set (ESS) configurations were described in the first chapter. A BSS network is composed of a base station or access point and a number of client devices. An ESS includes multiple access points that are interconnected by a wired LAN. The operation of those multiple access points might be coordinated by a central controller in a configuration called a wireless LAN switch. We will look at those WLAN switches more closely in a moment.

There are three major elements in this type of wireless LAN:

1. Wireless LAN NICs

2. Access Points (AP)—Possibly connected to a central controller

3. Antenna Systems

3.2 Wireless LAN NICs

To operate on a wireless LAN, each user device (PC, PDA, voice handset, etc.) must have a wireless Network Interface Card (NIC). Most laptop computers sold today come with a Wi-Fi card standard, however consumers have to ensure that the card that is provided has the features needed to support the overall design. If the device selected does not include the required features, it will be important to determine whether those additional capabilities can be added with a firmware upgrade. As a general rule, protocol features like 802.11e QoS can be upgraded with firmware, while radio links (e.g., 802.11n) and encryption functions (e.g., 802.11i/WPA2) require hardware upgrades. In all cases, consumers should be buying equipment that is *Wi-Fi Certified* to ensure that it will interoperate with any other devices on the network.

Here are the major features to be concerned with for Wi-Fi voice and data devices:

- **Radio Link:** Wi-Fi laptop cards typically support the 2.4 GHz 802.11b, g interfaces, though most early Wi-Fi voice handsets supported only the lower speed 802.11b interface. The second generation Wi-Fi handsets introduced in 2007 support the 802.11a, b, and g radio links. As the 2.4 GHz channels (i.e., 802.11b and g) become more congested, support for the 5 GHz 802.11a interface will grow in importance. The 802.11n standard has not yet been finalized, however in mid-2007, the Wi-Fi Alliance took the unprecedented step of developing a certification plan for devices built to the 802.11n draft specification; the Alliance claims that the devices they certify will be backwards compatible with the final standard when it is ratified. Normally enterprise buyers will avoid nonstandard devices, but with WLAN equipment, the assurance of interoperability comes not from adherence to the IEEE standard, but from the Wi-Fi Alliance's certification. For the moment this is a moot point, as there are no Wi-Fi voice handsets that support any version of 802.11n.

- **802.11e Quality of Service (QoS):** All Wi-Fi devices will support the standard 802.11 MAC, but if you intend to support voice and data devices on the same WLAN, you will want to use the 802.11e QoS extensions to prioritize the voice access. Pre-802.11e voice devices can coexist on the same network, however their transmissions will have a lower priority for channel access, which will lead to increased delay and jitter. Those legacy voice devices should be upgradeable to 802.11e with a firmware download. Given the limited memory available in Wi-Fi handsets, some legacy devices may not be upgradeable. This is a capability any buyer will want to confirm with the vendor.

- **Security:** Security has been a persistent concern with Wi-Fi devices given the rather meager protection provided by the original Wired Equivalent Privacy (WEP) encryption. Today the minimum security option for an enterprise network is Wi-Fi Protected Access (WPA) and most voice and data devices do support it. The preferred solution would be encryption based on the 802.11i standard, which the Wi-Fi Alliance refers to as WPA2 Certified. The 802.11i/WPA2 security uses the more powerful Advanced Encryption Standard (AES). Virtually all of the second-generation Wi-Fi handsets support WPA2.

3.2.1 Proprietary Extensions—Cisco Compatibility Extensions

Typically any Wi-Fi Certified WLAN NIC can interoperate with any access point or any other WLAN card (i.e., peer-to-peer). Some vendors, notable Cisco, have included additional functions in their wireless NICs that operate in conjunction with their own APs. Cisco's program is called Cisco Compatibility Extensions (CCX), and Cisco does offer the technical specifications to other vendors who can choose to implement them. The features deal with Cisco-defined security mechanisms, location capability, network management functions, and the ability to control transmit power in the client. Cisco NICs will still work with other vendors' APs, however, without the additional functions. To confirm whether a particular device supports CCX, visit: www.cisco.com/go/ciscocompatible/wireless.

3.3 Access Points (APs)

The 802.11 Basic Service Set and Extended Service Set configurations require an Access Point (AP) that acts as the base station for the wireless LAN. The AP includes a radio transceiver, the logic to generate the wireless LAN protocol, and an interface to a wired LAN (e.g., 10/100 BaseT); access points that support the higher speed 802.11n radio link will require a 1 Gbps wired interface (i.e., 1000 BaseT). Most APs support a single WLAN radio interface, but commercial models may support two independent WLANs and can use any of the standard radio link interfaces (i.e., 802.11b/g or 802.11a). Commercial models will also support multiple WLANs on the same unit, multiple wireless VLANs, and the full range of security protocols.

3.3.1 Four Major Types of APs

The four major types of APs are consumer, stand-alone enterprise, centrally controlled (or "thin"), and mesh capable.

- **Consumer:** These are the low-cost (e.g., $69) devices from Cisco/Linksys, D-Link, and Netgear that are used in consumer and small business installations and are typically packaged with a DSL/cable modem router. There is also free, third party firmware for the Linksys models, though it is recommended that you have a spare router available before you start altering the firmware. To give it a try, Google "Linksys Firmware" or try the Web site http://www.dd-wrt.com.

**Figure 3-1: Cisco Aironet 1200 Series Stand-Alone Access Points.
Courtesy of Cisco Systems, Inc. Unauthorized use not permitted.**

- **Stand-Alone Enterprise:** Cisco's Aironet 1100, 1200, and 1300 products
  represent the majority of this market, though there are others. Cisco's widely
  used Aironet 1200 Series stand-alone access points are pictured in Figure 3-1.
  Commercial access points include a number of important features that
  distinguish them from consumer models:
  - Adjustable Transmit Power: The ability to adjust the transmit power setting
    is very important in limiting co-channel interference in multi-cell
    installations. Typically that adjustment can be made remotely from a control
    station.
  - Multiple Radios: The ability to support 802.11b, g and 802.11a from the
    same unit. As there are typically only two antenna leads on the unit, one
    will be used for the 2.4 GHz network and one for the 5 GHz network.
  - Support for Multiple WLANs: The ability to support multiple WLANs from
    a single unit. This feature can be used to deploy separate WLANs for voice
    and data devices in the same area.
  - Support for Wireless VLANs: Wireless VLANs allow you to support
    different groups of users on the same shared radio channel. With this
    feature, the AP will support two different network names or SSIDs, and
    clients will associate with one or the other. This feature is often used to
    support internal users and visitor services while maintaining separate
    security policies for each. It can also be used to segregate legacy devices

that do not support the full range of security features without sacrificing the more advanced security options available on newer devices.

○ External Antennas: Access points typically come with two omnidirectional antennas. The AP determines which antenna to use in communicating with each client device based on the received signal strength; that feature is called *antenna diversity*. If we need to provide better coverage in a particular area, we can disconnect one of those antennas and attach an external antenna with a cable to cover the problem area.

○ Power over Ethernet (802.3af) Support: Access points are typically installed at ceiling height, and there are no electrical receptacles in the ceiling. Power over Ethernet (PoE) allows the AP to receive electrical power over the wired LAN interface, eliminating the need to install a separate power receptacle with each access point. That can mean a major cost savings in the deployment.

○ Network Management: Network management capabilities to record numbers of users, traffic loads, and other operational parameters.

- **Centrally Controlled or "Thin":** These are limited function commercial access points that can work only in conjunction with a central controller in a WLAN switch. The access point unit may contain little more than a radio, and the association, security, and management functions are provided by the central controller. The protocol for communicating with the central controller is proprietary, so the thin APs and the controller must be purchased from the same vendor. There are control link standards in development (e.g., LWAPP, CAPWAP, SLAPP), but multi-vendor WLAN switch configurations are at best a few years off. We will look more closely at WLAN switches later in this chapter.

- **Mesh Capable:** These are a special class of APs that use the standard 802.11 protocol to communicate with client devices, but employ a proprietary backhaul protocol to relay messages access point-to-access point to a central unit that is connected to a wired connection (e.g., an Internet access). These models cannot be used interchangeably between vendors nor can they be substituted for traditional APs. Over 100 US municipalities are deploying mesh-based Wi-Fi networks that cover entire metropolitan areas. While the idea seems to be popular with local politicians, initial reports on the service have not been enthusiastic. It takes a considerable number of access points to cover each square mile (e.g., 40 or more) and indoor coverage is still poor. Whether Muni Wi-Fi develops into a profitable business model remains to be seen, but the initial reports describe poor coverage and significant unanticipated support costs.

3.4 Antennas

One important though confusing topic in radio is the function of the antenna. An antenna is any structure or device used to collect or radiate electromagnetic waves. An oscillating *electric* field from the radio's transmitter is fed to the antenna that creates an oscillating *electromagnetic* field that is radiated through space. The radio waves (i.e., electromagnetic waves) create an electric current in the receiving antenna, which is then sent to the radio receiver.

3.4.1 Antenna Gain

Gain and directionality are the two main considerations in antenna selection. An antenna is a passive device; that means the antenna does not generate radio energy, it simply focuses it. The only way an antenna can increase the power of a radio signal is to concentrate the power into a narrower beam. By constraining the signal propagation in some directions, the antenna can concentrate the power in other directions. The level of concentration that is accomplished is referred to as gain.

Radio engineers use a light bulb as an analogy for what antennas do. A light bulb throws roughly equal amounts of light in every direction. However, if you put that light bulb in a narrow-beam flashlight with a reflector behind it, you can concentrate the light in a specific direction. When you point that flashlight, you provide greater brightness in a particular area.

Antenna gain is expressed in decibels (dB) relative to an isotropic source (dBi). Decibels are a logarithmic power measurement widely used in telecommunications systems as they allow a wide range of power variation to be represented with a small range of numbers. For example, a power increase or decrease of 3 dB is equivalent to doubling or halving the power. In the Appendix we have the formula for converting between milliwatts and dBm, the two power references that are typically used for describing WLAN transmit levels.

One advantage of using decibels to represent power is that we can use addition rather than multiplication to compute different power levels. Hence, a WLAN radio that is generating a transmit power level of 20 mW or 13 dBm coupled to a typical omnidirectional antenna with 2.14 dBi gain will radiate radio power at 15.14 dBm (i.e., 13 dBm transmit power +2.14 dBi antenna gain = 15.14 dBm). That radiated power is the fundamental input to the formula to determine radio path loss.

3.4.2 Directionality/Coverage Pattern

There are two main types of antennas, omnidirectional and directional. One example of each along with their RF coverage patterns is shown in Figure 3-2.

- **Omnidirectional:** This is the type of antenna built into WLAN APs and clients. These antennas are simple stick-like structures, whose length is determined by the frequency of the radio signal to radiate. The optimal electrical length of the antenna is one-fourth of a wavelength. The wavelength of a 2.4 GHz signal is about 12.5 cm, and a 5 GHz wavelength is about 6 cm; that is why the antenna for a 5 GHz radio is shorter than the one for a 2.4 GHz radio. Despite the name, omnidirectional antennas do not radiate power equally in all directions. An omnidirectional antenna does not radiate or receive in the direction in which the rod points; this region is called the antenna blind cone. As a result, omnidirectional antennas radiate power equally in all directions in the horizontal plane (assuming the antenna is aligned vertically). An omnidirectional antenna constrains the vertical radiation and concentrates the power horizontally. Omnidirectional antennas typically provide about 2.2 dBi gain in the horizontal plane.

- **Panel/Directional:** There are antennas that not only reduce vertical radiation, but also focus the energy in a particular horizontal direction. These antennas, called patch, panel, or Yagi antennas, concentrate the signal to a specific horizontal sector. These types of antennas come in various shapes, and it is critically important that the radiating surface be pointed in the correct direction. In many cases, the isolation or loss is seen in directions where we do not want to propagate the signal. The specifications for the antenna will include a diagram illustrating the coverage pattern viewed from the Azimuth (top) and Elevation (side). As the coverage angle is reduced, the gain that is produced in the desired direction is far greater than with an omnidirectional antenna. The radiation pattern is typically 30° to 120° wide and pie-shaped; this type of antenna can typically provide about 13 dBi of gain. Directional antennas can be used to improve signal coverage in difficult areas like hallways.

3.4.3 Limiting Signal Coverage for Security?

Some users have tried to use directional antennas as a security measure. The idea is to limit the amount of radio leakage outside the facility to discourage hackers from attempting to infiltrate the WLAN. As little energy is emitted from the back of the antenna (i.e., backscatter), they think that by installing directional antennas at the perimeter of the coverage area pointed inwards, they can limit the signal leakage. In practice, they typically wind up broadcasting the signal out the other side of the building! Unless it is possible to surround the entire building with a Faraday cage, it would be better to look at other techniques to secure a wireless LAN.

3.5 Distributed Antenna Systems

The other approach to ensure good signal coverage throughout a facility is to install a distributed antenna system (DAS). These are specially designed RF distribution systems that are custom designed for a particular facility. A DAS includes a signal source and a number of antennas, cables, and amplifiers that distribute the radio signal throughout the facility. Distributed antenna systems are rarely used for wireless LANs. If you distribute the access point's signal more widely, that channel would then be shared by a far larger number of users. Distributing the radio signal increases the coverage area, but it does not increase the network's traffic capacity.

Distributed antenna systems can be used to improve indoor cellular coverage in large public facilities (e.g., train stations, hotels, airports, etc.). In that case, the signal source is often an outside antenna that captures the cellular carrier's signal, which is then distributed throughout the facility. A DAS can represent a significant investment (hundreds of thousands of dollars). In some cases, cellular carriers install these systems at their cost to improve coverage in the headquarters of a large customer; those systems will be designed to distribute only the signal for that cellular carrier. In other cases, a venue owner will install the DAS and rent access to it to the various cellular carriers; any carrier who wants to improve their coverage in that facility can connect to the system for a fee.

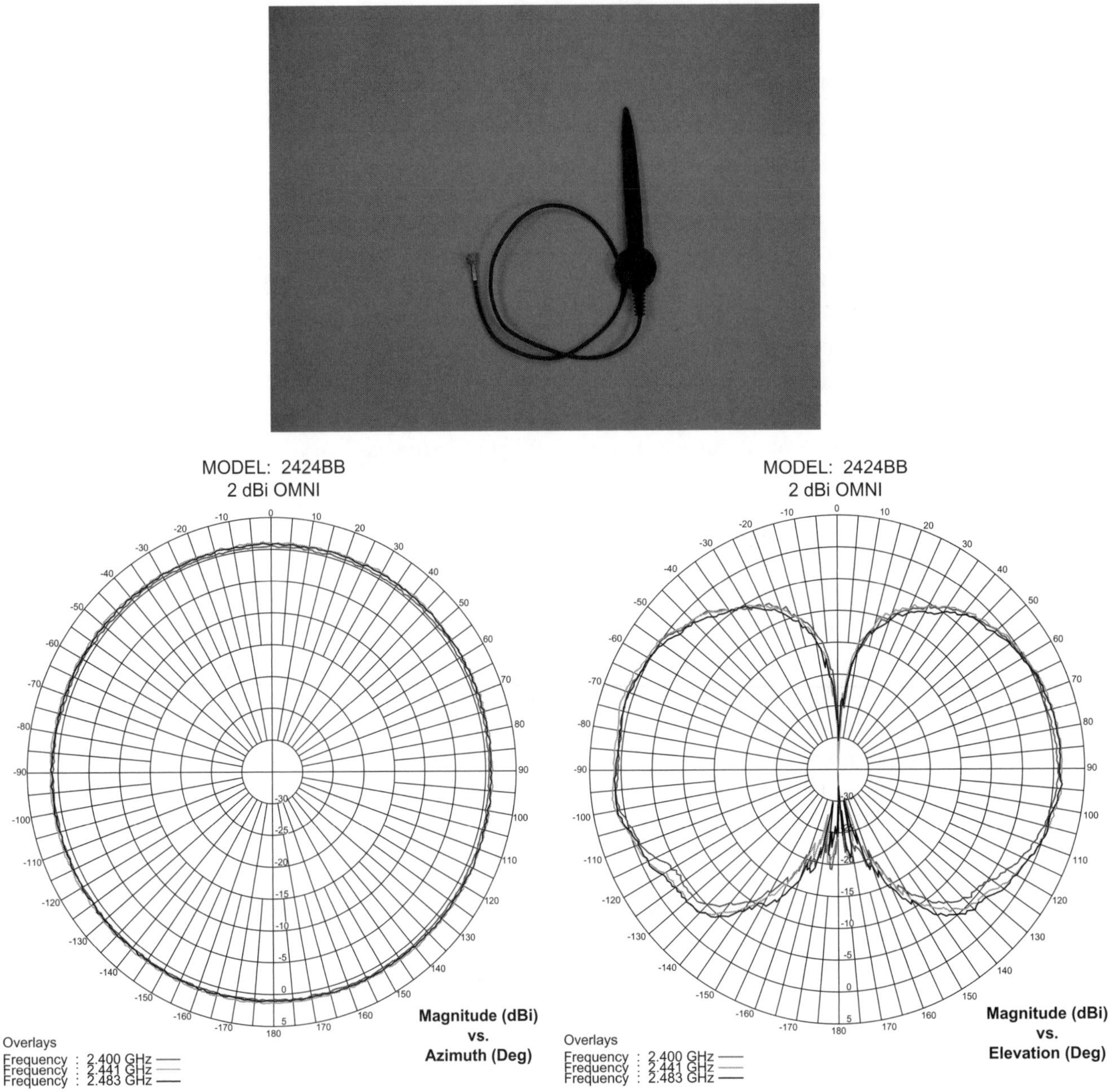

A

Figure 3-2: Antennas with Azimuth (Overhead) and Elevation (Side View) Coverage Patterns (a) Telex 2424BB 2 dBi Omnidirectional, and (b) 2443AA 12 dBi Outdoor Panel

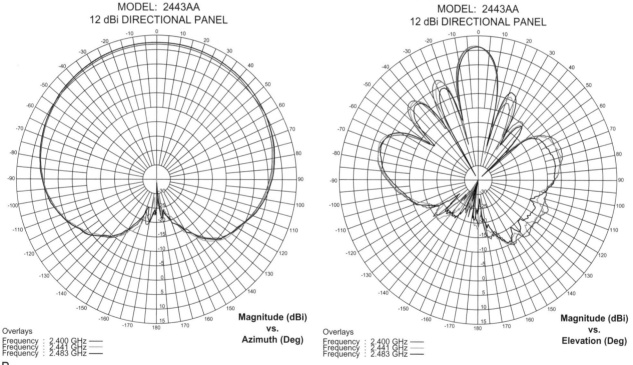

Figure 3-2 (*Continued*)

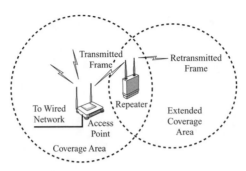

Figure 3-3: WLAN Repeater

3.6 WLAN Repeaters

Another way to extend the range of an access point is through the use of wireless repeaters. There are very few stand-alone 802.11 wireless repeaters on the market, but many access points can be optioned to operate in repeater mode. A radio repeater is a device that relays WLAN frames in order to extend the coverage range of an access point. A WLAN repeater does not physically connect by wire to any part of the network. Instead, it receives radio signals (i.e., 802.11 frames) from an access point or end user device and retransmits them on the same radio channel. A repeater is located between the access point and a distant user and can relay the frames traveling back and forth between them. See Figure 3-3.

Like external antennas, repeaters can provide connectivity to remote areas that cannot be reached by an access point connected to the wired network. The downside is that the repeater will reduce the throughput on the WLAN; every frame is sent twice over the radio link. Further, the relay function increases the transit delay, which is particularly important in voice applications. Given their impact on the network capacity, radio repeaters are not widely used in enterprise installations.

One new application for radio repeaters is improving indoor coverage in municipal Wi-Fi networks. As the operators have found problems in providing adequate signal coverage indoors, they now recommend installing a WLAN repeater near a window (where it will hear the signal from the outdoor access point) and then retransmit the frame for indoor devices.

3.7 Mesh Extension

The other type of radio repeater is a mesh extension. This solution is not intended to cover entire metropolitan areas, but rather, to extend WLAN coverage into areas where it is difficult to install wires (e.g., an outdoor storage lot). Unlike a radio repeater, which simply retransmits the frame on the same channel, a mesh device can have two radios: one for communicating with clients, and one for the backhaul transmission. Many WLAN switching systems support this capability and it is used to extend coverage outdoors or to areas where it is difficult to install LAN cabling.

3.8 Wireless LAN Switches

When you look at the process involved in setting up a network using stand-alone access points, you begin to recognize that this technology was developed for the requirements of small, residential applications rather than commercial installations. In a home network, only one AP and one radio channel are needed to cover the entire area. If there are coverage problems, they can generally be solved by relocating the AP or buying a better (i.e., higher gain) antenna for the access point.

Large-scale commercial customer installations require a wireless LAN infrastructure with multiple access points providing coverage in different parts of the facility. To build that type of WLAN, the customer must first determine where to locate the access points. Once the access points are in place, the customer must decide which channel to assign on each, while taking care not to assign the same channel in adjacent cells. When the channel layout is complete, the customer must determine the appropriate transmit power settings in each access point to ensure adequate coverage while minimizing co-channel interference.

Given that indoor radio propagation is difficult to predict precisely, after the preliminary installation is complete, the network will typically require additional "tuning." The tuning process involves relocating access points and adjusting power levels to ensure adequate signal coverage and network performance. As the network grows from a few access points to hundreds, the shortcomings of this manual, labor-intensive approach become painfully apparent. A subset of this process must be repeated every time a new cell is added to the network.

There are several problems associated with the traditional approach to WLAN design and implementation:

- **Network Layout:** The most obvious difficulty in deploying a large-scale network is the requirement of site planning to determine where APs should be placed, identifying the channels to use, and the optimal transmit power settings of each. With stand-alone access points, this process involves a time consuming process of trial and error that can takes weeks or months to resolve.

- **Security Exposure:** With stand-alone access points, the network name (called the System Services Identifier or SSID) and other security information is stored in the access point. A hacker who has access to the building could simply steal an access point, hack into it later, and discover much of the information needed to break into the network.

- **Roaming:** While there is a rudimentary handoff function defined in 802.11, it is slow (e.g., 5–10 seconds). While a 10 second handoff might meet the requirement of data users, it would certainly not support the real time requirements of a voice connection.

- **Detecting Rogue Access Points:** Even if the core wireless network is secure, the IT department must continually monitor the WLAN radio band to determine if any unauthorized "rogue" or "malicious" access points are present. They must also determine if there are user devices using unsecured Ad Hoc connections. Rogue APs are access points installed by users and connected to the wired LAN; as they are typically installed with no encryption activated, Rogue APs create a major security vulnerability. Malicious access points are installed by hackers in close proximity to the network as part of an attempt to gain unauthorized access. Traditionally, unauthorized networks were discovered by patrolling the area with a handheld test set, however this strategy was not effective in uncovering malicious APs or Ad Hoc networks that might be operating only sporadically. Comprehensive protection could only be provided by installing a continuous monitoring system from a company like AirMagnet, AirDefense, or AirTight Networks. Those systems included a number of sensors that would be installed throughout the facility that were connected through the wired LAN to a central controller.

3.9 Wireless LAN Switch Features

To address the requirements of large-scale enterprise wireless LAN installations, a number of established vendors and start-ups developed products called Wireless LAN

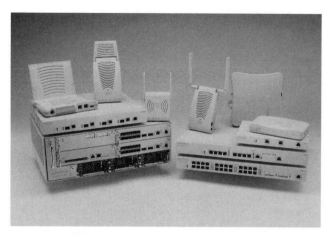

Figure 3-4: User-Centric Mobility Controllers and Access Points for the Aruba Networks WLAN Product Family. Copyright © 2007 Aruba Networks, Inc. Used by permission.

Switches. The essential concept of a WLAN switch is to use a central controller to manage a network of specially designed access points. The User-Centric Mobility Controllers and Access Points for the Aruba Networks WLAN Product Family are pictured in Figure 3-4. Network access, security, and network management functions are managed from the central controller. Most of these systems use relatively "dumb" or thin access points, and house the security, authentication, and network management functions in the central controller. In a thin AP environment, the user device will "think" it is associating with the access point, but in reality, the access point is merely relaying those control commands to the central controller. See Figure 3-5.

There are some standard capabilities we see supported across all vendor product lines: network design support, RF management, handoff capability, rogue/malicious access point detection/mitigation, and mesh extensions.

- **Network Design Support:** The first step in providing a voice-capable wireless LAN is to plan and build an infrastructure that provides adequate coverage and capacity. Some WLAN switches include design tools that allow the user to input a floor plan, identify critical attributes like building materials, which can affect signal propagation, and define the capacity required in each area. The tool then makes recommendations regarding the number and placement of access points to support those requirements. The tool also takes into account the

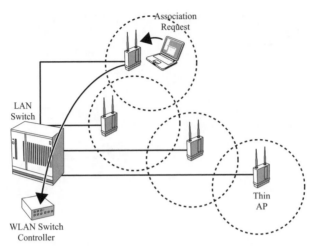

Figure 3-5: WLAN Switch Configuration

differing signal loss and wall penetration of 2.4 GHz versus 5 GHz signals. There are also third-party design tools available from companies like Motorola-Wireless Valley and Ekahau.

- **RF Management:** Once the basic layout is complete, the access points are installed but the network still needs to be tuned to ensure coverage while minimizing co-channel interference. The level of integration between the network design tool and the management system will be key in determining how difficult it will be to translate the design into a working configuration. The setup and tuning process is completely automated in a WLAN switch deployment, and the process takes a matter of seconds. Further, if an access point fails, the access points surrounding it will increase their transmit power to cover the affected area. Some vendors claim that the design phase can be minimized by simply installing APs in a dense grid (e.g., 50–100 foot spacing), and then letting the RF management system tune the configuration.

- **Handoff Capability:** As we have noted, voice users are far more mobile than data users and will require a handoff capability to transfer calls from one AP to another quickly. The IEEE is working on a roaming standard designated 802.11r that will provide for handing off connections between access points; the performance objective for the standard is to hand off a connection in less than 50 ms while maintaining the security association. Unfortunately, that standard will not be completed until some time in 2009. Virtually all WLAN switches will support fast secure handoffs today. The handoff can take anywhere from 10–150 ms; the longest handoff times are encountered when a user passes between APs on different IP subnets and reauthentication is required. The vendors are working on strategies like neighbor reporting and opportunistic key caching with adjacent APs to reduce that interval. However, even a 150 msec handoff interval results in a barely perceptible click in the connection path.

- **Rogue/Malicious Access Point Detection/Mitigation:** To enforce the organization's policy banning unauthorized/rogue access points and Ad Hoc networks while detecting malicious APs that are being used in a hacking attempt, the radio environment must be monitored continuously. As WLAN switches will "know" all of their own APs, they have the ability to identify any unauthorized WLAN transmissions within the coverage area. Many systems will also provide the ability to locate where the unauthorized device is operating, and determine whether or not it is connected to the wired LAN (i.e., differentiate between rogue and malicious APs). Many include the ability to disable the unauthorized AP until it can be located. Rather than investing in a separate dedicated monitoring system, that function can now be provided in the WLAN switch. One important feature that distinguishes this monitoring capability is whether it is done with APs dedicated to monitoring or is time-shared on APs that are handling live network traffic. Time-shared solutions are not monitored continuously, and reduce the amount of time the AP is handling live traffic.

- **Mesh Extensions:** Normally every AP is connected to the central controller over the wired LAN with an Ethernet interface. To support outdoor or other hard to wire areas, many WLAN switching systems support mesh APs to extend the coverage. The better designs include two radios, one for client access and a second for backhaul. The configuration is a simple mesh, and the vendors typically limit the number of radio hops a message might traverse. In short, they look at these capabilities as an extension of their basic product, not as a real competitor to mesh solutions for citywide applications.

3.10 Selecting WLAN Switches

WLAN switches were designed to provide users with the functionality needed to build and support large-scale, enterprise-grade wireless LANs. Any enterprise customer who is considering a WLAN with more than a few dozen access points would be well advised to explore a WLAN switch solution. You can still build a home or small office network using traditional stand-alone APs, but a large network requires a centrally controlled solution.

3.10.1 Standards for WLAN Switching—LWAPP, CAPWAP, and SLAPP

The first issue to recognize is that there is no standard for the control protocol that is used between the central controller and the thin access points. WLAN switch vendor Airespace (subsequently acquired by Cisco) had proposed a standard control protocol called the Light Weight Access Point Protocol (LWAPP). The protocol was submitted to the Internet Engineering Task Force (IETF) for consideration as an IETF standard. The IETF changed the name to the Control and Provisioning of Wireless Access Points (CAPWAP). A draft was published but it has not been accepted as a standard by the IETF. WLAN switch vendors Aruba and Trapeze have proposed the Secure Light Access Point Protocol (SLAPP), a simpler protocol that would allow vendors to download firmware to each others' APs, but it too has failed to catch on. So for the foreseeable future, customers must purchase the access points and controller from the same vendor.

There are a number of features that distinguish WLAN switches, the first being scalability. Virtually all vendors have products that will scale to hundreds of access points, but support for small office deployments remains a problem. Many organizations would like the ability to implement WLAN switch-based solutions in their smaller offices and home office/telecommuter configurations and still have the ability to monitor them remotely. The cost of deploying controllers in those small office and home office locations is often prohibitive. To address that requirement, vendors either offer small office controllers or provide the ability for their large controllers to also manage a number of smaller remote configurations. So if you are considering a WLAN switch deployment, be sure that you consider your small office requirements as well as the major facilities.

3.11 WLAN Switch Architectures

The development of higher speed radio links like 802.11n and the need to support voice services over the wireless LAN have given rise to a new set of concerns regarding the overall architecture of a WLAN switching system. At the most basic level, WLAN switches provide three primary sets of functions:

1. **Traffic Forwarding:** The basic job of picking up and delivering frames to WLAN clients and forwarding them over the wired LAN. This function would also include such functions as VLAN tagging and prioritizing voice frames (i.e., implementing QoS).

2. **Encryption and Key Management:** The task of generating encryption keys and encrypting/decrypting traffic that is sent over the radio link.

3. **RF Management:** The task of assigning channels to access points, adjusting transit power, rogue detection/location, and managing roaming.

The essential architecture decision is which functions should be handled in the central controller and which should be off-loaded to the access points?

3.11.1 Centralized Approach

In the centralized approach, pioneered by companies like Cisco/Airespace and Aruba, all of the network intelligence is housed in the central controller, and the access point contains little more than a radio. All traffic passing over the wireless LAN is routed through the controller. The centralized approach increases the traffic load on the wired LAN, as every frame must be forwarded from the access point to the controller, and then from the controller on to the server. The load on the wired network is less of a concern than the additional delay that might be introduced for voice frames. Further, as more WLAN stations upgrade from 802.11a/b/g to the higher speed 802.11n interface, the traffic volume being carried might require an upgrade of the central controller.

3.11.2 Distributed Approach

Some vendors are now moving to a distributed approach where traffic forwarding and encryption are handled in the access point, and the central controller is only involved in supporting the management functions. Besides lessening the traffic load on the central controller, this solution generally makes it easier to support remote access points, as

Figure 3-6: Xirrus WLAN Array with 12 Directional Antennas (1 Removed)

only the management traffic would have to be forwarded over the wide area network. Some solutions like those from AeroHive and Xirrus distribute all three functions to the access point and forego a central controller entirely; the intelligent access points exchange management messages with one another on a peer-to-peer basis. The Xirrus WLAN Array, which includes 16 individual radios and 12 sectionalized directional antennas, is pictured in Figure 3-6. See Figure 3-7.

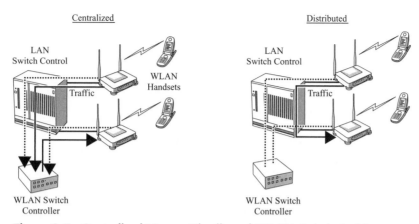

Figure 3-7: Centralized Versus Distributed WLAN Switch Architectures

The WLAN standards are also looking to push distributed traffic handling even further. One of the options in the 802.11e standard is Direct Link Protocol (DLP). If it were supported in the network, DLP would allow the access point to recognize when two of its voice clients were engaged in a conversation, and would allow them to communicate directly with one another rather than through the access point. In that case, none of that voice traffic would pass through the LAN switch or the WLAN switch controller. While we have not yet seen DLP supported in any actual products, it would mean that a voice conversation between two stations on the same access point would require only half of the radio link capacity of a traditional WLAN call. As the likelihood of two voice users being on the same access point is minimal, the overall impact of the DLP feature will be minor.

3.11.3 Meru Networks Approach

The one WLAN switch solution that stands out from the pack is Meru Networks. While all other solutions utilize the standard, contention-based, WLAN access protocol, the Meru design incorporates a proprietary scheduling technique called Air Traffic Control (ATC). In a traditional WLAN, all stations vie or contend for access to the same shared radio channel. With WLAN QoS (i.e., 802.11e), traffic can be categorized so that voice users are given preferred access to the shared channel, but all voice users are still vying for access on an equal basis.

Table 3-1: Major WLAN Switch Vendors

Vendor	Web Site
AeroHive	http://www.aerohive.com
Aruba Networks	http://www.arubanetworks.com
Cisco/Airespace	http://www.cisco.com
Colubris	http://www.colubris.com
Extricom	http://www.extricom.com
HP ProCurve	http://www.procurve.com
Meru	http://www.merunetworks.com
Motorola/Symbol Technologies	http://www.motorola.com
Siemens/Chantry Networks	http://www.siemens.com
Trapeze Networks	http://www.trapezenetworks.com
Xirrus	http://www.xirrus.com

Meru's ATC takes the randomness out of the 802.11 contention protocol. By scheduling when each client station can transmit, they minimize collisions and maximize utilization of the radio channel. While the company does not disclose the technical details of their solution, they are quick to point out that their system will work with any Wi-Fi Certified client device. They are apparently taking advantage of features that are inherent in any Wi-Fi client, but using their proprietary technology to schedule when each device will transmit. The result of that scheduled access is that they can support up to 30 G.729A (i.e., 8 Kbps) voice calls on an 11 Mbps WLAN, twice the number of any other solution.

The question with regard to Meru is always whether the user is willing to opt for a proprietary architecture that seems to be taking a different tack than the rest of the industry. That decision is made more difficult in that Meru is a privately held company and so is not required to release profit and loss information. That means we would be basing our WLAN infrastructure on a "dark horse" whose business prospects are difficult to assess. The Web sites for Meru and the other WLAN switch vendors can be found in Table 3-1.

3.12 Conclusion

The first step to a successful WLAN voice capability will be a sound infrastructure. In this chapter we have looked at the major elements that comprise that infrastructure and the major features customers should ensure are included. Clearly, WLAN switches are the biggest development, and they should be considered in any large-scale installation.

Many organizations deployed their first generation wireless LANs without any consideration for upgrades or future developments in the technology. Those companies are now spending good money replacing devices they should not have purchased in the first place! If we have learned anything in networking, it is that networks grow and change, so any plan that is not looking toward the future is probably an exercise at painting ourselves into a corner. Escaping from that corner is often an expensive proposition.

CHAPTER 4

Media Access Control Protocol

Dealing intelligently with WLAN voice requires a fundamental understanding of how the devices share the radio channel and transmit frames. The 802.11 protocols define the functions for Layers 1 and 2 of the OSI Reference Model. Layer 1 describes the radio transmission technique while Layer 2 defines the process for formatting and addressing frames and getting access to the shared channel.

While the convention in describing protocol options is to start at the bottom and work your way up, with WLANs it is easier to start at Layer 2 and work your way down. In this chapter we will describe the standards for the 802.11 media access control (MAC) protocol. The 802.11 MAC uses a technique called Carrier Sense Multiple Access with Collision Avoidance (CSMA/CA). CSMA/CA is a contention protocol, which means that all of the stations are vying or contending to access the channel. The result of that contention is the possibility that transmissions can collide in which case they must be resent.

The 802.11 MAC is based roughly on Ethernet's Carrier Sense Multiple Access with Collision Detection (CSMA/CD) protocol. Despite the similarity of the names, CSMA/CA defines a very different mode of operation. Early on, wireless LAN designers found that they could not simply take these wired LAN approaches and transplant them to the wireless domain. One basic reason is the problem of collision detection. In a wireless LAN, when a station is sending, it cannot "hear" any other transmitters. Technically, a station must turn off its receiver when it is sending. Even if it could leave the receiver on, the receiver is collocated with the transmitter, so when the transmitter is active, it would drown out all other signals.

Recognizing this and other difficulties, wireless LANs have their own protocols that are tailored to the peculiarities of the wireless environment. While CSMA/CA describes the overall protocol for wireless LANs, there are several different options included within it. To help avoid collisions, the protocol calls for waiting intervals that precede all transmissions. The operation also calls for message acknowledgments to confirm successful transmissions. If the transmission is unsuccessful due to a collision or radio link failure, the station retransmits after a random waiting interval. All of this additional overhead has an impact on the network's throughput. "Throughput," you will recall, is the amount of real data a network can deliver after factoring in all the delays and overhead.

In this section, we will look at the 802.11 MAC protocols, in particular, the fundamental design for data applications. First we will examine the characteristics of the wireless environment and how the characteristics affect the design of the protocol. Then we describe the overall processes for association and authentication. We will also look at the message formats and the major MAC options: the distributed control function (DCF), request-to-send/clear-to-send (RTS/CTS), and point control function (PCF). This description will help you to understand the challenges involved in supporting voice on wireless LANs. This chapter is the prelude to Chapter 11 where we will revisit these issues and describe the 802.11e standard for providing quality of service (QoS) in wireless LANs. To see how the two relate, Chapter 11 may be read immediately following this chapter. Finally, we will look at some of the other features described in the protocol like fragmentation and power save mode.*

4.1 Basic Characteristics and Peculiarities of Wireless LANs

To begin with, we should note a few important characteristics shared by all of the 802.11 wireless LANs:

- **Data Flow:** In a Basic Service Set or Extended Service Set (BSS/ESS) network, the access point sends to the clients, and the clients send to the access point. Even if the two clients are on the same WLAN, all transmissions between them will be relayed through the access point and be carried over the wireless link twice (i.e., inbound and outbound). Stations do communicate directly in an Ad Hoc network.

- **Half Duplex/Shared Media:** The access point and all of the users associated with it will share one half duplex transmission channel. Further, different users can operate at different bit rates on that shared channel. Lower bit rate users affect the performance of higher bit rate users, as those low bit rate users will keep the channel busy longer as they send a message. Management and control

*Note: You can download a free copy of the 802.11 Standard from the IEEE at: http://standards. ieee.org/getieee802/802.11.html.

messages must be sent at the lowest bit rate so that all devices on the network will be able to read them.

- **Capacity/Range Relationship:** The maximum data rate of the channel is based on the radio link interface used (i.e., 802.11a, b, or g). However, the actual data rate for each device is based on its distance from the access point, impairments in the radio path, and any radio interference encountered.

- **No Collision Detection:** As we noted above, as all stations send and receive on the same channel, a station cannot "hear" other transmitters when it is sending. If the stations cannot all hear each other, there is no way for them to detect collisions. Further, a WLAN station may not be within range all of the other stations with which it is sharing the channel. In that case the station might unwittingly cause collisions by sending when it "thinks" the channel is idle (i.e., the *hidden node* problem).

- **Message Acknowledgments:** Given the stations' inability to detect collisions on the wireless medium, all transmissions in a wireless LAN must be acknowledged. If no acknowledgment is received, the station schedules a retransmission. However, there are no negative acknowledgments (NAKs) or automatic retransmissions; the recipient simply discards frames with errors. Sending acknowledgments increases the time it takes to send each frame, and so reduces the network's overall throughput.

- **Total Network Capacity:** Wireless LANs deliver far less transmission capacity than what is typically found in a fully switched wired LAN environment. In a wired LAN, each user is provided with a dedicated, full-duplex 10 or 100 Mbps connection. The maximum capacity of a wireless LAN is a shared, half duplex 11 or 54 Mbps channel.

4.2 Media Access Control Protocol—CSMA/CA

There are several different radio link interfaces defined for 802.11 wireless LANs, but they all use the same MAC protocol. The basic 802.11 MAC protocol is called Carrier Sense Multiple Access with Collision Avoidance (CSMA/CA), but there are three major options defined.

- **Distributed Control Function (DCF):** The basic channel sharing protocol that is used in the vast majority of wireless LANs today.

- **Request-to-Send/Clear-to-Send (RTS/CTS):** An option developed to address the problem of hidden nodes.

- **Point Control Function (PCF):** An option to support time sensitive applications like voice and video, which is rarely (if ever) implemented.

A fourth option called the Hybrid Coordination Function was defined in 802.11e to provide Quality of Service, or the ability to prioritize access for voice devices over data devices. We will describe that option in Chapter 11.

4.3 IEEE 802.11 Message Format

The 802.11 MAC defines data, management, and control frames. In each case the basic message format involves four major parts (See Figure 4-1).

1. Physical Layer Convergence Protocol (PLCP)

2. MAC Header: ≤30 octets

3. Frame Body: ≤2,312 octets in 802.11b, ≤4,095 octets in 802.11a and g

4. Frame Check Sequence: 4 octets

4.4 Physical Layer Convergence Protocol (PLCP)

Technically, the Physical Layer Convergence Protocol (PLCP) Header is part of the Layer 1 protocol, and precedes all transmissions in a wireless LAN, including Data, Management, and Control frames. The total length of the PLCP header varies slightly based on the radio link protocol. All stations on the network must read the PLCP header, so it is always sent at the lowest data rate defined for that radio interface (i.e., 1 Mbps in 802.11b or 6 Mbps in 802.11a or g). The PLCP is different in 802.11b versus 802.11a and g networks.

4.4.1 802.11b PLCP

- **Synchronization Pattern/Start Frame Delimiter:** The PLCP begins with a synchronization pattern that allows the receiver to synchronize its clock with the transmitter. The 802.11b radio link defines a long (i.e., 128 bit) and a short (i.e., 56 bit) synchronization pattern (also called the synch pattern). The synch pattern is followed by the Start Frame Delimiter, a 2 byte field for establishing character synchronization.

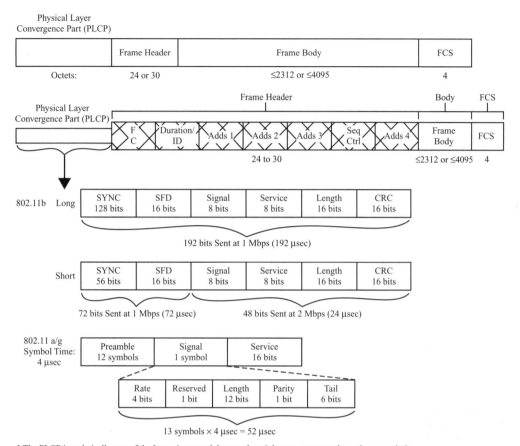

* The PLCP is technically part of the Layer 1 protocol, but we show it here to represent the entire transmission.

Figure 4-1: Basic 802.11 Frame Format

- **Signal (8 bits):** Indicates the data rate that will be used to send the MAC frame.

- **Service (8 bits):** Originally reserved, but subsequently used to signal high rate extensions in 802.11b.

- **Length (16 bits):** Time in microseconds (i.e., μsec) required to transmit the MAC frame (Note: The length is expressed in time, not in the number of octets in the frame).

- **Cyclic Redundancy Check (16 bits):** Header error check.

4.4.2 802.11a and g PLCP

The 802.11a and g radio links use an OFDM transmission format and the standards call for a different PLCP format.

- **Preamble (12 Symbols):** In the 802.11a and g interface, the preamble duration is defined in transmission symbols, where the symbol interval is 4 μsec. As with 802.11b, it is used to synchronize the transmitter and receiver.

- **Signal (24 bits):** The signal field includes five subfields:
  - Rate (4 bits): Data rate to be used to transmit the MAC frame.
  - Reserved (1 bit): Unused.
  - Length (12 bits): The number of bytes in the MAC frame.
  - Parity (1 bit): Used to protect against data errors.
  - Tail (6 bits): Used to terminate the forward error correction code.

- **Service (16 bits sent at the frame transmission rate):** Currently used to initiate the data scrambling function and is set to all 0 bits.

The transmission time for the PLCP formats is summarized in Table 4-1.

4.4.3 Start Slowly, then Speed Up

It is imperative that all stations be able to process the PLCP and frame headers, regardless of the quality of the signal they receive. As a result, this part of the frame is always sent at the lowest bit rate defined for that WLAN. If 802.11b and g users are present on the same network, the PLCP header is always sent at 1 Mbps. The signal

Table 4-1: PLCP Transmission Times

802.11b		Bits	Time
	Long	192	192 μsec
	Short	120	96 μsec
802.11a and g		Symbols	Time
		13	52 μsec*
*Plus a 16-bit service field sent at the frame transmission rate (6 to 54 Mbps).			

field in the PLCP header alerts the receiver to the full transmission rate that will be used for the frame, and will increase to that full bit rate for the data field.

4.5 MAC Frame Header

The 802.11 protocol defines data, management, and control frames; the frame type is identified in the Frame Control field in the header. The frame header includes four parts (see Figure 4-2):

1. **Frame Control:** 2 octets

2. **Duration/ID:** 2 octets

3. **MAC Addresses (Typically 3):** Total 18 octets (6 octets each)

4. **Sequence Control:** 2 octets

4.5.1 Frame Control (2 Octets)

All frames, data, management, and control begin with a 2-octet frame control (FC) field. While only 2 octets in length, the FC includes 11 subfields so it is used for a wide variety of functions.

- Protocol Version (2 bits): Current version is 00.

- Type (2 bits)/Subtype (4 bits): Identifies the type of frame being sent (i.e., Data, Management, or Control), and the specific subtype. See Table 4-3.

- To DS (1 bit): In a BSS or ESS network, this bit is set to 1 on frames being sent by a client to an access point to be forwarded over the wired LAN or distribution system (DS).

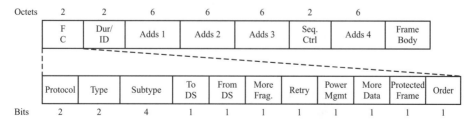

Figure 4-2: MAC Frame Header

- From DS (1 bit): In a BSS or ESS network, this bit is set to 1 on frames received by the access point from the wired LAN that are now being sent to a wireless client. The setting of the To/From DS bits define how the Address Fields are interpreted. See Table 4-2.

- More Fragments (1 bit): Set to 1 if the original message was fragmented and there are more fragments to follow. Set to 0 for the last fragment of a fragmented message or on non-fragmented frames. The fragmentation function is described later in this chapter.

- Retry (1 bit): Set to 1 if the frame is a retransmission; this helps in eliminating duplicate frames. This field can also be monitored to identify problems (i.e., excess frame retransmissions) on the network.

- Power Management (1 bit): Set to 1 if the station is in power save mode. In the power save or sleep state, the client turns off its radio to save power and switches it on to receive beacon frames from the AP. Beacon frames may be followed by Announcement Traffic Indication Map (ATIM) frames that alert stations to the fact that the AP is holding frames in buffer for them.

- More Data (1 bit): Used by the access point in communicating with stations in power save mode to indicate that there are additional frames buffered for them. This also serves as the indication that the station can return to sleep mode.

- Protected Frame (1 bit): Formerly called the "WEP bit," this bit is set to 1 if the data field is encrypted using any of the 802.11 encryption mechanisms.

- Order (1 bit): Indicates the frame is being sent using Strict Order service where the network guarantees all frames will be delivered in sequence.

4.5.2 Duration/ID (2 octets)

In normal operation, the field has two meanings. Its primary purpose is to set the Network Allocation Vector (NAV), which is a timer in each station that records the length of time it will take to send the frame. All stations record the NAV setting and will assume the channel is busy for that period even if they cannot sense the actual transmission (i.e., this is a command to force all stations to treat the channel as busy). In power save operation, a waking station sends a Power Save Poll message where the first 14 bits indicate the BSS that station belongs to, and the last 2 bits are set to 1s.

4.5.3 Sequence Control (2 octets)

This field is used to reassemble fragmented messages, and to discard duplicate frames. The field has two parts: Fragment Number and Sequence Number.

4.6 MAC Addresses (Address 1–4)

An 802.11 frame will include 3 or 4 addresses (normally 3). Like Ethernet frames, 802.11 devices are identified by a unique 48-bit MAC address that is burned onto the WLAN NIC. The access point will also have a MAC address called the Basic Service Set Identifier (BSSID). As the addresses are unique, the user does not have to assign or manage device addresses. Each 802.11 frame will include the source and destination MAC addresses along with the BSSID (i.e., the address of the access point) is also carried in the header. There are four defined address fields and the meaning of each address is based on the setting of the To/From DS bits in the Frame Control field.

Table 4-2 defines the meaning of the Address Fields based on the To/From DS bit settings.

4.6.1 Frame Body

This is the actual voice or data traffic that is being sent along with the higher level protocol headers. An 802.11b frame can carry 0 to 2312 octets, while 802.11a and g frames can carry 0–4095 octets. Most drivers will set the maximum frame size to 1500 octets to match the maximum Ethernet frame size and ensure that frames will not have to be fragmented in transit. Typically the frame body of a data frame will contain an IP packet and include an IP and TCP header. For voice packets the frame will include (see Figure 4-3):

1. The voice sample: Typically 20 msec of digitally encoded speech.

2. The Real Time Transport Protocol Header (12 octets): To identify voice coding used, and ensure sequence and timing continuity.

3. The UDP Header (8 octets): To provide a port address.

4. IP Header (20 octets in IP Version 4): To provide the IP address and other information for routing.

Table 4-2: 802.11 Address Assignments

To DS	From DS	Explanation	Address			
			1	2	3	4
0	0	Frame sent directly from one station to another station in an Ad Hoc or IBSS network	DA	SA	BSSID	Not used
0	1	Frame from an access point to a client in a BSS or ESS network	DA	BSSID	SA	Not used
1	0	Frame from a client to an access point in a BSS or ESS network	BSSID	SA	DA	Not used
1	1	Frame sent between two access points in a point-to-point bridge connection	RA	TA	DA	SA

DEFINITIONS: **DS, Distribution System:** The term used in 802.11 standards to describe a wired LAN or other network that interconnects APs; **DA, Destination Address:** The final destination of the frame, which is the station that will process the frame's contents; **SA, Source Address:** The originator of the frame, which is the station that created the content; **BSSID, Basic Service Set ID:** The address of the access point, which also identifies the network. In an Ad Hoc network, the BSSID is a random number assigned when the connection is established; **TA, Transmitter Address:** The MAC address of the station that is sending the frame onto the wireless LAN, but is not the originator of the content; **RA, Receiver Address:** The station that is receiving the frame over the wireless LAN, but is not the final destination for the content.

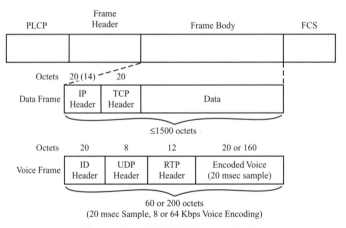

Figure 4-3: WLAN Frame Body

4.6.2 Frame Check Sequence (FCS)

A 4 octet cyclic redundancy check (CRC) is appended to the frame and used to test the received message for errors. An ACK message is returned for frames received without errors. Frames found to contain errors are simply discarded by the receiver (i.e., there is no Automatic Retransmission reQuest or ARQ process).

Table 4-3: Type and Subtype Field Identifiers

Frame Type	Subframe Type	Bits 5–8 in the Frame Control Field
Management Frames (00)	Association Request	0000
	Association Response	0001
	Reassociation Request	0010
	Reassociation Response	0011
	Probe Request	0100
	Probe Response	0101
	Beacon	1000
	Announcement Traffic Indication Map (ATIM)	1001
	Disassociation	1010
	Authentication	1011
	Deauthentication	1100
	Action (for spectrum management in 802.11h and for QoS)	1101
Control Frames (01)	Block Acknowledgment Request	1000
	Block Acknowledgment	1001
	Power Save Poll	1010
	RTS	1011
	CTS	1100
	ACK	1101
	Contention Free End	1110
	Contention Free End With ACK	1111

Table 4-3 (*Continued*)

Frame Type	Subframe Type	Bits 5–8 in the Frame Control Field
Data Frames (10)	Data	0000
	Data with Contention Free ACK	0001
	Data Poll with Contention Free Poll	0010
	Data Poll with Contention Free Poll and ACK	0011
	Null—No Data Transmitted	0100
	Contention Free ACK—No Data Transmitted	0101
	Contention Free Poll—No Data Transmitted	0101
	Contention Free Poll ACK—No Data Transmitted	0110
	QoS Data	1000
	QoS Data + CF ACK	1001
	QoS Data + CF Poll	1010
	QoS Data + CF ACK+ CF Poll	1011
	QoS Null—No Data Transmitted	1100
	QoS Data + CF ACK—No Data Transmitted	1101
	QoS Data + CF Poll—No Data Transmitted	1110
	QoS Data + CF ACK+ CF Poll—No Data Transmitted	1111

Note: Use of the various message types is described in connection with the protocol descriptions. CF stands for contention free.

4.7 Authentication and Association

The logical place to begin the protocol description is at the beginning. The first element in the protocol is the authentication and association processes.

The 802.11 Standards define three states for a station:

1. Initial State: Not authenticated and not associated

2. Authenticated but not yet associated

3. Authenticated and associated

During the initial state, only control frames that support network location and authentication can be sent.

4.8 Beacon Message

The beacon message is an important element in a wireless LAN. In BSS and ESS networks, access points transmit beacon messages to advertise their availability and provide a signal that clients can use to measure the quality of the radio link. The beacon message also announces the options and capabilities a station must support to join that WLAN. From an architecture standpoint, this means we can change the network features in the access point, and the access point will announce those new parameters in the beacon message. The clients know they must use those parameters to join that particular network.

Access points will typically send beacon messages at an interval of 100 msec, or 10 beacons per second. If virtual WLANs are supported on a network, the access point will send a beacon for each with their unique operating parameters. A beacon message is approximately 50 bytes long, with about half of that being the frame header and frame check sequence. The destination address in the beacon message is set to all 1s, the broadcast address. This forces all stations on the channel to receive and process each beacon frame. Each beacon frame carries the following information in the frame body:

- **Timestamp (8 octets):** The station uses the timestamp value to update its local clock to ensure synchronization with the access point. This is particularly important in power save operation.

- **Beacon Interval (2 octets):** This is the length of time between beacon transmissions. Before a station enters power save mode, it needs to know the beacon interval so it can wake up to receive beacons and discover whether there are frames buffered for it at the access point.

- **Capability Information (2 octets):** This field defines the options that are in use on the network. That would include such features as privacy, short preamble, and short time slot.

- **Service Set Identifier (up to 32 octets):** The SSID is a text name assigned by the user that identifies the wireless LAN; in an Extended Service Set (ESS) network, all of the access points will use the same name. By default, access points include the SSID in the beacon message to enable a sniffing function where the station will automatically configure its wireless network interface card (NIC). We have the option to disable the SSID broadcast to reduce security concerns. This is a rather feeble security measure, however, as the station must provide the correct SSID to join the network. A hacker need only monitor one successful association to determine the SSID.

- **Optional Fields:** The beacon message can also include a number of additional fields identifying parameters for frequency hopping, contention-free operation, and country information.

- **Announcement Traffic Indication Map (ATIM):** The access point can also send an Announcement Traffic Indication Map (ATIM) frame (type = 00, subtype = 1001) that identifies which stations in power save mode have data frames waiting in the access point's buffer. Those stations then poll the access point to retrieve their messages.

4.9 Authentication Process

Authentication is the first step in the process by which a station joins a WLAN. It provides a mechanism whereby the access point can determine if a particular station can legitimately join the wireless LAN. The 802.11 standard defines an Open Authentication and an optional Shared Key Authentication process. In networks with Open Authentication, there is no authentication, and the station proceeds directly to the association process; the beacon message informs the stations if Shared Key authentication is in use on that network.

4.9.1 Shared Key Authentication

The shared key authentication involves a four step process. See Figure 4-4.

1. The client sends an initial Authentication frame to initiate the process.

2. The access point responds with an Authentication frame that contains a random 128-byte message called the challenge text.

3. The station encrypts the challenge text using its encryption key and returns it to the access point.

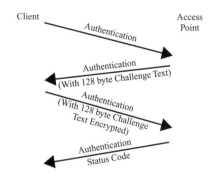

Figure 4-4: Shared Key Authentication Process

4. The access point decrypts the challenge text, and if the message comes out the same it means the station has the correct key, and the client is allowed to join the network.

In essence, the only credential the client must present to be authenticated on the network is a demonstration that it has the same encryption key as the access point. So the strength of the authentication process is based entirely on the possibility the encryption key could be compromised. The Shared Key authentication process used in conjunction with the WEP encryption key provides authentication in name only, given the ease with which the WEP key can be discovered. As a result, Shared Key Authentication based on the WEP key is inherently vulnerable to hackers.

4.9.2 IEEE 802.1x Extensible Authentication Protocol (EAP)

To address the potential deficiencies of Shared Key authentication, the IEEE defined a new and more functional authentication framework called the Extensible Authentication Protocol (EAP) in 802.1x. If 802.1x authentication is used, the WLAN uses Open authentication so the client associates with the network and then initiates the authentication process. During the Association process, the client can only communicate with the access point that in turn routes the authentication dialog to the authentication server (the authentication dialog is encrypted over the wireless link). Typically the authentication mechanism uses a user name/password sequence, and the access point will allow traffic onto the wired network only after the user is authenticated. We will look at the various 802.1x authentication options and tradeoffs in Chapter 6.

4.10 Association Options

The 802.11 protocol defines two Association options, but virtually all devices use the Passive Association option.

- **Passive Scanning (Mandatory):** In passive scanning mode, a station wishing to join a WLAN scans all channels looking for beacon messages from eligible access points. Clients use the access point's beacon message to gauge the received signal quality (i.e., Signal/Noise ratio) and select the best access point with which to associate. When the station selects an access point, it performs an authentication exchange (if necessary), then it sends an Association Request. The access point returns an Association Response if it elects to accept the station onto the network.

- **Active Scanning (Optional):** In active scanning, the station initiates the association process by sending Probe frames to all access points within range; available access points respond to those Probe frames with Probe Responses to initiate the process. The mobile station selects the AP whose Probe Response is received with the best signal. The problem with active scanning is the additional traffic it creates, and the battery drain from the client's probe transmissions.

4.11 Reassociation/Handoff

The 802.11 specification provides for a simple handoff in the event that a station moves from one AP area to another. In normal operation, clients continue to scan all channels even when they are associated with an AP. If a client finds a better signal from a different access point, it can automatically change its association. The process involves the client disassociating from one AP and associating with another.

4.11.1 Handoff Procedure

1. All APs send beacon messages periodically, and the clients monitor those messages to measure received signal strength (RSS) and the signal to noise (S/N) ratio.

2. The client compares the S/N on the channel it is using to that of any other APs that are within range.

3. If the station recognizes an access point with a better signal, it sends a Reassociation Request to the new AP that includes information about the client and the old AP.

4. The old and new APs now communicate over the wired LAN (i.e., the distribution system) to coordinate the handoff. There is an IEEE-defined Inter-Access Point Protocol (IAPP) designated 802.11f, though most APs use a vendor proprietary technique.

4.11.2 Other Handoff Options

The tricky part about the handoff is that the APs must ensure that any frames waiting to be delivered from the old AP are transferred to the new AP as part of the handoff. While the handoff process defined in the original 802.11 standards is suitable for data devices, it can still take several seconds to complete, particularly if the station must reauthenticate with the new access point. Voice users are far more mobile than data users, and this process is far too slow for voice applications; the goal in voice applications is to effect the handoff in less than 50 msec, though most short term approaches can take as long as 150 msec, which is still virtually unnoticeable to the voice users. There are two options to accomplish this.

*   **802.11r Fast, Secure Roaming:** The IEEE is developing a new standard called 802.11r that will provide for a fast, secure handoff in 50 msec or less. Barring any unforeseen difficulties, completion of the standard is expected in the late 2008 timeframe. While a standard had been anticipated earlier, the committee discovered a number of issues in ensuring that the security relationship was maintained during the handoff.

*   **Wireless LAN Switches:** With the advent of wireless LAN switches, the handoff procedure can be implemented in the WLAN switch architecture. You will recall that in a WLAN switch environment, the user associates with the central controller, not the access point. In that way, the switch can maintain the session as the users move throughout the coverage area. WLAN switches can already provide a fast, secure handoff in 10 to 150 msec. From a user standpoint, the tradeoff in opting for a proprietary protocol is that this may be limiting their choice of equipment further down the line. On the other hand, if a customer need a fast, secure handoff mechanism today, there are no standards-based options today.

4.12 CSMA/CA Distributed Control Function (DCF)

Once the user is associated and authenticated, he or she can begin sending and receiving traffic over that WLAN. The primary media access used with CSMA/CA is the Distributed Control Function (DCF). DCF describes the process by which stations transmit frames on the network and attempt to avoid collisions. There are other options defined in the 802.11 MAC, which we will describe later; however, DCF is the typical mode of operation. (Note: while a logical presentation calls for describing the authentication/association processes first, in actuality the beacon, authentication, and association messages are also sent using the CSMA/CA access protocol.)

4.12.1 Basic Definitions

To understand how the DCF operates, we must first introduce a few of the concepts that are employed.

- **Inter-Frame Spacing:** This is the period of time between frame transmissions. Even when the channel is idle, any station wishing to send must wait one of these time periods before initiating a transmission; the intervals vary depending on the radio link protocol and the type of frame being sent. There are three inter-frame spacing intervals defined in the original DCF.
  - Short Inter-Frame Spacing (SIFS): A fixed interval (i.e., 10 μsec) a station must wait before sending an acknowledgment (ACK), a clear-to-send (CTS) message, or subsequent fragments of a fragmented frame. This is the shortest inter-frame spacing, so these messages have the highest priority to access the channel.
  - DCF Inter-Frame Spacing (DIFS): This is the interval a station must wait before sending a data, management, or control frame. The DIFS interval is one SIFS (i.e., 10 μsec), plus 3 time slots. The time slot duration is 20 μsec for 802.11b networks, and 9 μsec for 802.11a and g networks, which yields intervals of 70 and 37 μsec, respectively. (Note: in mixed b/g networks, the longer time slot value must be used.) If the channel is idle, a frame is sent after the DIFS interval. However, if the channel is busy when a station attempts to send, it sets a back-off counter with a random number of time slots, and begins decrementing it after the channel has been idle for one DIFS interval. The standard also describes the range of values for the back-off counter (i.e., CW_{min} to CW_{max}).

○ Point Control Function Inter-Frame Spacing (PIFS): The point control function is a rarely used mechanism to support time sensitive traffic like voice and video. In PCF, the access point periodically takes control of the channel; that period of time is called the Contention-Free (CF) interval. The time interval for sending that control message is the PIFS interval. The PIFS interval is SIFS plus one time slot, so it is longer than the SIFS, but shorter than the DIFS. During the CF interval, the access point polls each station with time-sensitive traffic; at the end of the CF interval, the network returns to the DCF contention mode operation. See Table 4-4.

It is unlikely that the PCF operation will ever be implemented, as a more comprehensive approach, called Hybrid Controlled Channel Access, is now included in the 802.11e QoS standard. See Table 4-4 and Figure 4-5.

Table 4-4: 802.11 Inter-Frame Spacing Intervals

Interval	Computation	802.11b or b/g (µsec)	802.11a or g (µsec)
SIFS	10 µsec	10	10
PIFS	SIFS + 1 Time Slot	30	19
DIFS	SIFS + 3 Time Slots	70	37

- **Contention Window (CW)/Back-off Window:** This is the randomizing element in the DIFS. The contention window comes into play in two instances:
  ○ If the channel is busy when a station tries to access it.
  ○ If a station has tried unsuccessfully to send a frame.

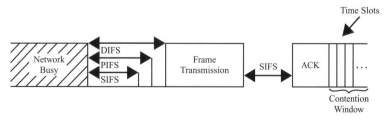

Figure 4-5: Inter-Frame Spacing

Table 4-5: 802.11 DCF Contention Window Ranges

802.11b or b/g		802.11a or g	
CW_{min}	CW_{max}	CW_{min}	CW_{max}
31	1023	15	1023

In either of these cases, the station wishing to send (or resend) a frame sets a back-off counter with a random number of time slots between CW_{min} and CW_{max}. The station then waits for the channel to be idle for one DIFS interval (SIFS plus 2 time slots) and then begins decrementing the counter. When the back-off counter expires, the station sends immediately. CW values are always a power of 2 less 1 (i.e., 31, 63, 127, 255, 511, and 1023). The counter advances to the next increment with each unsuccessful attempt (i.e., the station will "try harder" to avoid a collision).

If the channel becomes active during the back-off window (i.e., a station begins transmitting), all back-off timers are frozen and restart one DIFS interval after the channel becomes idle. If the timers were to remain running during a transmission, several timers might expire during that transmission and a collision would occur immediately after the channel became idle. The standard specifies a range of values for the back-off counter (i.e., CW_{min} to CW_{max}). See Table 4-5.

- **Network Allocation Vector (NAV):** The NAV is a timer stored in each terminal that is used for *virtual carrier sense.* The NAV is set by the duration value in each frame header, and it causes the station to assume the channel is busy even if it cannot hear the transmission. In the PCF, the AP reserves the channel for a period by sending a CTS frame to set the NAV timers in all stations on the network. The RTS/CTS option (described later in this chapter) also uses the NAV timers. When the NAV value is set, a station will not attempt to initiate any transmissions for that interval, and if any station is running a back-off counter, the counter will be frozen for that interval.

4.12.2 Distributed Control Function (DCF) Operation

In 802.11 WLANs, a single radio channel is used for both inbound (to the AP) and outbound (from the AP) transmissions. Even though the AP is the control station in a BSS or ESS WLAN, it vies for access to the channel on the same basis as all other stations. All transmissions are broadcast to all stations on that channel, and each station copies frames with its 48-bit MAC address in the destination address field. The protocol also supports the MAC broadcast address (i.e., all 1s), so that frames can be addressed to all stations. (Note: stations do not acknowledge broadcast frames.)

4.12.3 Distributed Control Function Transmission Process—The Channel is Clear

Distributed Control Function means that all stations will collaborate to avoid a collision. When the channel is idle, the basic process works as follows:

1. The sending station must first determine if the channel is idle before transmitting. It uses two metrics to make that determination:
   a. Carrier Sense: Senses for transmission activity on the radio channel.
   b. Virtual Carrier Sense: The station refers to its Network Allocation Vector (NAV). If the NAV is set to a greater value than 0, the station will assume the channel is busy for that duration regardless of whether it senses radio activity.

2. If the receiver senses no activity on the channel and the NAV is zero, the station assumes the channel is idle.

3. Once the station determines the channel is clear, it waits at least one DIFS interval and then transmits. If two stations sense the channel at exactly the same instant, they will collide. If a station hears a transmission begin during the DIFS interval, it will defer its transmission based on the rules for busy channel operation.

4. The receiving station copies the frame into a buffer and recomputes the frame check sequence (i.e., 32-bit cyclic redundancy check) to test for errors. If the frame is error-free, the receiving station transmits an ACK (14 bytes) after one SIFS. If the receiver detects an error (i.e., the frame is incomplete or the FCS is incorrect), it simply discards the frame and no response is returned (i.e., there is not such a thing as a "no acknowledge" or NAK message). See Figure 4-6.

5. If the transmitting station does not receive an ACK, it assumes a collision or some other failure has occurred. As the station cannot "hear" while it is sending, it must assume a collision or other failure occurred. In that case, the sending station sets its back-off counter, initiates it after one DIFS interval, and retransmits immediately after the counter expires. If any other station begins

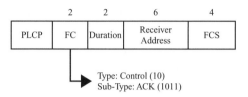

Figure 4-6: ACK Format

transmitting while the counter is running, the counter is frozen while that other message is being sent, and restarts one DIFS after the channel is idle.

The design of the timer system ensures that certain transmissions have priority over others. Any station wishing to send a frame must wait at least one DIFS interval while a station sending an ACK message waits one SIFS interval. So if one station is waiting to send a frame and another is waiting to send an ACK, the ACK message will always be sent first.

4.12.4 Distributed Control Function (DFC) Operation—Busy Channel

The previous section covered how frames are transmitted when a station finds the channel idle. When the channel is busy, stations wishing to send take additional steps to avoid creating a collision. That process works as follows.

1. If the channel is busy, the station wishing to send sets its back-off counter with a random number between 0 and CW_{min}, that indicates the number of fixed-duration time slots the station will wait before attempting to transmit. The timer does not begin to run until one DIFS after the channel is clear. The use of a variable length interval to access the channel after a transmission is one of the key mechanisms in avoiding collisions.

2. If another transmission begins while the back-off timer is running, the timer stops while the transmission is in progress and starts again one DIFS after the channel is clear. The station always waits one DIFS interval before restarting the timer so that the ACK message can also be sent.

3. When the counter expires, the station sends immediately and then waits one SIFS interval for an ACK.

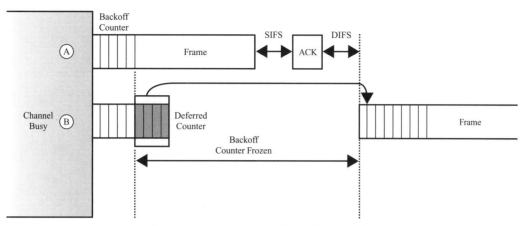

Figure 4-7: DCF Busy Channel Operation

4. If no ACK message is received, the station assumes a collision or some other failure has occurred. It then sets its back-off counter with the next increment and repeats the process; the increments are always a power of 2 less 1 (i.e., 31, 63, 127, 255, 511, and 1023). The back-off counter is moved to the next increment with each unsuccessful transmission attempt until it reaches the maximum (i.e., CW_{max}). The driver will determine how many attempts it will make to transmit the frame before abandoning the attempt. See Figure 4-7.

4.13 Request-To-Send/Clear-To-Send (RTS/CTS) Operation

The process of deferring transmissions while other stations are transmitting assumes that a station can hear all stations on their wireless LAN, and so defer transmissions when the channel is busy. However, stations may not be able to sense all transmissions. By definition, the access point must be able to hear all stations, but two stations might not be able to hear one another due to the distance between them or obstacles in the transmission path. The situation where two stations can both hear the access point but cannot hear each other is called the *hidden node* problem. The problem with hidden nodes is that they increase the likelihood of collisions, as those stations may send when a station they cannot hear is using the channel. To deal with hidden nodes, an optional form of CSMA/CA was developed called Request-To-Send/Clear-To-Send (RTS/CTS). See Figure 4-8.

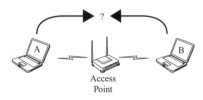

The access point can "hear" stations A + B,
but A and B cannot "hear" each other

Figure 4-8: Hidden Node Problem

4.13.1 Request-To-Send/Clear-To-Send (RTS/CTS) Operation

1. A station wishing to send in RTS/CTS mode (either a client or the access point) first sends a short RTS frame (20 bytes) one DIFS after it senses the channel is idle; if the channel is busy, the station sends the RTS frame according to the rules for busy channel operation. The RTS is a control message that includes the transmitter address (i.e., its own address), the receiver address, and the duration value, or length of time it will take to send the message. As a station operating in RTS/CTS mode may not "hear" all other transmissions, the RTS message could collide with another transmission, in which case the station uses the back-off process before trying to resend the RTS.

2. After a SIFS, the recipient returns a CTS control frame (16 bytes) that includes the receiver address and the duration value. As the CTS response is sent after a SIFS, it will have priority over any new transmissions. See Figure 4-9.

3. Note that both the RTS and CTS messages contain the duration value. As the access point is always involved in the RTS/CTS operation, it can be assumed that every station hears either the RTS or the CTS, and will therefore know the duration for which the channel will be busy. All stations use that duration value to set their NAVs, and during that period they will not try to access the channel. Also, any back-off counters that are running will be frozen until one DIFS after the NAV setting expires.

4. The station sends the frame a SIFS interval after it receives the CTS, and the recipient returns the ACK.

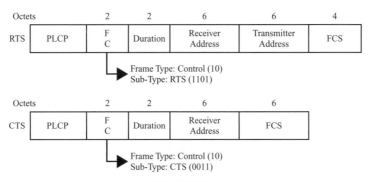

Figure 4-9: RTS/CTS Frame Formats

In effect, the RTS/CTS process provides a contentionless time window during which one frame and one ACK can be sent. However, as more overhead is involved, the RTS/CTS process reduces throughput for the network.

4.13.2 Protection Mode for Mixed 802.11b/g

RTS/CTS was designed to address the hidden node problem, however 802.11g devices switch to it automatically when a low-speed 802.11b device joins the same network; in 802.11g that operation is referred to as *protection mode*. The problem with a mixed 802.11b/g network is that the older 802.11b stations do not recognize 802.11g transmitters as legitimate users on their channel (i.e., the 802.11g transmitters are assumed to be noise). As a result, the 802.11b stations may transmit even when an 802.11g device is sending.

4.13.3 Protection Mode Operation

To avoid those difficulties, when operating in protection mode:

- The 802.11g stations will send RTS/CTS messages using the 802.11b radio link at 1 Mbps prior to the transmission of each frame.

- All transmission preambles and all management and control frames will be sent at 1 Mbps (i.e., the lowest transmission rate supported in 802.11b).

- All network timers and back-off counters will use the longer 802.11b intervals.

Given the additional overhead in a mixed network, the throughput for 802.11g stations in a mixed 802.11b/g network is about 15 Mbps! You can improve your 802.11g performance significantly by eliminating all 802.11b devices.

4.13.4 CTS-Only Mode

CTS-only mode is another option designed for mixed 802.11b/g networks. As all stations in the network can hear the AP, there is no reason for the AP to broadcast an RTS and then wait for a CTS to initiate a transmission. CTS-only means that when the AP needs to send a frame, it can do so after sending a CTS message (i.e., no RTS has to be sent).

4.14 Point Control Function (PCF)

The third option in the original 802.11 MAC is the PCF. PCF is a contentionless access method. Contentionless means that it guarantees that only one station can send during that period so there can be no collisions. PCF was designed to support time sensitive traffic like real time voice or video. Currently, almost no 802.11 products support PCF. There is a very similar option called Hybrid Controlled Channel Access (HCCA) that is included in the 802.11e QoS standard, and currently no vendors are looking at supporting that option either.

4.15 PCF Basic Concept

When PCF is used, the access point periodically assumes control of the network; the times when the access point takes control are called contention free (CF) periods. During the contention free periods, the access point polls stations with time sensitive traffic. The portion of time allocated to contention free access is variable, and the assignment is made by the AP based on the number of stations requesting contention free service, their transmission requirements, and data rates. At other times, data users access the channel on a contention basis using the DCF.

4.15.1 Point Control Function Operation

1. The AP broadcasts a control message after a PIFS (SIFS plus 1 time slot) interval causing all stations to set their NAVs to initiate a contention free period. As in the RTS/CTS operation, that NAV setting will inhibit stations from sending for that specific amount of time.

2. During the contention free period, the AP polls each station that has been assigned dedicated capacity. Stations can only transmit in response to those PCF polls. At the end of the contention free period, the network automatically returns to contention mode.

4.15.2 Throughput Impact

The use of the PCF would have a major impact on the throughput of the wireless LAN, as the NAV message essentially puts the entire wireless LAN on hold for a period of time. All of the contention-based traffic is now vying for a portion of the available channel capacity. With the development of the equally unpopular Hybrid Controlled Channel Access (HCCA) option in 802.11e, it is very likely we will never see PCF implemented.

4.16 Other Protocol Features

While the previous description covers the basic operation or the 802.11 MAC protocol, there are a number of other features that have been added to increase efficiency and reduce battery drain. Some of these were not included in the original 802.11 MAC, while others were added in later versions like 802.11e and 802.11n.

4.16.1 Streaming/Burst Mode

Streaming or Burst Mode transmission is a feature that is supported in 802.11e and with the MAC enhancements for the 8021.11n radio link. Streaming allows a station to download a series of frames without relinquishing the channel, and it is particularly important in supporting voice. In human speech, a burst of transmission may last for a several seconds, while a single voice frame typically carries 20 msec of digitally encoded speech. With streaming support, a device can get access to the channel after waiting the required IFS interval, but it will then send subsequent frames after a SIFS interval. As the SIFS is shorter than any other waiting interval, no other station will be able to seize the channel until all packets in the buffer have been sent. See Figure 4-10.

Some have expressed a concern that a streaming device would essentially lock out all other users on the channel. However, it is important to put the issue in context. A single 64 Kbps voice transmission with the associated protocol overhead and pre-transmission timers requires the equivalent of at least 250 Kbps on a channel with the effective

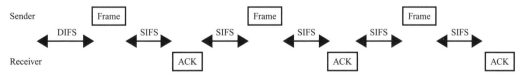

Figure 4-10: Streaming Mode

capacity of about 5.5 to 30 Mbps (i.e., 50 percent of the line rate). Therefore, that streaming transmission will not consume the entire capacity of the channel for the duration of a speech burst. On the other hand, a multi-megabit streaming video source may have that impact.

4.16.2 No Acknowledgment Operation

Transmission acknowledgments are important for data transmissions on a wireless LAN, as they provide the only mechanism by which the sender can be assured that their message was received. If no ACK is received, the sender assumes the transmission failed, sets a back-off counter, and tries again when the back-off expires. Given the time sensitivity of voice traffic, the 802.11e MAC allows an option for stations to operate without ACKs to improve the overall efficiency of the network.

4.16.3 Fragmentation Mode

Fragmentation is a transmission function of the original 802.11 MAC that allows a station to divide larger messages into a series of smaller frames. The idea is that longer transmissions take longer to send, and are hence more likely to experience errors. To avoid the requirement to retransmit those large frames, in fragmentation mode the sender will divide large frames and send them as a series of fragments. When a fragmented frame is sent, the sender waits the required DIFS interval before beginning the transmission. Once the first fragment is sent, the sender waits an SIFS interval after each ACK to send each subsequent fragment (i.e., similar to the streaming mode operation described above). The idea is that you want to clear all parts of the fragmented message before another user can initiate a transmission. If an error occurs, only that fragment will have to be resent. To implement the fragmentation option, the user sets a maximum frame length threshold; frames longer than that will automatically be fragmented.

4.17 Power Save Features

One persistent problem we have discovered with WLAN stations is that they are power hogs. Access points will be powered with commercial power either through an AC adapter or a Power over Ethernet capability, but client devices are powered by batteries. Battery life is becoming one of the major limiting factors in wireless LANs, and that is particularly important with voice devices.

In dealing with power issues, it is important to recognize that both the WLAN transmitter and receiver draw power, but the transmitter creates the greater power draw. The 802.11 standards now have two mechanisms for improving battery life: Power Save mode, and Automatic Power Save Delivery (APSD)/WMM Power Save.

4.17.1 Power Save Mode

The original 802.11 standards included a rudimentary data-oriented power save feature. If a station is inactive, it can alert the AP that it is entering the sleep state or Power Save mode by setting the power save bit in the frame control field of the last frame it sends. The access point records which stations are sleeping and buffers packets addressed to them. The sleeping stations must wake up every 100 msec to receive the beacon messages, which could be followed by an Announcement Traffic Indication MAP (ATIM) message announcing which stations the AP is holding traffic for.

The stations having frames in buffer will remain awake, and send a Power Save Poll control frame to the access point to retrieve each frame. After all the frames are downloaded, the station goes back to sleep. The beacon message also provides the beacon interval, so the stations know when it is time to wake up again to check for incoming traffic.

4.17.2 Automatic Power Save Delivery (APSD)/WMM Power Save

There are a few problems with the original power save feature. First, all stations have to wake up (i.e., turn on their receiver) 10 times per second to hear the beacon message and send Power Save Poll (PS Poll) messages, thus draining the client's battery. Further, if the network is supporting voice traffic, a 100 msec beacon interval will add significant latency and jitter to voice frames. In voice, the objective is to keep the transit delay below 150 msec, and the wake-up interval will add 100 msec of delay by itself.

To address these difficulties, a new power save feature was included with the 802.11e QoS standard called Automatic Power Save Delivery (APSD); the Wi-Fi Alliance identifies compliance with that capability as Wi-Fi Multimedia Power Save (WMM Power Save). In this operation, the client wakes up on its own schedule and sends a "trigger frame" to the AP to initiate a burst-mode download of all buffered frames; the trigger frame can be a data frame with a bit set in the header. In response, the AP downloads all of the frames in the buffer, and uses a bit to indicate the last frame so the client can return to the sleep state. The original power save feature required that the station send a Power Save Poll message to retrieve each individual frame.

For voice traffic, the recommended wake-up interval is 20 msec, which is also the standard interval for generating voice frames. So the station is not waking up solely to retrieve frames. Further, as the station does not have to poll for each frame, it generates fewer transmissions, and is awake for a shorter period of time. The Wi-Fi Alliance claims that APSD/WMM Power Save should improve power consumption 15 to 40 percent, while also improving latency for voice traffic (i.e., the client is looking for traffic every 20 msec versus 100 msec).

Beyond these basic protocol modifications, the equipment vendors are moving to better battery technologies and incorporating other features to further increase battery life. We will look at those other options in Chapter 11.

4.18 Throughput Considerations

Now that we have looked at the protocol, it is evident that there is a lot of waiting, acknowledging, and potentially retransmitting, that will have to go on to send data over the wireless medium. This is particularly important when sending voice frames where the data field must be kept short to minimize delay. Further, using different access protocols (DCF, RTS/CTS, PCF, etc.) and mixing different speeds will increase the overhead and decrease the throughput.

4.18.1 Performance Models?

As yet, we have not seen many good performance models that will estimate the volume of traffic the network will really be able to support in these different configurations. For the time being, we are working with rough estimates for capacity planning based on rather simple rules of thumb.

4.18.2 Determining Throughput Capacity

You will recall the term throughput is used in data communications to indicate the total amount of real traffic that can be sent over a channel in a given period of time. Throughput is the raw transmission rate or *line rate* (e.g., up to 54 Mbps in 802.11a or g), less all of the overhead elements (e.g., waiting intervals, acknowledgments, and retransmissions). In determining the amount of capacity available per user, we take the available throughput of the channel and divide it by the number of users. In making these estimates, we use the estimated throughput of the channel, not the line rate. The estimated throughput of our different configurations is:

- 50 percent of line rate for 802.11b or 5.5 Mbps

- 55 percent of line rate for 802.11a/g or 30 Mbps

- 25 percent of line rate or 15 Mbps for 802.11g devices in mixed 802.11b/g networks

In general, the 802.11a and g protocols are slightly more efficient because the waiting elements (i.e., back-off time slots and inter-frame spacing intervals) are shorter. The bottom line is that with WLANs we start with a relatively low-capacity, shared media channel, and then use it with considerable inefficiency.

4.19 Conclusion

In this chapter, we have taken a detailed look at the protocols that are used in 802.11 wireless LANs. You can see some similarities to Ethernet's CSMA/CD, but everything gets a special twist for the wireless environment. If you know how Ethernet works, you are only 20 percent along the path to understanding CSMA/CA.

While we have described a number of options, the basic DCF is far and away the primary mechanism used on wireless LANs today. The big lesson to take away from this is that wireless LANs were designed for the requirements of data traffic, not voice. The basic contention operation will increase latency and jitter, two of the major factors that degrade the quality of voice service. So while we can carry voice traffic on a DCF-based wireless LAN, ensuring adequate voice quality will require special attention in the design. Fortunately the standards committees have recognized this and defined two new protocol options in IEEE 802.11e that will provide mechanisms to prioritize voice or guarantee capacity for voice users.

Wireless LANs are still a fairly new technology, and developments are coming fast and furious. One of the most critical tasks in assessing WLAN equipment is understanding what the new options are, and determining which of these the customer will require. The customer will want to ensure that those required functions can be implemented with software upgrades rather than swapping out hardware.

CHAPTER 5

802.11 Radio Link Specifications

The key element in the wireless LAN is the radio interface that is used to transmit the users' information over the radio channel. In Chapter 2, we introduced the fundamental issues that govern radio link capacity: bandwidth, received power, and interference. This is where that background will come into play. The purpose is to provide a fuller understanding of the radio technologies being used, and to develop an appreciation of the capabilities and constraints.

In this chapter, we will begin with an overview of the major radio link interfaces defined in the 802.11 specifications. Then we will also take a closer look at the 2.4 GHz ISM and the 5 GHz U-NII unlicensed radio bands and how they are used with 802.11 wireless LANs. We will also review the advantages and disadvantages of moving from 2.4 GHz to 5 GHz. and discuss the tradeoffs involved. We will describe each of the radio link interfaces in detail including the bit rates supported and the modulation techniques employed. Finally, we will look at the impact of pre-standard and non-standard radio interfaces in enterprise environments.

5.1 Defined Radio Link Interfaces

The IEEE 802.11 standards currently describe three major radio link interfaces with a fourth in development. In reality, each of those radio link interfaces supports a range of bit rates, and each of those fallback rates defines a different transmission system. We will begin with a quick overview of the options in Table 5-1, which was also presented as Table 2-7, but is reprinted in Chapter 5 for convenience.

Radio transmission systems can be divided into two broad categories, analog and digital. In an analog radio system, a carrier signal is modulated (i.e., combined with) an analog source signal. That modulation is based on either the amplitude or the frequency of an analog source signal, that is, the basic definition of AM and FM radio. Wireless LANs on the other hand carry information that is digital in nature, and so a different set of concepts applies.

Table 5-1: IEEE 802.11 Radio Link Interfaces

Standard	Maximum Bit Rate (Mbps)	Fallback Rates	Channel Bandwidth	Non-Interfering Channels	Transmission Band	Licensed
802.11b	11	5.5, 2, or 1 Mbps	22 MHz	3	2.4 GHz	No
802.11g	54	48, 36, 24, 18, 12, 11, 9, 6, 5.5, 2, or 1 Mbps	20 MHz	3	2.4 GHz	No
802.11a	54	48, 36, 24, 18, 12, 9, or 6 Mbps	20 MHz	23	5 GHz	No
802.11n (Draft)	≤289 (at 20 MHz) ≤600 (at 40 MHz)	Down to 6.5 Mbps (at 20 MHz)	20 or 40 MHz	2 in 2.4 GHz 11 in 5 GHz (40 MHz)	2.4 or 5 GHz	No

Digital radio is a somewhat deceptive term, because all radio transmissions are essentially analog; we are sending a continuously varying wave of electromagnetic energy. The idea of a digital radio is to take a digital bit stream, modulate it (i.e., represent it in an analog form), and then introduce that modulated signal onto the radio carrier. The methods used to convert the digital bit stream in an analog format are very similar to what we do with voice band modems. The advantage of digital radio is that it can produce a transmission signal that is more bandwidth efficient and robust, and it can incorporate forward error correction coding so the receiver can actually correct errors that are introduced in the radio link. As a result, digital systems can provide acceptable performance even in fairly hostile radio conditions.

The basic capacity measure for a digital transmission system is bits per second (bps). As we noted in Chapter 2, analog capacity is measured in bandwidth, or the range of frequencies that can be carried on the channel. In a wireless LAN, the bandwidth is fixed, and higher digital transmission rates are achieved by improving the *bandwidth efficiency* or sending more bits on each cycle of bandwidth. In digital radio systems, group of bits are represented as an analog *symbol*. In years past we used the term *baud rate* to identify the number of symbols per second we sent. Unfortunately, the term *baud* was so often confused with *bits per second* that now it has largely been abandoned. When we describe a digital radio system, we identify the symbol rate (i.e., the number of symbols transmitted per second) and multiply it by the number of bits per symbol to get to the digital transmission rate (i.e., bits per second).

5.2 Signal Modulation

To represent the digital bit stream in an analog format, we begin with a carrier signal and alter one of three basic characteristics: frequency, amplitude, or phase. Some modulation systems use a combination of two of these characteristics, typically phase and amplitude. See Figure 5-1.

- **Frequency:** In frequency modulation or Frequency Shift Keying (FSK), different frequencies are used to represent the 1 bit and the 0 bit. This is a fairly

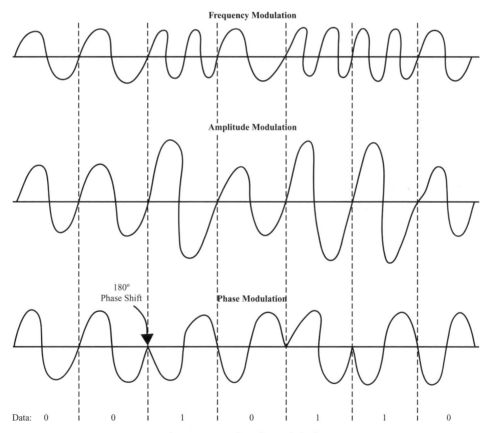

Figure 5-1: Signal Modulation

simple but inefficient mechanism for signal encoding. The original 1 and 2 Mbps 802.11 frequency hopping radio links used an FSK technique called Gaussian Frequency Shift Keying (GFSK) where the frequency shift is held to a minimum and a Gaussian filter is used to limit the frequency band.

- **Amplitude:** Amplitude means power, and in the simplest case we can transmit the carrier wave at two different power levels, one to represent a 1 bit and a second to represent a 0 bit. More complicated systems might define four amplitude levels (i.e., 2^2), and use each of them to represent a 2-bit pattern (i.e., 00, 01, 10, and 11); that technique is called Quadrature Amplitude Modulation (QAM). If sixteen different amplitude levels are used (i.e., 2^4), the technique is called 16-QAM. Each of those 16 states could be used to represent a 4-bit pattern (i.e., 0000, 0001, 0010 . . . 1111). Some systems go as high as 64-QAM or 256-QAM (i.e., 2^6 or 2^8), which encodes 6 or 8 bits per symbol. While the QAM term only references "amplitude," in reality these techniques combine amplitude and phase changes. The IEEE 802.11a and g radio link protocols use 16-QAM and 64-QAM for their highest transmission rates.

- **Phase:** Phase changes are shifts from the expected direction of the signal. If a signal is sent with no changes in its phase relationship, each cycle will look just like the one before it. In Figure 5-1, we show what the signal would look like if

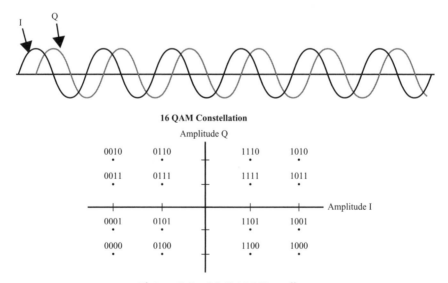

Figure 5-2: 16-QAM Encoding

we introduced a 180° phase shift. This technique is called Binary Phase Shift Keying (BPSK). We can also introduce four different phase shifts (i.e., 0°, 90°, 180°, and 270°), and each of those signals could represent 2 bits of information. That technique is called Quadrature Phase Shift Keying (QPSK). The 802.11b radio link uses QPSK modulation, and it is also used for some transmission rates for 802.11a and g.

The higher-capacity 802.11a and g modulation schemes use a combination of phase and amplitude. The transmitter generates two carrier waves that are 90° out of phase (i.e., the second is following a quarter of a cycle behind the first). The two are designated I (for in-phase) and Q (for quadrature). The transmitter varies the amplitude on each of the two carriers. Bit encoding is based on the amplitude of the first carrier and the amplitude of the second. The shorthand way for representing this type of coding system is called a constellation.

5.2.1 Basic Tradeoff: Robust and Inefficient Versus Feeble but Efficient

The basic point to stress is that as we add more phase and amplitude levels to the encoding system, we can carry more bits on each symbol. The problem with using a more efficient coding system (i.e., more bits/sec/Hertz) is that the receiver must be able to read that signal with greater precision or it will make mistakes (i.e., generate errors). The basic goal in the system design is to optimize the efficiency, but still create a signal that can be read accurately in the presence of noise. So the challenge is to design a system that is both efficient and robust. See Table 5-2.

5.3 Spread Spectrum Transmission

Wireless LANs employ a particular type of digital radio technology called spread spectrum, which simply means that the signal is encoded or spread over a wide band

Table 5-2: Bandwidth Efficiency of QAM Coding Systems

Encoding System	Number of States	Bandwidth Efficiency (bps/Hz)
QAM	4	2
16-QAM	16	4
64-QAM	64	8
256-QAM	256	16

of frequencies. This is a critical technology as it is used in 3G cellular systems, WiMAX, Bluetooth, and a variety of other applications as well as wireless LANs.

There are three basic types of spread spectrum transmission, and all of them are used in 802.11 WLANs:

- Frequency Hopping Spread Spectrum (FHSS)

- Direct Sequence Spread Spectrum (DSSS)

- Orthogonal Frequency Division Multiplexing (OFDM)

5.4 Frequency Hopping Spread Spectrum (FHSS)

The actress Hedy Lamarr is credited with developing the Frequency Hopping Spread Spectrum (FHSS) technique during World War II. The allies required a jam-proof radio guidance system for torpedoes, and Ms. Lemarr, who was married to an arms merchant at the time, came up with the idea of a system that would switch or hop quickly among a number of frequencies or channels in a seemingly random pattern. The pattern would appear to be random (i.e., pseudorandom), but would be known by the transmitter and the receiver. The transmitter would send on each channel for a short period of time and then hop to a different channel; knowing the pattern, the receiver could hop at the same instant to receive the next transmission. The amount of time the transmitter stays on a particular channel is called the *dwell interval*.

As the enemy would not know the hopping pattern, it would be impossible for them to intercept or override the communication. Further, as the transmitter could jump over a wide band of frequencies, traditional radio jamming to disrupt the communication would be impractical because the jammer would have to cover all of the channels. Frequency hopping also protects against power loss or *fading* at a particular part of the frequency spectrum (i.e., *frequency selective fading*) as the transmitter would not send on any one channel for very long. See Figure 5-3.

The downside for FHSS is that if 100 hopping channels were to be used, the system required 100 times the radio bandwidth! However, in modern FHSS systems, a number of users could also share one set of hopping channels. The hopping patterns can be designed to limit the likelihood of *collisions*, or the instance where two different transmitters send on the same channel at the same time. One of the original 802.11 radio links used a FHSS technique, though it is rarely found today.

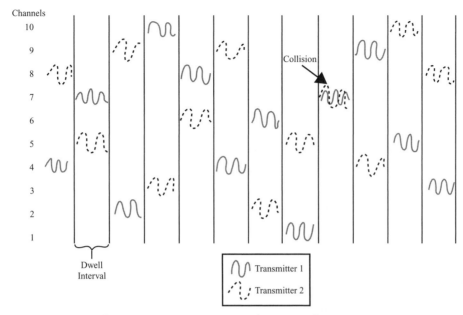

Figure 5-3: Frequency Hopping Spread Spectrum

5.5 Direct Sequence Spread Spectrum (DSSS)

Direct Sequence Spread Spectrum (DSSS) is the type of spread spectrum technology used in 802.11b wireless LANs 2.4 and 5 GHz cordless telephones, CDMA-based second generation digital cellular telephone networks, and all third generation cellular services.

DSSS transmission involves two primary steps:

1. Each bit of the transmission signal is spread or mapped into a series of shorter duration bits called *chips*. The number of bits in the spreading code is called the *bandwidth expansion factor*.

2. The chips are then transmitted through a traditional modulator (e.g., QAM or DPSK). As a result of the spreading, the signal has a higher transmission rate and therefore requires a much wider bandwidth channel.

At face value, the idea of increasing the transmission rate appears counterproductive, but we will see there are a number of offsetting advantages. High on that list is the fact that the transmission is tolerant of noise or fading that occurs in just part of the frequency spectrum.

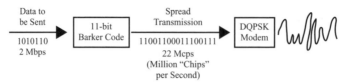

Figure 5-4: 802.11 Direct Sequence Spread Spectrum

In wireless LANs, the 802.11b standard uses a form of direct sequence spread spectrum transmission. The 1 and 2 Mbps 802.11 radio links use an 11-bit Barker code for spreading. With an 11-bit spreading code, a 1 Mbps transmission is sent at a rate of 11 Mcps (i.e., million chips per second) and a 2 Mbps transmission is sent at a rate of 22 Mcps (see Figure 5-4). The receiver correlates the received signal using a duplicate of the spreading code to correlate and recover the original bit. The 5.5 and 11 Mbps data rates in the 802.11b radio link standard use a technique called Complementary Coded Keying (CCK), where the data pattern is sent using a set of code words.

5.5.1 DSSS in CDMA Cellular Systems

Qualcomm has developed a more enhanced form of direct sequence spread spectrum technology that is used in CDMA cellular systems. That technology provides two additional features not found in WLAN systems.

1. In a CDMA cellular system, multiple users can transmit on the same transmission channel at the same time. The individual transmissions are kept separate by using different 64-bit spreading codes called Walsh codes. At the base station, the individual channels can be reconstructed by subtracting the codes; that is where the name *Code Division* Multiple Access comes from.

2. In CDMA-based cellular telephone networks, frequency channels can also be reused in adjacent cells. A pseudorandom number code (PN Code) is used to scramble the transmissions from each base station, and this allows the base station to identify which callers are connected through it.

5.6 Orthogonal Frequency Division Multiplexing (OFDM)

Orthogonal Frequency Division Multiplexing (OFDM) is a newer version of spread spectrum transmission that is designed to counteract the effects of multipath. Multipath, you will recall, is created when multiple reflected images of the signal are detected at the receiver. As those signals are at the same frequency but are out of phase, they will tend to cancel one another out. The IEEE 802.11a and g wireless LANs use OFDM, and the new 802.11n radio link will use it as well. OFDM is also used in the 802.16 or WiMAX standards for broadband wireless access systems.

The OFDM technique combines three basic ideas:

1. Divide the digital transmission signal into *n* subchannels

2. Divide the channel bandwidth into *n* orthogonal subcarriers or "tones"

3. Encode each subchannel (i.e., each part of the bit stream) onto one of the subcarriers and send them in parallel on the channel.

An OFDM system can be thought of as using a number of narrowband modems, each of which carries some of the bits and occupies some part of the total channel bandwidth. In 802.11a and g, there are 48 data subchannels and four pilot tones. The subcarriers are numbered −26 to −1 and 1 to 26; subcarrier 0 is not used. The pilot tones are sent on subcarriers −21, −7, 1, and 21. The transmission interface also includes FEC coding on the data carriers. The subcarriers are 312.5 KHz wide and overlap one another, but they are arranged in a manner that eliminates interference (see Figure 5-5). In mathematics, *orthogonal* means non-interfering. The 802.11n standard will use 52 rather than 48 data subcarriers, plus the four pilot tones in a 20 MHz channel and the 40 MHz implementation will use 108 data subcarriers and six pilot tones.

5.6.1 Multipath Mitigation

While the 802.11b standard launched the WLAN market, as time went on the designers determined that multipath was the major difficulty to be addressed in indoor radio systems. The measure that is used to quantify the multipath characteristics is *delay spread*, or the difference in time between the arrival of the first and last signal image; higher transmission rates require a shorter delay spread. Multipath is particularly problematic when the delay spread is so great that some of the echoes arrive during the next symbol interval, a problem called Inter-Symbol Interference (ISI). With a higher transmission rate the symbol interval will typically be shorter, and so those higher rates are less tolerant of long delay spread. With the 802.11a or g implementations of ODFM, 1/48th of the bits are carried on each subcarrier, and so the

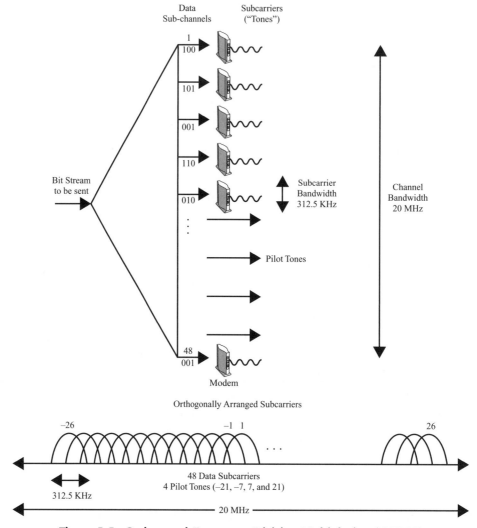

Data Sub-channels

Subcarriers ("Tones")

Bit Stream to be sent

Subcarrier Bandwidth 312.5 KHz

Channel Bandwidth 20 MHz

Modem

Orthogonally Arranged Subcarriers

−26   −1 1   26

48 Data Subcarriers
4 Pilot Tones (−21, −7, 7, and 21)

312.5 KHz

20 MHz

Figure 5-5: Orthogonal Frequency Division Multiplexing (OFDM)

symbol interval could be 48 times as long. In that way, the receiver can tolerate a far longer delay spread before the images cross over into the next symbol period. Addressing the multipath problem is what allowed the OFDM-based 802.11a and g radio links to carry almost five times the bit rate in a channel that is about 10 percent smaller (i.e., 20 MHz versus 22 MHz).

5.7 Forward Error Correction (FEC)

Before we look at the radio links that are defined for 802.11, there is one more essential building block. As we have noted, the key tradeoff in a radio transmission system is robustness versus bandwidth efficiency. Carrying more bits per cycle requires more and subtler distinctions to be introduced on the carrier wave, the result being that even minor distortions in the channel will result in transmission errors.

To improve reliability, many digital radio systems employ some form of FEC. The idea of an FEC system is to combine redundant information with the transmission, and then allow the receiver to detect and correct a certain percentage of the errors using a probability technique. The downside of FEC is that it increases the number of bits that must be transmitted over the channel. Increasing the bit rate requires a more efficient (and hence less reliable) transmission, however, that is more than offset by the error performance gain provided by the FEC. Both 802.11a and g radio link interfaces use FEC coding; the original 802.11 and 802.11b radio links do not.

There are a number of different FEC techniques that can be employed, and the 802.11a, g, and n wireless LANs use a version called *convolutional coding*. Convolutional coding works by taking the transmitted bit sequence, selecting a group of the most recently occurring bits, distributing them into two or more sets, and performing a mathematical operation (convolution) to generate one coded bit from each set as the output. The decoder in the receiver runs a comparison of the coded bit sequence and retains the best matches. With that information, it can make a maximum likelihood estimate of the correct decoded bit sequence.

5.7.1 FEC Overhead: 3/4 or 1/2 Coding

The coding rate of an FEC system is expressed as a fraction that represents the number of uncoded bits input at the FEC encoder and the number of coded bits output to the transmitter. The 802.11a and g interfaces use either 3/4 (where each 3 uncoded bits are sent as 4 coded bits, with 33 percent overhead) or 1/2 (where 1 uncoded bit is sent as 2 coded bits, with 100 percent overhead).

5.8 The 2.4 GHz Radio Links

The 802.11b and g radio links use the 2.4 GHz unlicensed Industrial, Scientific, Medical (ISM) band. The national radio regulatory authority (e.g., the FCC) defines the range of frequencies available and the rules for their use; the 802.11 standards define how the available bandwidth should be split into channels and how the signal should be modulated. The FCC regulations for the ISM band allow transmission power up to 100 mW, but to aid in heat dissipation, most WLAN products limit their transmission power to 30 mW.

The 2.4 GHz ISM band provides 83.5 MHz of continuous radio spectrum between 2.400 GHz and 2.4835 GHz. The 802.11 standards divide that radio band into 11 channels that are spaced at 5 MHz intervals. As an 802.11b radio channel requires 22 MHz of bandwidth and the channels are spaced at 5 MHz intervals, it takes five of those 5 MHz channels to support one 802.11b transmission. As a result, only three of the available channels can be used without creating interference with one another; those three channels are 1, 6, and 11. Using any of the other channels (i.e., 2, 3, 4, 5, 7, 8, 9, or 10) will create interference with one or more of those three. In Japan where more spectrum is included in the 2.4 GHz band, the channels extend up to 14, and four non-interfering channels are available. See Table 5-3 and Figure 5-6.

Table 5-3: IEEE 803.22b Channels

Channel	Center Frequency (GHz)	North America	Most of Europe	Spain	France	Japan
1	2.412	X	X			X
2	2.417	X	X			X
3	2.422	X	X			X
4	2.427	X	X			X
5	2.432	X	X			X
6	2.437	X	X			X
7	2.442	X	X			X
8	2.447	X	X			X
9	2.452	X	X			X
10	2.457	X	X	X	X	X
11	2.462	X	X	X	X	X
12	2.467		**X**		X	X
13	2.472		X		X	X
14	2.484					X

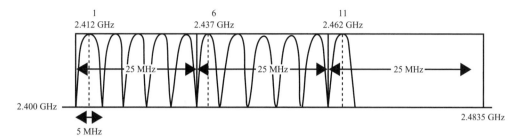

Figure 5-6: IEEE 802.11 2.4 GHz Channels

5.8.1 802.11G: 20 MHz Channel

As we have noted, an 802.11g channel is only 20 MHz wide, so if they had started with 802.11g, the IEEE could have squeezed four non-interfering channels into the 83.5 MHz of spectrum available in the 2.4 GHz band. However, when 802.11g was ratified, networks based on 802.11b was already widely deployed, so the same three channels are used allowing those channels to be shared between 802.11b and g users.

5.9 802.11 Radio Link Options: 1 and 2 Mbps

The original 802.11 radio link interface defined two different radio transmission formats each supporting data rates of 1 or 2 Mbps. There was also an infrared optical interface called Diffused IR (DIFR), which was never deployed. We will focus on the radio based alternatives: Frequency Hopping and Direct Sequence Spread Spectrum (FHSS and DSSS).

5.10 Frequency Hopping Spread Spectrum

The 802.11 frequency hopping radio link was adopted from Proxim's early RangeLAN wireless LAN product. The FHSS specification for the US defines 78 radio channels spaced at 1 MHz intervals in the band between 2.402 and 2.480 GHz. Countries with different allocations in the 2.4 GHz band will have different numbers of hopping channels.

In the US allocation, the hopping channels are divided into three groups with 26 channels in each.

Group 1: Channels 0 (2.402 GHz), 3 (2.405 GHz), 6, 9, . . . , 75 (2.478 GHz)

Group 2: Channels 1 (2.403 GHz), 4 (2.406 GHz), 7, 10, . . . , 76 (2.479 GHz)

Group 3: Channels 2 (2.404 GHz), 5 (2.407 GHz), 8, 11, . . . , 77 (2.480 GHz)

The channel hopping normally occurs every 20 msec, and that interval we referred to as the dwell time. The radio link used either two- or four-level Gaussian Frequency Shift Keying (GFSK) to deliver the 1 or 2 Mbps data rate. The transmitters hop only between channels in their group, so three separate FHSS WLANs can be supported in the same area.

The problem with the frequency hopping plan is that the channels in each group extend across the entire 2.4 GHz band, so they will overlap with channels 1, 6, and 11. The result is that using of any of the three FHSS groups would create interference with any 802.11b or g network operating in the same area. Very few products were ever built with the FHSS option. See Figure 5-7.

5.11 802.11 DSSS Radio Link

The other low-speed 802.11 radio link uses Direct Sequence Spread Spectrum technology, and is based on NCR's WaveLAN product. Like the frequency hopping option, it supports data rates of 1 or 2 Mbps, but it is designed to operate in the same 22 MHz channels as 802.11b. As a result, it has the same three non-interfering channels as 802.11b and g. While no products are being built that use this interface exclusively, the 802.11 DSSS radio link is used for the 1 and 2 Mbps fallback rates for 802.11b and g devices.

5.11.1 Direct Sequence Spread Spectrum: Bits to Chips

You will recall that the basic idea of a DSSS system is to take an input bit stream, spread or multiply it with a special code, and then modulate the signal for transmission. The 802.11 DSSS system uses an 11-bit Barker code, so the input bit rate is multiplied by a factor of 11. Each bit of user data is represented as 11 shorter transmission bits called *chips*. The spreading code is designed to allow the receiver to easily correlate the 11-chips as either a 1 bit or a 0 bit.

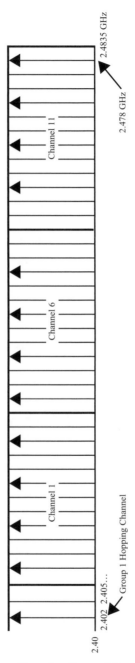

Figure 5-7: 802.11 Frequency Hopping Spread Spectrum Bandwidth Utilization

5.11.2 Signal Modulation: DBPSK/DQPSK

The actual data rate sent over the wireless channel is 11 or 22 Mcps (i.e., million chips per second). To provide the two different data rates, two different forms of phase shift keying are used. The 1 Mbps transmission rate uses a Differential Binary Phase Shift Keying (DBPSK) where each symbol represents one bit. The 2 Mbps transmission rate uses a Differential Quadrature Phase Shift Keying (DQPSK) where each symbol represents 2 bits.

5.11.3 PLCP Header Always Sent at 1 Mbps

One problem with supporting multiple transmission rates is that the receiver must know what modulation scheme the transmitter is using. In the 802.11 radio link, the transmission frame is preceded by a 192-bit Preamble and Physical Layer Convergence Protocol (PLCP) header (transmission time: 192 μsec). To accommodate the two transmission rates, the PLCP header is always sent at 1 Mbps (i.e., using the DBPSK modulation) and the Signal field in the header indicates whether the rest of the frame will be sent at 1 or 2 Mbps.

5.12 802.11b Radio Link Interface: DSSS

The 802.11b specification defined a High Rate-Direct Sequence Spread Spectrum (HR-DSSS) transmission format with two higher-speed (5.5 and 11 Mbps), transmission rates. The two lower-speed (1 and 2 Mbps) fallback rates were adopted from the original 802.11 DSSS scheme. Users operating at different bit rates can share the same radio channel. See Table 5-4.

Table 5-4: IEEE 802.11b Radio Link Interfaces

Data Rate (Mbps)	Modulation	Spreading Code
11	Quadrature Phase Shift Keying 8 bits/symbol, 1.375 Msps (I = 2.75 Mbps, Q = 8.25 Mbps)	Complementary Coded Keying (CCK)
5.5	Quadrature Phase Shift Keying 4 bits/symbol, 1.375 Msps (I = 2.75 Mbps, Q = 2.75 Mbps)	Complementary Coded Keying (CCK)
2	Differential Quadrature Phase Shift Keying (802.11)	11 bit Barker Code
1	Differential Binary Phase Shift Keying (802.11)	11 bit Barker Code

The 5.5 and 11 Mbps 802.11b modulation schemes use the CCK encoding technique. The transmission process begins by taking a group of 4 or 8 bits and dividing them into two groups (2 and 2 for 5.5 Mbps or 2 and 6 for 11 Mbps). The 2-bit group is sent on the I carrier, and the second 2- or 6-bit group is sent on the Q carrier. The Q carrier is generated by taking the 2 or 6 bits and representing them as a 2- or 6-bit code word. The transmitter uses a symbol rate of 1.375 Msps so the I carrier encodes the 2 bits per symbol (1.375 Msps × 2 = 2.75 Mbps), and the Q carrier encodes 2 or 6 bits per symbol (1.375 Msps × 2 = 2.75 Mbps or 1.375 Msps × 6 = 8.25 Mbps).

5.12.1 Long and Short Preambles

As with the original 802.11, the 192-bit PLCP Preamble/Header is sent using the 1 Mbps Differential Binary Phase Shift Keying modulation; the Signal field in the PLCP header identifies the transmission rate at which the MAC frame will be sent (i.e., 1, 2, 5.5, or 11 Mbps). However, that represents a considerable amount of overhead relative to the actual data that is being transmitted. As the PLCP Preamble/Header is sent at 1 Mbps, that will take 192 μsec to send. A 1500-octet data frame (plus the 28 octets of MAC frame header and trailer) takes 1.11 msec to send at 11 Mbps, so the PLCP Preamble/Header increases that by 17 percent.

The high rate 802.11b specification also defined an optional short Preamble. The Short Preamble is 120 bits long and the first 72 bits are sent at 1 Mbps while the last 48 bits are sent at 2 Mbps. The Short PLCP Preamble/Header takes 96 μsec to send, reducing that PLCP overhead to 8.6 percent for a 1500-octet frame sent at 11 Mbps.

5.13 IEEE 802.11g Radio Link Interface: OFDM

The other 2.4 GHz transmission interface is 802.11g or Extended Rate: PHY, which uses Orthogonal Frequency Division Multiplexing (OFDM). As we described earlier, a 20 MHz channel is divided into 52 orthogonally arranged subcarriers, each with a bandwidth of 312.5 KHz. Within the 52 subcarriers there are:

48   Data subcarriers

4   Pilot tones

The data stream is divided into 48 substreams that are coded onto the 48 data subcarriers; the coding is done so that sequential bits are sent on carrier channels that are far apart. There are eight different OFDM coding formats defined, and they use different modulation schemes (BPSK, QPSK, 16-QAM, and 64-QAM). The coding schemes also include varying levels of FEC coding (1/2, 2/3, or 3/4). Some non-standard 802.11g devices support data rates up to 72 Mbps by eliminating the FEC coding altogether. Four of the coding formats are mandatory (i.e., 54, 24, 12, and 6 Mbps) while the rest are optional. For backwards compatibility with 802.11b networks, 802.11g devices must also support the four data rates defined for 802.11b (11, 5.5, 2, and 1 Mbps). The data rates, modulation, and FEC coding options are summarized in Table 5-5.

5.13.1 Interoperability with 802.11b: Protection Mode

One major advantage of 802.11g is that it was designed to be backwards compatible with 802.11b. That means that 802.11b users can share channels with 802.11g users. However, that does create problems for the legacy 802.11b devices. As an 802.11b device may have been designed before 802.11g was created, it might not recognize 802.11g signals as valid transmissions in its network (i.e., the 802.11b device would

Table 5-5: 802.11g High Speed Modulation Systems

User Rate (Mbps)	FEC Coding	Line Rate (Mbps)	Total Bits per Subcarrier	Modulation	Bandwidth Efficiency (bps/Hz)	Symbols/ Subcarrier (Ksps)
54*	3/4	72	1.5 Mbps	64-QAM	2.7	250
48	2/3	72	1.5 Mbps	64-QAM	2.4	250
36	3/4	48	1 Mbps	16-QAM	1.8	250
24*	1/2	48	1 Mbps	16-QAM	1.2	250
18	3/4	24	500 Kbps	QPSK	0.9	250
12*	1/2	24	500 Kbps	QPSK	0.6	250
9	3/4	12	250 Kbps	BPSK	0.45	250
6*	1/2	12	250 Kbps	BPSK	0.3	250
11	None	11	—	CCK	0.5	—
5.5	None	5.5	—	CCK	0.25	—
2	None	2 M	—	DPSK	0.18	—
1	None	1 M	—	BPSK	0.09	—
*Mandatory data rates. Others are optional.						

interpret the 802.11g transmission as noise). The result is that the 802.11b device might initiate a transmission when an 802.11g device is transmitting.

To allow the two generations of devices to operate on the same channel, as soon as one 802.11b device joins the network, the 802.11g devices must shift into Protection Mode, which places three requirements on them:

1. The 802.11g users automatically shift into the RTS/CTS option where each frame transmission is preceded by an RTS/CTS exchange. The RTS/CTS messages are sent at the 1 Mbps transmission rate so they can be read by the 802.11b devices.

2. As all devices on the network must be able to understand the control traffic, all transmission Preambles and all Management and Control frames must be sent at 1 Mbps (i.e., the lowest data rate supported in 802.11b).

3. The 802.11g MAC protocol defines shorter pre-transmission waiting intervals (DIFS and SIFS) and shorter back-off counters (CW_{MIN} and CW_{MAX}) than 802.11b. However, when the two are used together on the same network, the 802.11g devices must use the longer 802.11b intervals.

The 802.11b users will be operating at a lower data rate, so it will take them longer to send frames, causing the higher-speed 802.11g users to wait longer to get access to the channel. The end result is that when 802.11b and g users share a channel, the maximum throughput for the 802.11g users drops from around 30 Mbps to about 15 Mbps. Given that performance degradation, WLAN users would do well to eliminate all 802.11b devices if at all possible. The one potential problem for WLAN voice is that many early wireless LAN handsets only supported the 802.11b radio link. So having even one of those 802.11b handsets place a call on a shared 802.11b/g network will have a significant impact on the capacity for all voice and data devices that use the 802.11g interface.

5.14 802.11a 5 GHz Radio Link Interface

Currently, 802.11a is the only WLAN interface that operates in the 5 GHz U-NII band. It uses the same Orthogonal Frequency Division Multiplexing (OFDM) coding options as 802.11g. As 802.11a does not have to deal with legacy installations in the 5 GHz band, it does not include the 1, 2, 5.5, and 11 Mbps modulation options we have in 802.11g. See Table 5-6.

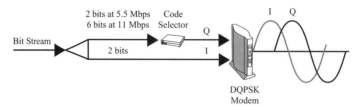

Figure 5-8: High Rate Direct Sequence Spread Spectrum (5.5 and 11 Mbps)

Table 5-6: 802.11a High Speed Modulation Systems

User Rate (bps)	FEC Coding	Line Rate (bps)	Total Bits per Subcarrier	Modulation	Bandwidth Efficiency (bps/Hz)	Symbols/ Subcarrier
54 M*	3/4	72 M	1.5 Mbps	64-QAM	2.7	250 Ksps
48 M	2/3	72 M	1.5 Mbps	64-QAM	2.4	250 Ksps
36 M	3/4	48 M	1 Mbps	16-QAM	1.8	250 Ksps
24 M*	1/2	48 M	1 Mbps	16-QAM	1.2	250 Ksps
18 M	3/4	24 M	500 Kbps	QPSK	0.9	250 Ksps
12 M*	1/2	24 M	500 Kbps	QPSK	0.6	250 Ksps
9 M	3/4	12 M	250 Kbps	BPSK	0.45	250 Ksps
6 M*	1/2	12 M	250 Kbps	BPSK	0.3	250 Ksps
*Mandatory data rates. Others are optional.						

One advantage of using the 5 GHz band is that for the time being, there are fewer other users on those channels and hence less interference. As they are unlicensed, however, there is no reason to expect that advantage to persist. Further, as 5 GHz signals lose power more rapidly than 2.4 GHz signals, the transmission rate/range relationship will be different. The 5 GHz devices typically transmit at a higher maximum power rate (e.g., 40 mW versus 30 mW for 2.4 GHz devices) but they will still shift to a lower transmission rate at a shorter range.

Like the 2.4 GHz channel assignments, the 5 GHz channels are spaced at 5 MHz intervals and are numbered sequentially. In the definition of the 802.11a radio link, the IEEE specified that devices would only use every fourth channel (e.g., 36, 40, 44, etc.), and hence there are no overlapping channels in the 5 GHz band.

5.15 Additional 5 MHz Spectrum: 11 Additional Channels

The original 5 GHz allocation in the US included 300 MHz of non-contiguous spectrum, arranged in Low (5.15 to 5.25 GHz), Middle (5.25 to 5.35 GHz), and Upper (5.725 to 5.825 GHz) frequency bands of 100 MHz each. The Low band was designated for indoor operation only, the Upper band for outdoor operation only, and the Middle band could be used for either. Based on their different applications, the maximum transmit power was different for each. Each of those bands could accommodate four non-interfering 20 MHz channels, for a total of 12 802.11a channels.

In November 2003, the FCC added 255 MHz of spectrum at the low end of the Upper band (i.e., 5.470 and 5.725 GHz). The FCC did not define rules for the use of those channels until January 2006. The IEEE has elected to define 11 additional channels (105 to 145) in that new band for a total of 23. See Table 5-7.

5.16 Tradeoffs with 802.11a

Up until this point in time, few enterprise networks have migrated to the 5 GHz band, though that may change with the advent of WLAN voice. There are a number of factors that impact the decision to use 802.11a. First, the plus side:

- **Number of Channels:** The biggest factor in favor of 802.11a is the number of channels that are available in the 5 GHz band. With only three channels to work with, it is difficult to limit co-channel interference in a large-scale 2.4 GHz network.

- **Interference:** At the moment there are far more interference sources in the 2.4 GHz than in the 5 GHz band. However, as the 5 GHz band is also unlicensed, that advantage will likely disappear over time. We are already seeing 5 GHz cordless phones, and unlicensed WiMAX implementations will likely operate in the 5 GHz band.

- **802.11a Devices Do Not Interfere with 802.11b/g Devices:** As they operate on a completely different set of channels, it is possible to install an 802.11a and an 802.11b/g network in the same area. This dual-overlay approach can be used:
  - To support voice users on an 802.11b network while supporting data users on an 802.11a network.
  - To support internal users on a secure 802.11a network while supporting visitors on an open 802.11b/g network that only provides Internet access.

Table 5-7: 802.11a Transmission Channels (US)

Channel Number	Center Frequency (GHz)	Max Transmit Power	U-NII Band
36	5.180	**Indoor Only** 50 mW + 6dBi Antenna Gain	**Low** (5.150–5.250 GHz)
40	5.200		
44	5.220		
48	5.240		
52	5.260	**Indoor/Outdoor** 250 mW + 6dBi Antenna Gain	**Middle** (5.250–5.350 GHz)
56	5.280		
60	5.300		
64	5.320		
104	5.520	**Indoor/Outdoor** 250 mW + 6dBi Antenna Gain	**Approved Nov. 2003** (5.470–5.725 GHz)
108	5.540		
112	5.560		
116	5.580		
120	5.600		
124	5.620		
128	5.640		
132	5.660		
136	5.680		
140	5.700		
144	5.720		
149	5.745	**Outdoor** 1 W + 6dBi Antenna Gain	**Upper** (5.725–5.825 GHz)
153	5.765		
157	5.785		
161	5.850		

- **Security:** As there is greater signal loss at 5 GHz, hackers may have a more difficult time monitoring 802.11a networks.

There are also a number of negative factors to consider:

- **Signal Loss:** The basic problem with 802.11a is that it uses higher frequency 5 GHz radio channels where the signal attenuation is greater than at 2.4 GHz transmissions; the difference is roughly a factor of 2. Despite the higher transmit power, 802.11a NICs typically operate at the maximum bit rate (i.e., 54 Mbps) for a range of 40 feet, while an 802.11g radio can operate at the maximum rate at a range of 100 feet. The bottom line is that more access points are probably needed to cover an area for 802.11a than for 802.11g. It also means that to deploy both a 2.4 GHz and a 5 GHz network in the same facility, a separate radio coverage design will need to be done for each.

- **Supporting the Installed Base:** Many of the newer WLAN NICs support 802.11a, b, and g, but the vast majority of the installed base uses 802.11b or g. If we are purchasing cards for internal users, we can begin buying 802.11a components. However, if we need to provide "public" access, most of the public will have 802.11b/g, so focusing on 2.4 GHz support will be the obvious choice.

5.17 The Developing IEEE 802.11n Radio Link

The newest WLAN radio link will be defined in IEEE 802.11n. The goal for that standard was to provide a throughput (i.e., not a raw data rate, actual *throughput!*) of at least 100 Mbps. Accomplishing that will require both a higher capacity radio link and a more efficient MAC protocol. After a long, drawn out battle, a draft standard was accepted in early 2006 and final ratification was expected by September 2007. However, over 1000 comments were received on the draft specification, and so the final ratification will likely occur some time in 2009. In the interim, the Wi-Fi Alliance published a certification plan in mid-2007 to ensure compatibility among devices built to the *Draft-n* specification.

The members of the 802.11n committee were originally divided between two alternate designs:

- One plan looked at delivering the higher bit rate by increasing the channel bandwidth from 20 to 40 MHz. That would reduce the number of available channels, and complicate the problem of backwards compatibility, as a 40 MHz transmitter would impact 2–20 MHz channels.

- The other solution sought to maintain the 20 MHz channel and incorporate Multiple Input–Multiple Output (MIMO, pronounced MY-mo) antenna technology to achieve the higher data rate. Opponents said that MIMO was too new, and still too expensive.

The final draft standard called the Joint Proposal incorporated elements from both designs.

5.18 The IEEE 802.11n Draft Specification

The draft 802.11n specification includes both 20 MHz and 40 MHz implementations operating in either the 2.4 GHz or 5 GHz bands. Using an optional MIMO antenna system, it can support a maximum raw data rate of 288.9 Mbps in a 20 MHz channel and 600 Mbps in a 40 MHz channel. The draft specification is available on the Enhanced Wireless Consortium's Web site (http://www.enhancedwirelessconsortium. org).

5.18.1 Backwards Compatibility

The most important attribute about 802.11n is that it will be backwards compatible with existing 802.11a/b/g networks. However, operating in what is referred to as Mixed Mode, there will result a significant reduction in throughput for the 802.11n devices. The Mixed Mode operation must use additional preamble information so that legacy (i.e., 802.11a/b/g) devices can recognize 802.11n transmissions. The standard also defines a Greenfield Mode, which assumes all devices utilize the 802.11n High Throughput (HT) link.

5.18.2 Multiple Input–Multiple Output (MIMO) Antennas

The key technology element that is responsible for the vastly higher data rates provided in 802.11n is the use of a MIMO antenna system. MIMO represents one of the most significant developments in radio technology and holds the promise of improving bandwidth efficiency, range, and reliability. The basic idea of a MIMO system is to send multiple simultaneous transmission signals in the same frequency band, but to transmit them from multiple antennas spaced some distance apart. Separating the antennas will create a difference in the multipath images for each signal allowing for spatial multiplexing. The receiver will be able to distinguish the individual signals because each will have a different and unique set of multipath images. See Figure 5-9.

5.18.3 MIMO Transmission

On the transmit or output side, a MIMO system divides the bit stream into two or more substreams or *transmission chains*; the 802.11n draft standard can support up to four

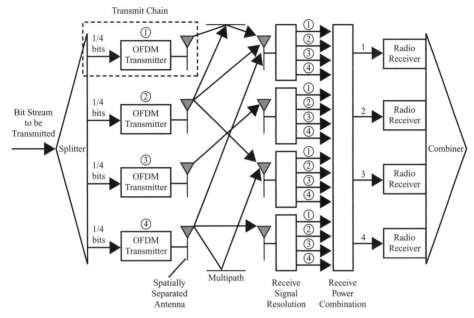

Figure 5-9: Multiple Input-Multiple Output Antenna System

streams. The data streams are then modulated using separate OFDM radio transmitters but sent using separate antennas that are spaced some distance apart; to achieve the special diversity effect, the antennas have to be spaced at least a half-wavelength apart. Again, all of those radios will be transmitting in the same frequency band.

5.18.4 MIMO Receivers

The tricky part of a MIMO system is making the receiver work. As all of the transmit chains will be sending in the same frequency band, the receiver must be able to resolve and reconstruct the different transmission chains independently. The input side also uses multiple receiving antennas that are also spaced some distance apart. As the transmission chains are being sent from different points in space, the receiving antennas will receive a different set of multipath images or a radio fingerprint for each transmitted signal. Where one receiver may get a weak copy of a particular transmission, another could get a stronger copy. The key to a MIMO system is that all of the transmit chains are received on all of the receive antennas. The receive processor resolves the separate transmission signals received at each antenna and adds their power

Table 5-8: 802.11n Data Rates in 20 MHz Channel

Modulation	FEC	Data Rate (Mbps)							
		One Stream		Two Streams		Three Streams		Four Streams	
		Guard Interval		Guard Interval		Guard Interval		Guard Interval	
		800 ns	400 ns	800 ns	400 ns	800 ns	400 ns	800 ns	400 ns
BPSK	—	6.5	7.2	13.0	14.44	19.5	21.7	26.0	28.9
QPSK	—	13.0	14.4	26.0	28.89	39.0	43.3	52.0	57.8
QPSK	—	19.5	21.7	39.0	43.33	58.5	65.0	78.0	86.7
16-QAM	—	26.0	28.9	52.0	57.78	78.0	86.7	104.0	115.6
16-QAM	—	39.0	43.3	78.0	86.67	117.0	130.0	156.0	173.3
64-QAM	2/3	52.0	57.8	104.0	115.56	156.0	173.3	208.0	231.1
64-QAM	—	58.5	65.0	117.0	130.00	175.5	195.0	234.0	260.0
64-QAM	5/6	65.0	72.2	130.0	144.44	195.0	216.7	260.0	288.9

Note: The 400 ns guard interval will not likely be practical initially.

together. MIMO systems can also support a different number of transmit chains in each direction, so we might have three transmit chains generated by an access point and two transmit chains generated by a client.

5.18.5 OFDM Radio Link Features

The radio link for each 802.11n transmit chain uses OFDM modulation. The interface can be configured to operate in either a 20 MHz or a 40 MHz channel. In a 20 MHz channel, the 802.11n OFDM system uses 52 data subcarriers and four pilot tones. The 40 MHz 802.11n interfaces 108 subcarriers and six pilot tones. As with 802.11a and g, there will be multiple modulation and FEC combinations supported with adaptive modulation based on the path characteristics. The modulation, FEC coding, and data rates for one to four data streams in 20 MHz and 40 MHz channels are summarized in Tables 5-8 and 5-9.

5.18.6 Other Radio Link Features

There are a number of other optional features that are included in the specification that can further improve the range and reliability.

Table 5-9: 802.11n Data Rates in 40 MHz Channel

Modulation	FEC	Data Rate (Mbps)							
		One Stream		Two Streams		Three Streams		Four Streams	
		Guard Interval		Guard Interval		Guard Interval		Guard Interval	
		800 ns	400 ns	800 ns	400 ns	800 ns	400 ns	800 ns	400 ns
BPSK	—	13.5	15.0	27.0	30.0	40.5	45.0	54.0	60.0
QPSK	—	27.0	30.0	54.0	60.0	81.0	90.0	108.0	120.0
QPSK	—	40.5	45.0	81.0	90.0	121.5	135.0	162.0	180.0
16-QAM	—	54.0	60.0	108.0	120.0	162.0	180.0	216.0	240.0
16-QAM	—	81.0	90.0	162.0	180.0	243.0	270.0	324.0	360.0
64-QAM	2/3	108.0	120.0	216.0	240.0	324.0	360.0	432.0	480.0
64-QAM	—	121.5	135.0	243.0	270.0	364.5	405.0	486.0	540.0
64-QAM	5/6	135.0	150.0	270.0	300.0	405.0	450.0	540.0	600.0

Note: The 400 ns guard interval will not likely be practical initially.

- **Space-Time Block Coding/Spatial Diversity:** This is another spatial effect where one bit stream can be encoded into two spatial streams using block coding and sent from spatially separated transmitters. This technique is used to improve reliability rather than to increase the capacity of the channel.

- **Beam Forming:** A phased-array type antenna technology can be used to steer the radio signal in a particular direction.

5.18.7 MAC Layer Enhancements

Along with the high performance radio link, the 802.11n standards also define enhancements for the MAC protocol. Among the major features are frame aggregation, multiple traffic ID block acknowledgment, reduced inter-frame spacing, spatial multiplexing power save, and power save multi-poll.

- **Frame Aggregation:** As each transmission will require a preamble, the more data we send on each channel access, the more efficiently we use the channel, and the higher the throughput. Also, at higher transmission rates, those frames will take less time to send. To provide that efficiency, the 802.11n MAC provides the ability to carry several higher level protocol units into a single

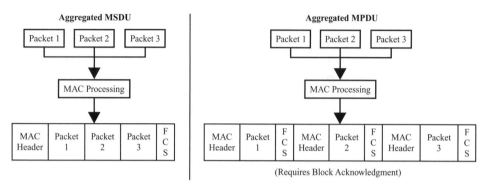

Figure 5-10: 802.11n Frame Aggregation

WLAN frame with one MAC header (Aggregated MSDU Format) or to take several higher level messages, each with its own MAC header, and send them on a single network access (Aggregated MPDU Format). For 802.11e compliance, all of the PDUs/SDUs must be in the same access category (i.e., voice, video, data, or background data). See Figure 5-10.

- **Multiple Traffic ID Block Acknowledgment (MTBA):** When the Aggregated MPDU format is used, a single block acknowledgment must be sent. When Aggregated MSDU is used, there is one MAC header and one MAC trailer (i.e., frame check sequence) so only one acknowledgment is needed. For Aggregated MPDU, a block acknowledgment is used and a bit map in that message specifies which subframes within that transmission were received with errors.

- **Reduced Inter-Frame Spacing (RIFS):** The 802.11n specification includes an expedited form of streaming transmission using a shorter Inter-Frame Spacing interval or RIFS. Where the original SIFS interval was 10 μsec, RIFS is 2 μsec. The RIFS option can only be used in all-n (i.e., not Mixed-Mode) networks.

- **Spatial Multiplexing Power Save (SM Power Save):** As a MIMO transmitter will require power to drive multiple spatial streams, its power draw could be considerable. To reduce that, in SM Power Save all but one transmit chain is shut down. The station can wake up on its own (static mode) or it can be woken up by the access point with an RTS/CTS exchange (dynamic mode).

- **Power Save Multi-Poll (PSMP):** PSMP combines ideas from 802.11e Automatic Power Save Deliver and Hybrid Controlled Channel Access options.

An 802.11 client under a continuous transmission requirement would first send a transmission specification (T-Spec) message to the access point, detailing its data rate, frame size, frame interval, and access category (i.e., priority level). The access point then defines a PSMP Service Period during which it would download buffered data for all PSMP stations, and then allows upstream transmissions from those stations.

While the final 802.11n specification will not be completed until 2009, WLAN products that incorporate MIMO technology are already appearing on the market. A number of consumer-oriented access point vendors introducing non-standard or *Pre-n* devices that offered higher bit rates. While these will be a boon for home users, enterprise customers will be well advised to steer clear of non-standard Pre-n devices. While a user may get a higher bit rate in the short term, long-term compatibility will be a bigger factor in an enterprise environment.

However, in mid-2007, the Wi-Fi Alliance introduced a certification plan for devices built to the Draft-n specification, and ensuring enterprise users that those devices will be backwards compatible with the standard when it is finally released. Given its track record on backwards compatibility up to this point, enterprise users may consider the Certified Draft-n option if they have a pressing need for the improved rate and range capabilities of 802.11n.

5.18.8 802.11n Deployment Options

The advent of 802.11n does make our deployment planning somewhat more complex. As we noted, 802.11n devices can share channels with legacy 802.11a/b/g devices, but the throughput for the 802.11n devices will drop substantially. The preferred approach will be to deploy 802.11n only networks, what is referred to in the standard as Greenfield Mode. The question is, do we use the 2.4 GHz or the 5 GHz band for those 802.11n only networks? Users who had been considering deploying voice devices in the 5 GHz band will now have to choose between 5 GHz for voice or 5 GHz for 802.11n. At the moment there are no 802.11n voice handsets, but those could certainly appear in the future. Fortunately, with 23 channels available in the 5 GHz band, some could be reserved for voice while others are used for 802.11n data devices.

The 802.11n standard also calls for other changes in the network. The signal propagation characteristics will be different for 802.11n transmissions, which will call for a different layout of APs. Also, 802.11n access points will require a 1 Gbps connection to the wired LAN rather than the 10/100 Mbps interfaces that are common today. In short, 802.11n is a major development in WLANs, but its direct impact on voice deployments is still a few years out. However, any decision that is being made with regard to WLAN deployments today must consider the long-term impact on 802.11n deployments.

5.19 Non-Standard Radio Links: Pre-n and Super G

While the battle over the 802.11n standard raged, chip vendors began developing their own proprietary radio links to provide higher data rates in advance of the standards.

5.19.1 Atheros: Super G and Turbo Mode

WLAN chipmaker Atheros developed a 108 Mbps radio link for the 2.4 GHz band called Super G that ran in a 40 MHz channel. They also developed a similar specification for the 5 GHz band called Turbo Mode. In each case, the specification supports transmission rates up to 108 Mbps, twice the maximum rate of 802.11a or g. Some implementations push that data rate to 154 Mbps by eliminating the FEC coding. The Atheros trick is to double the bandwidth of the radio channel so that additional OFDM subcarriers can be supported. Rather than using 48 subcarriers arranged in a 20 MHz radio channel, the Super G and Turbo Mode transmitters use 96 subcarriers and spread over a 40 MHz band. There are two major difficulties introduced with the use of these non-standard radio links:

1. The number of available channels in the 5 GHz band is reduced from 23 to 12. The 2.4 GHz Super G modulation can only run in Channel 6, so the number of available channels is reduced from three to one.

2. Tests have found that using the 2.4 GHz Super G radio link on Channel 6 causes significant interference and reduced throughput with standards-compliant radio links on Channels 1 and 11. With a Super G transmitter operating on Channel 6, the data rate for users on Channels 1 and 11 is reduced by about 50 percent.

5.20 Conclusion

The radio link is the Layer 1 for the wireless LAN, and there are a number of options available and in development. The range of options and inherent difficulty of radio transmission makes the WLAN radio link an area of particular concern. Beyond recognizing the basic tradeoff between efficiency and reliability, it is also important to note the longer-term impact of opting for a non-standard radio link. While a higher bit rate may be available initially, the use of non-standard products can greatly limit the choice of products further down the line.

The initial wave of WLAN voice handsets supported the 11 Mbps 802.11b interface only. While that did allow those devices to be intermixed with data devices on a 2.4 GHz network, the low data rate limited the number of simultaneous voice calls that could be supported. The introduction of 802.11a and g handsets should more than double the number of simultaneous calls, though we will still have the problem of decreased throughput for 802.11g devices on mixed 802.11b/g networks. As devices can rarely be upgraded from one radio link to another, it is important for consumers and businesses to know what they are buying before signing the contract.

CHAPTER 6

Privacy and Security Issues in WLANs

Countless surveys of enterprise users have shown that the biggest concern regarding the use of wireless LANs is security. Access to the wired LAN can be secured by posting a security guard at the front door, and assuming that anyone who is in the building is authorized to have network access. In actuality, the focus on wireless security alerted many organizations to the vulnerabilities that existed in their wired systems. Unescorted visitors can easily connect laptops to wired Ethernet ports, and a simple USB thumb drive can provide a mechanism to pilfer gigabytes of sensitive information from an unattended PC. However, the free-space propagation of radio signals in a wireless LAN changes the exposure dramatically, as a hacker can now access the network while sitting in a car in the parking lot. A directional antenna could allow him or her to monitor WLAN transmissions from over a mile away. The limited security features of the original 802.11 wireless LANs brought this issue to the forefront.

The badly flawed security features provided in the original Wired Equivalent Privacy (WEP) created an exposure to both eavesdropping and unauthorized network access. A further security exposure came from users implementing their own wireless LANs by simply connecting an unauthorized (and unsecured) "rogue" access point to their wired Ethernet connection. Those users failed to realize that the WLAN signals did not stop when they got to the outside wall, but continued well into the parking lot and to the street beyond. Rogue access points are a major security concern because they are seen as wired (and assumed trusted) connections to the wired LAN.

Fortunately, help is on the way. The anemic security features of the original 802.11 protocols were not designed for the requirements of commercial customers. Now the IEEE has responded with enterprise-grade security features for wireless LANs, and those tools should address all but the most extreme security requirements. We do need to give special attention to voice devices, as unauthorized access exposes the organization to toll fraud or theft of service. In the public VoIP arena, we have already seen carriers victimized by hackers.

In this section, we will begin with an overview of the major areas that must be addressed in network security and identify the major security exposure introduced through the use of wireless LANs. Then we will examine the basic capabilities of WEP and describe why it is considered inadequate for any commercial installation. Our main focus will be on the newer IEEE security protocols like 802.11i and 802.1x, how they work, and the level of protection they will afford. In particular, we will describe the security mechanisms that will be most appropriate for voice applications.

6.1 Security Requirements: Authentication, Privacy, and Availability

Security remains the number one reason that enterprise customers cite for deferring plans to implement wireless LANs. However, when we say "security" in a network context, it is a general term to address four major functions: authentication, privacy, non-repudiation, and network availability.

- **Authentication:** The task of ensuring that only legitimate users get access to the network. For a client device, that means that only legitimate users are allowed to join the network.

- **Privacy:** The process of ensuring that all transmissions are seen only by the legitimate sending and receiving parties.

- **Non-Repudiation:** Confirmation that the message is coming from the expected source, and has not been altered in any way.

- **Network Availability:** Protection from denial of service, viruses, worms, and other strategies that could render the network impaired or unusable.

To further complicate matters, the network may have to enforce different levels of security for different groups of users. Internal users will need secure access to internal resources, while the network may also have to support unsecured Internet access for visitors. Voice adds yet another dimension, as an unauthorized voice handset might be able to place phone calls that would be charged to the company's telephone bill.

6.2 WLAN Security Policy Recommendations

Security has been a weak point in WLANs, but tools are now available to build and maintain a secure wireless infrastructure. The first step will be to develop an adjunct to the organization's security policy that specifically addresses WLANs. As every organization's security philosophy will be different, we can only make general recommendations regarding what that policy should cover.

- WLAN security should be an additional responsibility for the corporate security team, not something separate and distinct. The security team will need extra training to support this new responsibility.

- It is now critically important to ensure compliance with legal mandates like Sarbanes-Oxley (i.e., the Public Company Accounting Reform and Investor Protection Act of 2002), the Health Insurance Portability and Accountability Act (HIPAA), and the other regulations that impact IT systems. Unfortunately, these regulations do not spell out required techniques for securing information but instead outline overall goals and guidelines like "current best practices."

- No wireless devices should be installed without the approval of the IT department.

- The IT department should also require that all wireless clients be configured by them to ensure they have personal firewalls and system integrity checks that are up to date. Systems that provide network access control (i.e., a server that automatically confirms the client has the required firewalls and latest security updates before it is allowed to access the network) can also be considered, although many organizations have found these systems to be too costly.

- To enforce the wireless security policy, some form of continuous RF monitoring is required to ensure that no unauthorized WLAN devices are being used by employees or contractors. That RF monitoring can be done through a WLAN switch or with a stand-alone monitoring system like those from AirMagnet, AirDefense, or AirTight Networks. Manual or walk around scanning on a periodic basis is not effective in detecting WLAN security vulnerabilities.

- All WLAN devices (e.g., access points and switches) should be hardened and tested to the same degree as any Internet-facing equipment.

- The Intrusion Detection System (IDS) should be evaluated to ensure that it recognizes WLAN attack signatures.

- All WLAN communications should support as a minimum 802.1x authentication (with a mutual or two-way authentication protocol) and Wi-Fi

Protected Access (WPA) based encryption. The new 802.11i defined encryption that incorporates the Advanced Encryption Standard is stronger, but may require that older wireless NICs be upgraded.

6.3 WLAN Security Exposure

In reality, all network-connected systems have some degree of security exposure. In a wired network, our primary defense will be physical security of the facility and a policy that requires any visitors be accompanied during their entire time within the facility.

To prevent unauthorized access from the Internet, firewalls are installed on the Internet access connection, and the firewalls inspect and filter incoming and outgoing packets according to a set of rules. Intrusion detection systems can also be used to monitor for known attack signatures, and application proxies can shield internal network addresses from the public Internet. As hacking strategies are being refined continuously (hackers are a crafty lot), the security personnel must update the software in all of those devices continuously to counteract new attack strategies.

Besides the use of free-space radio propagation, there are a few particular vulnerabilities to note with WLANs in general.

- **Radio Leakage:** WLAN networks often provide excellent coverage in areas surrounding the building. Some organizations have tried to constrain radio coverage to the desired coverage area by using directional antennas and carefully adjusting the access point's transmit power. There is even a company that sells paint that when applied to walls will attenuate radio signals. Unfortunately, unless we build a Faraday cage around the entire facility, strategies to constrain radio signals have been found to be only marginally effective and can adversely impact coverage inside of the facility.

- **Clear Text Headers and Control Messages:** While the content of a WLAN frame may be encrypted, the header fields are sent in the clear. That gives the hackers a head start for strategies that depend on MAC spoofing. Management frames are also sent "in the clear." Early on, users thought they could keep the network "secret" by optioning the access points to exclude the network name or SSID in their beacon messages. Unfortunately, any device that is associating includes the SSID in an unencrypted control message, so a hacker can determine the network name by monitoring a single association exchange.

- **Denial of Service:** WLAN protocols add new options for denial of service attacks. The most typical is a Disassociation attack where a hacker monitors users associating with the network, and immediately sends a Disassociation message to disconnect them. A flood of CTS messages can fool stations into thinking that the network is continuously busy. Further, radio jamming can also be added to the list of DoS strategies, and 2.4 GHz radio jammers are available for sale on the Internet for a few hundred dollars.

- **Infected Clients:** If a client device does not have a functional firewall and security software, it can pick up any number of worms and viruses when connected to a public Hot Spot. That virus-laden device can then infect the internal network when it is connected to the wired LAN.

- **Bridging Through a Client:** Windows Internet Connection Sharing (ICS) allows bridging between network interfaces on a laptop, allowing a user's Wi-Fi interface to provide an open portal to the wired LAN. A study by Network Chemistry in 2006 found 37 percent of sampled devices to be so configured.

- **WLAN Hacking Tools Are Readily Available**
  - Locate Networks: NetStumbler (http://www.netstumbler.com)
  - Crack WEP Keys: AirSnort (www.airsnort.shmoo.com), WEPCrack (http://wepcrack.sourceforge.net/)
  - Improve Signal Reception: Pringles Antenna (www.arwain.net/evan/pringles.htm): improves reception
  - Voice Decoding: VOMIT (http://vomit.xtdnet.nl/) and IPCrack (http://www.crack.ms/cracks/i_1.shtml)

6.4 Recognized Security Vulnerabilities

There are a number of potential security exposures introduced with the implementation of wireless LANs.

- Eavesdropping on unencrypted wireless transmissions or cracking the encryption key to allow eavesdropping: This is the most obvious and widely reported hack, but can now be prevented with more functional encryption options like those described in 802.11i/WPA2.

- Eavesdropping or gaining network access through user deployed rogue access points or Ad Hoc connections on bridge enabled laptops: This is more an administrative problem, but continuous RF (radio frequency) monitoring is required to ensure compliance. Users must be informed repeatedly that it is against company policy to install or use WLAN technology without the approval of the IT department.

- Introducing malicious access points to intercept association exchanges and user credentials: A malicious AP is an AP (typically a Soft AP, or a program installed in a laptop that allows it to mimic the functions of an access point) operated by a hacker in close proximity to the facility. The idea is to get legitimate client devices to associate with the malicious AP as part of a strategy to gain unauthorized access. This type of attack is called a man-in-the-middle or evil twin attack, and works by tricking users into providing legitimate user names and passwords to a hacker. This type of attack can be thwarted with mutual authentication, an authentication system where the network authenticates the user and the user then authenticates the network.

- Visitors disabling security features in APs: APs should be installed in areas where they are hard to reach, as physical access may allow users to disable security features. The simplest method is to press the Reset button to restore the factory default settings.

- Stealing APs with encryption keys and MAC address lists stored in them: These can be hacked later, off-line.

- Access through lost or stolen client devices: Users are always the weak link in a security system. Users must be instructed regarding the vulnerability created when a corporate device falls into the hands of a hacker.

- Radio Link Disruption: Intentional Denial of Service attacks targeting the RF environment (i.e., radio jamming) or unintentional radio interference from other devices using the unlicensed radio bands.

- Denial of Service attacks targeting the WLAN protocols.

6.5 Three Generations of WLAN Security

While wireless LAN security may have gotten off on the wrong foot, there are effective security measures available today. It is important to recognize, however, that WLAN security is the user's responsibility. The user must recognize the vulnerability, investigate the alternatives, and then ensure that they are implemented throughout the network. As it turns out, implementing the best security features may require expensive hardware upgrades and service disruptions. Further, your range of options may be restricted if you must also support older legacy devices that cannot be upgraded to the newer security features.

To organize the options, we can look at WLAN security solutions in three generations:

1. **Pre-Security:** WEP's anemic security features required the development of security work arounds using overlay techniques like a VPN/VLAN configuration (described below).

2. **Short-Term Improved Security (Acceptable Practice):** Wi-Fi Protected Access (WPA) is an intermediate term security enhancement developed by the Wi-Fi Alliance that provides significantly improved features. The primary advantage of WPA is that it can be implemented with a software upgrade rather than a hardware upgrade. WPA can be implemented using a pre-shared key that is manually entered in the device, or with 802.1x authentication that provides session-based key distribution and key refresh.

3. **Recommended Security Solutions (Best Practice):** The IEEE has developed a more comprehensive solution to WLAN encryption called 802.11i; the Wi-Fi Alliance refers to 802.11i compatibility as WPA/2. This option uses a far more powerful AES-based encryption, however, AES generally requires different hardware than WEP or WPA. That means that implementing 802.11i may involve replacing older NICs and access points. Like WPA, 802.11 can be implemented using a pre-shared key that is manually entered in the device, or with 802.1x authentication that provides session-based key distribution and key refresh. See Table 6-1.

6.6 Wired Equivalent Privacy (WEP)

The WEP described the original security features for 802.11 WLANs, and it is equivalent to very little security. The name Wired Equivalent Privacy refers to the goal of providing privacy that is equivalent to a wired connection with no encryption. The purpose of WEP was to protect transmissions from casual eavesdropping, not from a determined hacking attempt. WEP provides three basic security features:

Table 6-1: WLAN Security Options

	Description	Encryption	Authentication
Wired Equivalent Privacy (WEP)	WEP	40 or 104 bit RC4	Shared Key
	Dynamic WEP	40 or 104 bit RC4	802.1x with Key Distribution for Per-Session Keys
	VPN/VLAN Configuration	168 bit 3DES	802.1x
WI-Fi Protected Access (WPA)	Intermediate Term Option from the Wi-Fi Alliance	128 bit RC4 with TKIP	Personal: User Implemented Pre-Shared Key
		128 bit RC4 with TKIP	Enterprise: 802.1x with Key Distribution
802.11i/WPA2	IEEE Standard for WLAN Security	AES	Personal: User Implemented Pre-Shared Key
	IEEE Standard for WLAN Security	AES	Enterprise: 802.1x with Key Distribution

1. RC-4 based encryption using a static 40 or 104 bit key.

2. Shared Key Authentication based on the WEP encryption key.

3. MAC Address Filtering.

6.6.1 RC-4 Encryption/Dynamic WEP

The WEP encryption uses a technique called Rivest Ciper-4 (RC4) with a 40 bit or 104 bit key.* There was no key distribution protocol, so the key would be input manually in the access point and the client devices and would last for the life of the network. Recognizing the vulnerability of a static key, some enterprise networks implemented authentication systems that would provide a new key for each session; that capability was called Dynamic WEP. Dynamic WEP was not defined in the standards.

6.6.2 Shared Key Authentication

The 802.11 protocol defines an optional Shared Key Authentication procedure for joining a network. In Shared Key Authentication, the AP responds to an Authentication frame from a client by sending a message with a 128-byte Challenge Text. The user

*Note: The WEP key is typically identified as 64 or 128 bit, but the key includes a 24 bit initialization vector, so the actual key is 40 or 104 bits.

device encrypts the Challenge Text with its WEP key and returns it to the AP. If the AP can correctly decipher the Challenge Text (i.e., the client and the AP are using the same WEP key), the user is authenticated. In essence, authentication is based on the same static WEP key, so if a hacker cracks the WEP key, he or she can eavesdrop on all WLAN transmissions and can also authenticate on the network.

6.6.3 MAC Address Filtering

While not specifically defined in the protocol, MAC Address Filtering is supported on most APs. With this feature, the network administrator can input a list of MAC addresses on the access point, and only those addresses will be allowed to associate. As the MAC address is a unique identifier, the theory is that this filter can be used to exclude unauthorized stations. The problem is that WLAN header fields are sent unencrypted, so it is easy for a hacker to discover and spoof a legitimate MAC address. Further, the list must be input on each access point, and it must be updated every time a user is added to the network.

6.7 How Vulnerable is WEP?

In WEP, both authentication and privacy are based on the same static key. It was quickly determined that the key was extremely vulnerable to brute force (i.e., trial and error) attacks. There are free software tools available on the Web that will allow a hacker to crack the encryption key in a matter of hours using a sample of a few million packets. Some techniques can discover the key in a matter of minutes. Under no circumstances should a commercial customer consider using static WEP (i.e., WEP with a key that does not change) as their sole means of securing the network.

Why is WEP So Bad?

A number of analysts have attributed WEP's deficiencies to incompetent developers. However, that is not an accurate assessment. At the time WEP was being developed, the US Government had a requirement that no security mechanism could be exported if it included an encryption algorithm that used a key length greater than 56 bits. As WLANs were intended for worldwide deployment, the developers' hands were tied.

6.7.1 Tools of the Trade: AirSnort, WEPCrack

AirSnort is a Linux-based program that is designed to discover WEP encryption keys; AirSnort and other programs with similar capabilities (e.g., WEPCrack and dweputils) are available for free on the Web. Anyone can just Google those names to find the sites. With the program loaded on a laptop, all the hacker needs to do is find a location to monitor the WLAN and collect five to ten million packets. To hack the WLAN more surreptitiously, the hacker could use a directional antenna to monitor the network from a greater distance. Those directional antennas are often called *Pringles Antennas*, as some of the early homemade models were fashioned out of discarded Pringles potato chip cans. In actuality, coffee cans are a better size for 2.4 GHz signals.

Using that sample, the tool uses a trial and error mechanism to crack the WEP key. Collecting that number of packets from a small WLAN might take weeks, but larger networks generate that volume of traffic in hours. In 2007, a group of German researchers demonstrated a more efficient cracking mechanism that required a sample of fewer than 100,000 packets.

6.7.2 Longer Keys?

Many users thought that they could provide better protection by using the longer 104 bit key. In actuality, the cracking techniques work equally well regardless of the key length; a 104 bit key is just as vulnerable as a 40 bit key. The key is composed of a 24 bit initialization vector (IV) and either a 40 bit or a 104 bit WEP key making the total length is 64 or 128 bits. The initialization vector (IV) is actually a hint to use to crack the key. The 24 bit IV provides 2^{24} or 16,777,216 different RC4 encryption streams for every key, regardless of how long the rest of the key is. Further, not all of those IV numbers work equally well. There are approximately 9,000 weak IVs, and the snooping programs can recognize those to shortcut the cracking process. Some later WEP implementations avoid using those weak IVs.

6.7.3 The Three-Minute WEP Crack

AirSnort and similar programs crack the encryption key by a brute force technique. At an Information Systems Security Association (ISSA) meeting in 2004, the FBI demonstrated a much faster technique for cracking WEP keys. While the complete details were not disclosed, the technique apparently involved sending a message that would cause a WLAN station to send a known message in response. Even though the message was encrypted, the hacker knew exactly what the clear text version of the message would be. Working backwards from the encrypted version, they could determine the key that was used to encrypt it!

WEP Cloaking

One interesting technique to thwart brute force attacks against WEP is called WEP Cloaking. Developed by the wireless security company AirDefense, WEP Cloaking uses a relatively simple strategy to confound key crackers. AirDefense makes wireless intrusion detection monitors, but they can have those monitors periodically transmit fraudulent WLAN frames encrypted with a different WEP key. As the brute force system is attempting to discover the encryption key by trying different possibilities to determine if they produce readable text, these spurious frames will confuse the cracking program.

6.8 The VLAN/VPN Stopgap

Given the inherent weakness of WEP-based security, many early WLAN users deployed a stopgap security configuration using virtual LANs (VLANs) and secure tunnel Virtual Private Network (VPN) access. The ability to define virtual LANs is a standard feature in enterprise-oriented LAN switching equipment. In practice, each Virtual LAN corresponds to an IP Subnet; traffic passing between VLANs/IP Subnets must go through a routing function where security screens can be inserted.

The basic idea of the VLAN/VPN configuration is to configure a separate virtual LAN for all of the wireless access points; by default, all WLAN clients will become part of that VLAN when they associate with any of those access points. Users accessing the network over the WLAN are treated as untrusted, and there would be no servers in that virtual LAN. Before any wireless LAN clients would be provided access to any other virtual LAN, they would first be authenticated, and a secure tunnel connection created between the wireless client and a firewall. The authentication would deter unauthorized access, and the secure tunnel encryption would ensure privacy even though the transmissions are sent over the radio link. Rather than the relatively weak security provided in WEP, VPNs typically use IPsec that provides session-based keys and 168

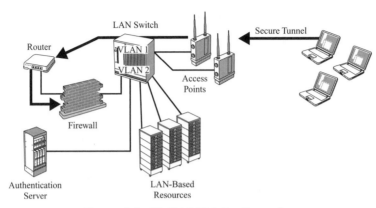

Figure 6-1: VLAN/VPN Configuration

bit 3DES encryption. In essence, WLAN users are treated in the same way as remote users accessing the network through the public Internet. See Figure 6-1.

6.8.1 Security in Spite of the WLAN

While the VPN solution is ingenious, it is essentially providing security *in spite of the wireless LAN*. In this configuration, neither WEP's RC4-based encryption nor the shared key authentication is used. Rather, authentication and encryption are done around the wireless LAN, using tools that are implemented in the client device, the authentication server, and the firewall on the wired LAN. One concern with this strategy is the performance of the firewall. Stations accessing the firewall over the Internet operate at relatively low speeds compared to WLAN stations. Supporting a large number of concurrent WLAN clients may require an upgrade to the firewall. Finally, voice handsets typically lack the software to support secure tunnel connections and so this strategy would not support voice devices.

6.9 New Wireless LAN Security Protocols: 802.11i and 802.1x

There are now two important enhancements that address most of the current security concerns we have identified. The IEEE has defined two standards that address the requirements for privacy and authentication:

- **802.11i:** An encryption standard based on the Advanced Encryption Standard that provides far better security than WEP's RC-4 algorithm.

- **802.1x:** An authentication framework for wired and wireless LANs called the Extensible Authentication Protocol (EAP).

6.10 802.11i/WPA2: Advanced Encryption Standard (AES)

The enhanced encryption capabilities for 802.11 wireless LANs are described in 802.11i; the Wi-Fi Alliance identifies compatibility with 802.11i with the designation *WPA2 Certified*. The key feature is the use of the Advanced Encryption Standard (AES). With computer technology getting better and faster, encryption schemes like WEP's RC-4 and even the Digital Encryption Standard (DES) and 3DES are becoming more susceptible to brute force attacks. In 1997, the National Institute of Standards and Technology (NIST) initiated a study to choose a replacement for DES. After consulting with cryptographic experts from around the world, the committee chose a new algorithm for what they termed the Advanced Encryption Standard (AES). The 802.11i implementation of AES is called Counter Mode with CBC-MAC Protocol or CCMP.

For AES, NIST chose a version of the *Rijndael* algorithm developed by Vincent Rijmen and Joan Daemen (the name Rijndael, pronounced RIN-dall, is a contraction of their two last names). Rijndael is a symmetric block cipher, which means that it operates by breaking the unencrypted data into blocks and applying the Rijndael algorithm. *Symmetric* means that the block size is the same for the input and the output. The data block is 128 bits, and the encryption key can be 128, 192, or 256 bits long. The 802.11i implementation uses only 128 bit keys. The algorithm itself is insanely complicated, but the bottom line is that a key length of 128 bits allows AES to create up to 3.4×10^{38} possible combinations. In short it is superior to the 3DES encryption used in VPN secure tunnel communications and is vastly superior to RC-4.

6.10.1 Federal Information Processing Standards (FIPS) 197

In May of 2002, AES became an official standard for use by the US Federal government as outlined in the Federal Information Processing Standards (FIPS) document 197, published by NIST. FIPS publication 197 outlines how AES should be used to secure classified information, when it should not be used, as well as other procedures pertaining to how the Federal Government handles information. For more information see: http://csrc.nist.gov/encryption/aes/rijndael/, or search Wikipedia's excellent description of AES (http://www.wikipedia.org).

6.11 Implementation Difficulties

One significant problem with 802.11i is that most older access points and NICs do not include the hardware to generate the AES algorithm. To ensure adequate performance, data encryption is typically performed in hardware rather than software; any access point or client NIC would require that chip to generate the AES encryption. In advance of the standard's final ratification, some access points began to advertise that they were upgradeable to 802.11i. In that case, the access point included the required hardware and needed only a firmware upgrade to implement 802.11i. However, to implement that capability, the NICs would also require 802.11i compatibility. Few NICs sold prior to 2006 were 802.11i capable; further, few Wi-Fi voice handsets included 802.11i capability.

Starting in mid-2006, the Wi-Fi Alliance is requiring 802.11i compatibility in any new Wi-Fi devices it certifies. All of the second generation Wi-Fi handsets introduced in 2007 included WPA2 capability, and we can expect that on all new handsets. Even without its inclusion in the certification requirements, Wi-Fi equipment vendors recognize the need to include 802.11i capability to remain competitive in the market. However, as implementation involves a hardware upgrade, 802.11i poses a challenge for the base of installed/legacy devices.

6.12 Wi-Fi Alliance's Wi-Fi Protected Access (WPA Certified)

Recognizing user concerns regarding security and the potential costs of upgrading to 802.11i, the Wi-Fi Alliance developed an interim plan for improved wireless LAN security called Wi-Fi Protected Access (WPA). The important feature of WPA is that it can generally be implemented on existing 802.11 networks with a software upgrade. Devices that have been tested and found to be in compliance with Wi-Fi Protected Access are identified as WPA Certified. Most Wi-Fi voice handsets will support WPA.

There are two major elements included in WPA.

- Temporal Key Integrity Protocol (TKIP).

- Message Integrity Check (MIC).

6.12.1 Temporal Key Integrity Protocol

To address the encryption deficiencies of WEP, WPA uses the Temporal Key Integrity Protocol (TKIP). TKIP uses the same RC-4 encryption as WEP, so it does not require a hardware upgrade. WPA uses the RC-4 encryption algorithm with a 128 bit key (plus a 24 bit initialization vector). TKIP means that the encryption key is changed on every packet. As it takes a few million packets with the same key to crack the encryption key, it is impossible to collect enough packets with the same key. Further, as WPA changes the key on every packet, a hacker will not be able to copy and repeat the same packet (i.e., a message replication hack).

6.12.2 Message Integrity Check (MIC)

WPA also provides a CRC-based message integrity check (MIC) called *Michael*. This check is intended to ensure the receiver that the message was not altered in any way. Unfortunately, the MIC does not provide the integrity it promises. It is particularly vulnerable to bit-flipping, where a hacker changes a bit in the message and then changes the integrity check to compensate.

6.13 WPA and WPA2 Implementation: Personal Versus Enterprise

There are two defined implementations for WPA and WPA2 that use different mechanisms to generate the initial key:

- **Personal:** This version uses a Pre-Shared Key (PSK) approach. That means the user inputs the initial key in the access point and the client device as they did with earlier WEP devices.

- **Enterprise:** The Enterprise implementation requires a server using one of the 802.1x authentication protocols. The user inputs credentials (e.g., user name and password), the server authenticates them, and provides the encryption key as part of that exchange. See Figure 6-2.

There is one widely reported vulnerability involving the Personal or Pre-Shared Key implementation, and the exposure involves the selection of the initial key. With the PSK implementation, the user selects the initial key, and the subsequent keys are generated by a published algorithm. As the algorithm for changing the key is known, if a hacker can guess the initial key he or she can determine all of the subsequent keys; that process of guessing the initial key using a list of typical selections is called a *dictionary attack*. To thwart this type of attack, the user must select an initial key that is difficult to guess (i.e., not the user's birthday or child's name). The Personal or Pre-Shared Key option is

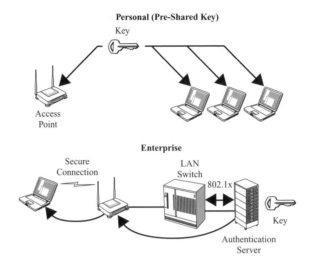

Figure 6-2: WPA, WPA2 Implementation Options

available with both WPA and WPA2, but it is used primarily with the former. PSK is not recommended with WPA2.

As of early 2007, WPA and WPA2 are supported on roughly an equal number of networks, though the VPN/VLAN configuration has a slightly higher percentage of implementations. Moving forward it does appear that WPA2 will be the preferred solution, particularly if voice devices are supported. Enterprise users are still uncomfortable with WPA's dependence on RC-4 encryption, and that discomfort has caused many users to pass it by. WPA does provide a useful security upgrade for home networks, however.

6.14 Extensible Authentication Protocol (EAP): 802.1x

WPA and 802.11i address privacy or protection from eavesdropping. The other major concern in security is authentication or ensuring that only legitimate users get access to the network. The protocol that addresses authentication for wired and wireless LANs is IEEE 802.1x, the Extensible Authentication Protocol or EAP ("Extensible" means "the ability to be extended"). That standard defines a generic framework for port-based authentication for both wireless and wired local networks. The standard specifies a framework that will accommodate various authentication methods including certificate-based authentication, smart cards, and traditional passwords.

Given the nature of wireless LANs, there are three critical elements that must be included in the authentication process:

1. **Strong Protection of User Credentials:** As the user credentials are being sent over a radio link that can be monitored, it is essential that the user's credentials be protected from eavesdropping. Most implementations use Transport Layer Security (TLS) to protect the user credentials.

2. **Mutual Authentication:** To thwart man-in-the-middle/evil twin attacks, it is important that the user be able to authenticate the network to which he or she is connecting. In that way the user device is ensured to not be delivering his or her authentication credentials to a hacker.

3. **Key Distribution:** The weakness of static keys has been adequately demonstrated, so the authentication system must also have the ability to provide a new encryption key for each session.

6.14.1 802.1x Authentication Process

When 802.1x authentication is used, the access point initially allows the client to communicate solely with the authentication server. Once the authentication process is completed successfully, the access point will then allow the authenticated user to communicate over the network with other entities. That authentication process involves a number of EAP Request and EAP Response messages to be exchanged between the client or *supplicant* and the authentication server or *authenticator*. The actual exchange will vary based on the authentication protocol that is used. The authentication method being used is identified by a Type Code that is included in the initial exchange. See Figure 6-3.

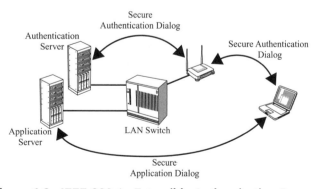

Figure 6-3: IEEE 802.1x Extensible Authentication Protocol

6.15 802.1x Authentication Protocols

IEEE 802.1x does not require a specific protocol for authentication. EAP is essentially an encapsulation protocol that allows different authentication protocols to be used with it. Effectively, EAP serves as a conduit for the authentication protocol.

There are several authentication protocols supported with 802.1x, which provides flexibility with regard to implementation. However, it also introduces an array of potential compatibility issues. This text is not intended as a definitive source for evaluating the options, but rather as a guide to assist in initiating such an investigation.

- **Cisco's Lightweight Authentication Extension Protocol (LEAP):** This is Cisco's proprietary username-based authentication, and it was the first widely used authentication protocol for WLANs. LEAP uses two MS-CHAP (Microsoft Challenged Handshake Protocol) V2 exchanges to provide mutual authentication. While it does provide for mutual authentication and session-based encryption keys, it has been found to contain numerous vulnerabilities. The weakness in LEAP stems from its dependence on the antiquated MS-CHAP V2, and its vulnerability to dictionary attacks in environments where strong password policies cannot be enforced. While it did provide greater protection than WEP, Cisco no longer recommends LEAP as an authentication alternative.

- **Cisco's EAP FAST:** Cisco's recommendation is now PEAP (described below) or EAP FAST; FAST stands for Flexible Authentication via Secure Tunneling. EAP-FAST uses symmetric key algorithms to achieve a tunneled authentication process. The tunnel establishment relies on a Protected Access Credential (PAC). As it uses a mutually authenticated tunnel, EAP-FAST protects against dictionary attacks and man-in-the-middle strategies. The process operates as follows:
  - Phase 0 (Optional): This phase can be used to enable the client to be dynamically and securely provisioned with a PAC.
  - Phase 1: A mutually authenticated secure tunnel is established between the client and authentication server using PAC.
  - Phase 2: Client authentication is performed over the tunnel by the client sending username and password to authenticate and establish client authorization policy.

- **Transport Layer Security (EAP-TLS):** EAP-TLS is an IETF-defined authentication mechanism that uses a Public Key Infrastructure (PKI) certificate-based mutual authentication. TLS is nearly identical to the Secure Sockets Layer (SSL) protocol used to secure Web transactions. Once the authentication is complete, 802.1x enables key generation. The biggest drawback with TLS is that it requires digital certificates in all user devices. If the organization does not currently employ client certificates, EAP-TLS will involve a major upgrade.

- **Tunneled TLS (EAP-TTLS) and Protected EAP (PEAP):** TTLS and PEAP are similar approaches that are based on TLS extensions. These techniques can be used with higher-layer authentication protocols (e.g., MS-CHAPv2) and require certificates only in the authentication servers. Both techniques operate in the same fashion. In the first phase, they establish a TLS tunnel. In the second phase, the authentication exchange is sent over the encrypted connection. That first phase is often referred to as the *outer* authentication and the second phase is called the *inner* authentication.

TTLS is required in organizations that wish to retain a non-EAP RADIUS infrastructure, such as those running Microsoft Active Directory. Such enterprises must front-end the RADIUS server with a TTLS server, which will convert EAP requests to legacy authentication methods. PEAP is an extension to EAP that is also designed to overcome the problems associated with certificate management under EAP-TLS. The main difference is that PEAP supports just Windows XP and 2000 operating systems natively, while TTLS supports many more (including several Microsoft handheld platforms). TLS, TTLS, and PEAP methods offer the strongest standards-based authentication and are recommended for 802.1x implementations.

6.15.1 Other EAP Authentication Methods

There are other authentication mechanisms supported with EAP, however they do not provide the strong cryptographic protection required for wireless networks. They might be used as inner authentication protocols with TTLS or PEAP.

- **MD5:** This is a one-way user authentication using a password. MD5 is the simplest authentication method, but it is also the weakest. With MD5, user passwords are stored in a way that allows the authentication server access to the plain-text password. This arrangement makes it possible for entities other than the authentication server to gain access to the password file. Further, MD5 does not provide for key distribution.

- **Generic Token Card:** This protocol was developed to support token cards like RSA's SecureID that provide one-time passwords.

- **EAP-SIM:** EAP-SIM provides authentication for GSM-based mobile phones using the Subscriber Identity Module (SIM) card.

6.16 Wireless Intrusion Detection and Network Access Control Systems

A corporate wireless security policy is essentially meaningless if there is no mechanism to ensure it is actually being followed. As the "honor system" has not proved to be an effective means to secure the enterprise IT infrastructure, organizations are starting to turn to automated systems to ensure compliance. Two of the solutions that are coming into widespread use are wireless intrusion detection systems (WIDS) and network access control (NAC).

6.16.1 RF Monitoring and Wireless Intrusion Detection Systems (WIDS)

Companies like AirMagnet, AirDefense, and AirTight provide RF Monitoring and Intrusion Detection Systems. These are stand-alone systems for monitoring the RF environment for Rogue APs, RF Hacking Attempts, and the use of Ad Hoc/peer-to-peer networks, which represent security vulnerabilities. Many also include the ability to monitor the 2.4 GHz and 5 GHz bands for intentional (e.g., radio jamming) or unintentional RF interference (e.g., microwave ovens or other devices that use the unlicensed frequency bands).

The configuration of a WIDS involves a number of passive RF monitoring devices connected over the wired LAN to a controller; the monitors continuously sample or "sniff" the RF environment. The typical configuration will have one monitor for every three to five access points. In the initial system configuration, the network technicians identify all legitimate access points (e.g., access points in neighboring WLANs) so the monitor will ignore them. Those legitimate access points are also recorded as benign to protect against false alarms. Once the initial configuration is set, if an unrecognized access point or Ad Hoc network is detected, the system immediately issues an alarm. See Figure 6-4.

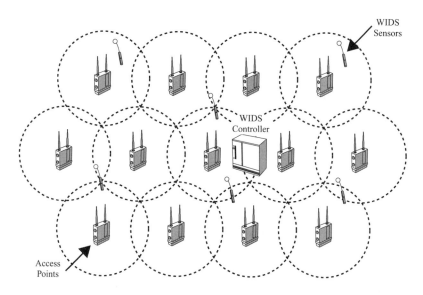

Figure 6-4: Wireless Intrusion Detection System

When an alarm is reported, the network engineers must determine if it is hostile and what should be done about it. Many of these alarms are not threats; if your neighbor adds an access point to his or her WLAN, the RF monitoring system will detect it. A number of attributes of the report can allow the network engineers to characterize it.

- **Signal Strength:** Provides a rough indicator of proximity.

- **Network Name/SSID:** If the network name is the same as other known benign neighbors, it is probably a new access point that has been added to that network.

- **Continuous or Intermittent Contact:** Weak, intermittent contact is typically a distant network that is detected occasionally. Strong intermittent contact could be a hacker.

- **Whether the Unknown Access Point Is Connected to the Wired Network:** An authorized AP on the wired network is typically a rogue that was installed by a user, a department, or by a contractor working in the facility. If the unknown access point is not connected to the wired network, it might be a benign neighbor or an actual hacking attempt.

There should be procedures in place to respond to alarms immediately, and address typical scenarios. For example, if the AP has a network name similar to a network known to be benign, contact the network administrator for that network to confirm they have added a new AP to their network. While they might be somewhat unsettled about someone monitoring their network, if that is not one of their APs, they will probably want to know about it too. These systems now include tools that allow you to locate the unknown AP; the precision of those location capabilities varies widely, and the best attempt to identify location down to a room number.

The Mystery Access Point

One organization that was diligent about monitoring their RF environment got a scare when they found an unknown access point with a strong signal that appeared at roughly the same time every afternoon. As they determined the device was not connected to their wired network, they assumed a hacker was sizing up their defenses, though they did not detect any attack signatures. After staking out the parking lot for several days and scanning the surrounding streets with binoculars, they discovered there was a delivery van with a WLAN access point that pulled up to their loading dock at roughly the same time every afternoon. Case closed.

If the unknown device cannot be characterized as benign and the situation warrants it, the network engineers can take steps to disable the unknown access point; this is a recommended procedure if the access point is connected to the wired LAN. The most typical attack strategy is called a Disassociation attack. The idea is to monitor stations associating with the unknown access point and immediately send a disassociation message to disconnect them. This type of defense is very effective, so you must be very sure that the unknown AP is truly malicious. If you attack one of your neighbor's legitimate wireless LANs, and you could be exposing your company to legal action from the company whose network you are attacking.

As we noted earlier, WLAN switches now incorporate many of these same monitoring capabilities. The vendors of stand-alone systems maintain that their systems are more flexible and comprehensive. One important feature to note with the WLAN switch-based monitoring capabilities is whether the monitoring is continuous or a time-shared function. For continuous monitoring it is essential to deploy additional access points on the WLAN switching system and dedicate them to the monitoring function. If a company depends on monitoring through access points that are handling live traffic (i.e., the time-shared approach), those access points cannot be monitoring at the same time that they are handling network traffic. The WLAN switch vendors contend that even if one access point is handling live traffic, the surrounding access points will be monitoring, so essentially there is uninterrupted monitoring. Most systems will allow the installation of extra access points dedicated to full-time scanning.

6.17 Network Access Control (NAC)

The other automated security compliance system that is coming into widespread use is Network Access Control. NAC systems utilize servers that work in conjunction with the authentication process to confirm the security capabilities of client devices accessing a network. Before a client is allowed to connect to the network, the NAC server confirms that the client device has the latest operating system patches, the anti-virus software is up-to-date, and that the firewall is activated. More advanced systems can confirm a variety of settings in the client including whether internal bridging is disabled.

The concern is that a laptop that is not adequately protected may pick up a worm, virus, or other malware while connected to a public Hot Spot, and then spread it to other stations when that device is connected to the wired network. If a client device does not pass the screening, it can be quarantined or connected to a remediation server that updates the device's software. Sygate, Cisco, Symantec, and Microsoft all have products that perform these functions.

6.18 Special Issues in WLAN Voice Security

Up to this point we have focused primarily on security for data devices, but voice adds an additional set of concerns. Confidential information is routinely passed over phone calls, and users are preconditioned to expect that their phone calls are private and secure.

6.18.1 WLAN Voice Exposure

With many Wi-Fi voice handsets, that perception of security may be false. A number of Wi-Fi phones still use the rudimentary WEP-based security. Further, voice traffic has specific patterns and port addresses that can be readily recognized on a wireless LAN. The other difficulty is that there is limited processing power in many handsets, so they may not have the capacity to add even WPA-based encryption.

6.18.2 Threat: Eavesdropping, Not Theft of Service

The concern with WLAN voice security is primarily centered on privacy. Virtually all enterprise Wi-Fi handsets do 802.1x-based authentication, and depend on additional proprietary authentication mechanisms. A handset must be registered with the telephony server before it can make or receive calls, and the server telephony often provides a code that must be input on the handset to thwart registration hijacking. As a result, it is unlikely that a hacker would be able to break into the network and make free phone calls. SIP-based Wi-Fi phones are a different matter, given that the signaling is based on an open standard, and the security features are still being developed. However, steps can be taken in the IP PBX system to defeat SIP phones that attach to the network over the WLAN.

Monitoring WLAN phone calls that are encrypted with the minimal WEP security is another matter. Once the WEP key has been cracked, voice decoders like VOMIT (for G.711) or IP Crack (for G.729A) can be used to listen in on the conversation. As we have seen, a voice system depending on WEP will offer little assurance of privacy. There is an encrypted voice protocol called Secure RTP (SRTP) that uses AES-based encryption, but we have not seen it deployed on any voice handsets as yet.

The good news is that there have been no reported hacks on WLAN voice. However, that may be due to the fact that there are relatively few WLAN voice networks out there and the hackers have not found them yet. In the meantime, many of the early Wi-Fi handsets have been upgraded to support WPA at least. The newer Wi-Fi handsets that are coming on the market should offer the full range of recommended security features including 802.11i /WPA2 and 802.1x authentication. However, most of these voice security solutions have not been put to the test by determined hackers, so enterprise security personnel should be particularly watchful when it comes to Wi-Fi voice security.

6.19 Solutions for Mixed Security Environments

Security solutions become especially challenging when there are different groups of users or devices that require different security policies. Among the typical scenarios for mixed environments would be:

- **Guest Services:** Installations where WLAN-based Internet access is provided for visitors in the facility.

- **Legacy Devices:** The requirement to support fixed-function devices like bar code scanners that can only support WEP and have no potential for upgrades.

- **Voice Systems:** The requirement to support legacy voice devices that cannot be upgraded to the security solution used for data devices (e.g., VPN/VLAN based data security).

Initially WLANs could support only one security mechanism for all devices on the network. The result was security by least common denominator; the security solution would be dictated by the options available on the least functional device. There are a number of solutions to the mixed environment today, but it is important for the network designer to recognize the capabilities and limitations of each.

6.19.1 Guest Services in a VPN/VLAN Configuration

If secure internal access is provided using a VPN secure tunnel connection from a separate VLAN for wireless devices, normally there are no servers available in the VLAN supporting the access points. To provide Internet access for visitors, an Internet connection can be added to that VLAN. However, if the WLAN now uses open access, anyone can get access to that Internet connection, not just recognized visitors. Further, the guests might be accessing inappropriate material over the organization's Internet access connection. The preferred solution would be to have an authentication server involved in that process and an employee that provides a unique password to legitimate guests, allowing the guests' Web activities to be traced back to them.

6.19.2 Dual Network/Overlay Network

A more comprehensive, though potentially more expensive, solution would be to maintain two separate WLANs, one for internal users and the other for visitors, and use different channels for each. The visitor network would most certainly have to use a 2.4 GHz channel, as most visitors will have 802.11b or g devices. That solution might work in a small office where the entire location can be served with a single channel, but in large facilities where the coverage area requires multiple access points in a cellular arrangement, the three channels available in the 2.4 GHz band will not allow for two separate, non-interfering networks. In that case, the solution might be to support internal users on a 5 GHz 802.11a network and leave the 2.4 GHz band for visitors. Many enterprise access points now have the ability to support two networks and dual radios to enable this type of configuration.

6.19.3 Wireless VLAN/Virtual APs

The other approach would be to support a single WLAN, but define multiple virtual WLANs over it. In this case the APs would broadcast beacon messages for multiple SSIDs on the same channel, and those different networks would be associated with different VLANs on the wired network. The network administrator can then define different security policies for each VLAN and allow guest access to only one of them. While an elegant way to support differing sets of requirements, it is important to remember that all of these different groups of users are still sharing one relatively low-capacity radio channel. If all user traffic is given equal priority (i.e., no QoS) a guest's multi-gigabyte download could affect the performance for the internal users.

6.20 Conclusion

While security has clearly been a weak link in wireless LANs, the situation has improved dramatically with the new security options. Unfortunately, hackers are a crafty lot, and there is a never-ending stream of strategies for attacking network security. Security is an ongoing effort, and our tools and techniques must be continuously evaluated in light of the new threats. Voice devices are a particular concern as most of the first generation WLAN voice handsets have fallen a generation behind in their security features.

The key issue to recognize with WLAN security is that implementing these capabilities is the user's responsibility. That means the network designer must understand the risks and the capabilities of the various solutions and ensure that they are implemented throughout the network. Further, from a practical standpoint, a company may be required to support older devices that are not capable of supporting the newest security features. In that case the company can either partition the network and provide different levels of protection for different classes of devices, or simply depend on the security features that are universally available (i.e., the least common denominator approach).

Despite the range of tools and techniques available, some security environments will continue to avoid wireless LANs entirely. Military, national security, and intelligence systems immediately come to mind. For the vast majority of commercial users, however, we should have acceptable levels of security available in the not too distant future.

CHAPTER 7

IP Routing for Voice

For the last several chapters we have looked at the radio links, protocols, and security attributes of wireless LANs. Now it is time to address the other major technology that supports WLAN voice, Voice over IP (VoIP). Traditionally, organizations have built and maintained separate networks for their voice and data traffic, but we now have the network tools to build a single network that would combine both types of traffic and carry video as well. The underlying technology would be packet switching based on the concepts of IP routing.

The obvious advantage of this convergence would be cost savings. Rather than leasing separate equipment and facilities to support voice, video, and data requirements, an organization could invest in a single shared infrastructure. Should we elect to use it, the dynamic allocation capability of packet switching could reduce the total wide area network capacity leased from the carriers. Further, new additions to IP routing like Multi-Protocol Label Switching can provide preferred handling for voice and video packets, thereby ensuring better service quality for those time and loss sensitive applications.

To understand the challenges involved in carrying voice on an IP network, some background in TCP/IP is essential. We will introduce the overall concept of a layered model and the Open Systems Interconnection (OSI) reference model, and then examine in detail how those concepts apply to the TCP/IP protocol suite. In particular we will describe the functions of the Real-time Transport Protocol (RTP), the IP protocol for voice transport, and how Multi-Protocol Label Switching (MPLS) can be used to provide Quality of Service (QoS) for voice applications.

7.1 Protocols, Compatibility, and The OSI Reference Model

Protocols are the basic building blocks of a communications network. A protocol is a set of rules describing how some part of the information exchange should be conducted. In order for two devices to interoperate and exchange data successfully, they must employ compatible protocols.

One of the central themes in modern networking has been the adoption of open standards so that products from different vendors will communicate with one another using a set of standard interfaces and protocols. Those standards are developed by industry bodies like the International Telecommunications Union (ITU), the Internet Engineering Task Force (IETF), the Institute of Electrical and Electronics Engineers (IEEE), and the Electronics Industries Association/Telecommunications Industries Association (EIA/TIA).

In dealing with IP voice, there are two major sets of protocols that must be addressed:

1. **Voice Transport:** These are the protocols describing how packets of voice information are carried between the two parties in the conversation. The RTP, User Datagram Protocol (UDP), and Internet Protocol (IP) are the major elements involved in IP voice transport.

2. **Signaling Protocols:** The exchange of voice packets cannot begin until a connection is established between the two parties. In a telephone network, signaling deals with the processes of setting up/tearing down connections and activating features. In IP telephony, the signaling dialog is typically carried on between the calling and called stations and a telephony server. There are also peer-to-peer IP telephone systems where the calling and called stations communicate directly without a telephony server. There have been a number of different signaling protocols used in IP telephony including H.248/Megaco, H.323, and the Session Initiation Protocol (SIP). While early IP telephone systems employed some version of H.248 or H.323, the current trend is toward SIP as the standard mechanism for VoIP signaling.

7.1.1 The Development of Protocol Standards

In the early days of data processing, each computer vendor developed a unique set of network protocols. The vendors used incompatibility as a marketing strategy that can best be expressed as: *ensuring customer loyalty through incompatibility*. The result was that if a customer had computer systems from two or more vendors, they typically had to build and operate a separate network with separate lines and modems for each system. While incompatibility may have been in the best interest of the individual vendors, it certainly was not in the best interest of the user. The most widely-used proprietary architecture was IBM's Systems Network Architecture (SNA); users groused that the letters actually stood for "Stifles Non-IBM Attachments." IBM has since abandoned this idea and now is a major supporter of standards-based networking.

Starting in the mid-1970s, there was a major push for the development of multi-vendor standards for data networking. The argument for the adoption of standards in network technology is based on three major economic benefits.

1. Standards reduce overall network cost by allowing the user to share network resources among computer systems from different vendors.

2. Standards foster competition that in turn leads to lower prices and larger markets.

3. Standards also lead to the development of chip-level components that result in a far lower cost of production.

The reluctance to endorse standards was one the major features that characterized the early stages of the computer industry, and it sprung from a *zero sum mentality* on the part of the vendors. The view among the vendors was that there was a finite amount of money customers would be willing to spend on IT, which meant the market pie was only so big. The large vendors like IBM were intent on holding on to the biggest slice, and the smaller vendors feared losing their small slice to the bigger vendors.

The Internet has proven that the economic case for standards was sound. The widespread adoption of standards has led to lower prices. Further, as the cost of the technology came down, people found more places where they could use it. The end result is that the overall market expands, and even if a dominant vendor gets a smaller slice of the pie, it's an exponentially larger pie.

7.2 The Open Systems Interconnection (OSI) Reference Model

The first major step toward computer system compatibility was the development of the OSI Reference Model in the early 1980s. Developed by the International Organization for Standardization (ISO), the OSI Reference Model was the first step towards organizing the problem of end-to-end network compatibility. While it did not provide a very useful mechanism for developing actual products, OSI provided a standard framework and vocabulary for defining and addressing compatibility issues.

The first step in establishing compatibility is to organize and define the overall task; in OSI, this is done by means of a layered model. To construct the model, the entire job of data networking is broken down into a series of tasks that are then assembled into logical groupings, each logical grouping becoming a layer in the model. The OSI Reference Model defines seven layers or groups of tasks that must be addressed. In order for two devices to communicate, they must be compatible at all seven layers.

7.2.1 Role of a Layered Model: The What not the How!

It is important to note that the OSI Model is not a protocol, but rather a definition of functions to be provided by protocols. The OSI Reference Model sets out what functions must be accomplished at each layer so that protocols could be developed to define exactly how those tasks are performed. So rather than actually making things compatible, the OSI model describes how the job of arranging compatibility should be organized. In short, OSI tells you what has to be done, but you still need to develop the protocols that tell you how to do it.

7.2.2 Key Idea: Modular Flexibility

As we review the various functions, it is important to keep the overall purpose in mind. The goal of the layered model is to provide modular flexibility. Rather than having one rigid set of protocols (i.e., an all or nothing approach), a layered model allows an environment where the communications structure is assembled out of a series of interchangeable elements. If someone comes up with a new way of performing the tasks for Layer 2 (e.g., a wireless LAN rather than a wired LAN), we can substitute that one element in the structure without changing everything else. The idea of defining an organized process that incorporates the type of modular flexibility was one of the key ideas adopted in TCP/IP.

7.3 OSI Layers and Functions

OSI provides a standard organization and vocabulary to define compatibility issues. To understand how the structure works, we begin with a brief description of the various layers and the functions they encompass. By convention, we begin at the bottom layer and work our way up. See Figure 7-1.

7.3.1 Layer 1: Physical Layer

Layer 1 describes the transmission of bits over a physical medium. In the simplest terms, Layer 1 is the "bit pipe" over which our digital information will be sent. Layer 1

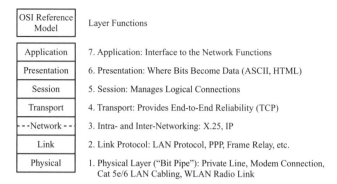

Figure 7-1: OSI Layers and Functions

addresses such issues as physical connectors and the signal encoding. Depending on the media to be used, that encoding might be electrical, optical, or radio. In a wired LAN, the primary physical medium for desktop access is the Category 5e or Category 6 unshielded twisted pair (UTP) cabling. For example, a Layer 1 protocol for wired LANs would be 10 Base-T or 100 Base-T, which define how we encode the 1 bits and 0 bits as electrical signals to be transported over the cable path. In a wireless LAN, the radio link protocols, 802.11a, b, g, or n define the Layer 1 options.

7.3.2 Layer 2: Link Layer

The link layer provides rules for the transmission of data between devices over a single transmission link or through a single network. The link layer describes how transmission elements called frames are formatted, addressed, and how they are introduced onto the transmission link. LAN protocols like Ethernet and Wi-Fi provide definitions for Layers 1 and 2 of the OSI model. Wi-Fi's Carrier Sense Multiple Access with Collision Avoidance (CSMA/CA) and Ethernet's CSMA/CD address the Layer 2 functions. Wide area Layer 2 protocols would include the frame relay interface and the Point-to-Point Protocol (PPP). Early link layer protocols like IBM's Synchronous Data Link Control (SDLC) defined techniques for the Layer 2 entity to detect and correct transmission errors. Most newer link layer protocols like Ethernet and PPP provide a mechanism to detect errors, but if an error is found, the frame is simply discarded. These newer protocols assume the incidence of error will be rare and that some higher-level protocol (e.g., TCP) will provide reliability if it is required.

7.3.3 Layer 3: Network Layer/Inter-Network Layer

The network layer is now divided into upper and lower sublayers. An example of an early network layer protocol would be the X.25 packet network interface. The attention today is shifted to the upper sublayer that addresses inter-networking, or the task of connecting networks together. IP is the primary inter-networking protocol we use today. IP provides an end-to-end address that will carry a packet router-to-router through a series of networks on the way to its destination. We will have more to say about IP and its cohort TCP in a moment.

7.3.4 Layer 4: Transport Layer

The functions of the network and transport layers are closely linked. In TCP/IP, the network layer is responsible for connecting networks together and delivering packets to the end device on a best effort basis. That means that IP is allowed to lose packets, and can deliver packets with errors. The transport layer ensures that all of the inter-network messages are delivered, in the right order, and provides end-to-end error detection and correction. The reliability complement to IP at the transport layer is the Transmission Control Protocol (TCP). IP is inherently unreliable, but TCP can recognize and recover from those failures and ensures overall data integrity. The TCP/IP model also includes an unreliable transport, the UDP, which is used to transport voice packets.

7.3.5 Layer 5: Session Layer

There are few recognizable protocols designed specifically to provide the functions defined for Layer 5, the session layer. The session layer deals with the structure of communications between applications. Layer 5 establishes and maintains logical connections or sessions between applications. For example, a user device could dial in to a host (i.e., a Layer 1 connection), but during that one connection he or she could log on and log off several different application sessions. Layer 5 defines that log on/log off function.

7.3.6 Layer 6: Presentation Layer

The presentation layer handles the format of the information so that it can be presented to the user or to the application. At the layers below Layer 6, the data is treated simply as bits. Initially, the presentation layer dealt with simple data transmission codes like ASCII or EBCDIC. These codes supported only alphanumeric information. The primary presentation format for data is the Hypertext Mark-up Language (HTML), which provides color and graphics for Web or Intranet applications.

7.3.7 Layer 7: Application Layer

The term application is used differently in data processing and data communications. In data processing, the term application has traditionally meant the application program itself; for example, the software that processes the order and prints the invoice. Application has a different meaning in OSI and other networking environments. In networking, application protocols are utilities that the computer applications use to get access to particular network services. Multiple application protocols are required to address different communication requirements such as terminal access, file transfer, electronic mail exchange, and network management. In IP voice, the RTP provides the application function, and its major task is to ensure sequence and timing consistency of packets within the voice stream.

7.4 The Internet and TCP/IP

While OSI introduced the idea a framework for compatibility and modular flexibility, those ideas came to fruition in the Transmission Control Protocol/Internet Protocols (TCP/IP). Vinton Cerf and Robert Kahn originally developed TCP/IP in the early 1970s for use in ARPANET, the first packet switching network. While the first major applications on ARPANET were time sharing access and email, even in those early years the developers were discussing the potential of using packet switching technology to support voice. Today, TCP/IP has become the dominant protocol set for all data communications. For more information on the early development of the Internet and TCP/IP, go to http://www.isoc.org/internet/history/brief.shtml.

TCP/IP is not simply two protocols, but an entire set of inter-networking protocols and functional definitions. TCP/IP describes its own layered model akin to OSI, but it uses four rather than seven layers. TCP/IP does not replace protocols like Ethernet and Wi-Fi, but rather defines their roles within its overall model. The bottom line is that TCP/IP delivers the type of multi-vendor solution that the OSI planners had envisioned.

OSI provided a standard vocabulary for addressing compatibility problems, but practicality demanded a simpler model for implementation, and TCP/IP was just the thing. The convenient term TCP/IP is simply the names of two of over a hundred protocols that comprise the TCP/IP architecture. However, when we implement TCP/IP,

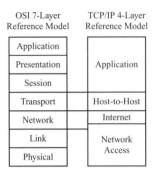

Figure 7-2: Transmission Control Protocol/Internet Protocol Model

we actually implement the entire architecture and the various protocols it is made of. One way to think of TCP/IP is an implementable version of the OSI concept. The important distinction is that TCP/IP includes fully defined protocols for all of the layers in the model. TCP/IP not only identifies the functions (i.e., the what), it also defines the protocols to implement them (i.e., the how). Figure 7-2 shows the correlation between OSI and TCP/IP's four layer model.

The major capability of TCP/IP is its ability to interconnect networks, and deliver packets from a source on one network, through a variety of intervening networks, to a destination on another network. In the local area, IP can be sent over Ethernet or Wi-Fi LANs. In the wide area, those IP messages can be forwarded through wireless, private line, or frame relay network services. A single IP datagram could be sent over any combination of these networks on its trip between two end devices.

7.5 Overall Organization of TCP/IP

Besides defining the protocol specifications, TCP/IP also defines the overall division of responsibility within the various network layers. The primary division is between TCP and IP. IP defines a connectionless best effort packet forwarding service. Connectionless means that each packet finds its way router-to-router to its destination but does not follow a predetermined path. That means that routers do not guarantee to deliver every packet to its destination. That means that IP can potentially deliver packets out of sequence.

Best effort means that IP can:

- Lose packets.

- Deliver packets with errors.

- Provide no mechanism to ensure timing consistency.

The reason that IP is unreliable is tied to the basic design of routers. Routers are the packet switches in an IP network, and like other packet switches, routers utilize buffers to compensate for momentary bursts of transmission. That buffering will introduce jitter, the variation in delay that is seen packet-to-packet. Further, if the buffer capacity is exceeded, the router will drop some of the packets.

7.5.1 Role of TCP

The reason that IP can operate in such a freewheeling fashion is that applications that require reliable delivery can use TCP. Where routers implement IP, TCP runs in the end devices, typically the user's PC and the server it is communicating with. TCP's basic job to recover lost, out-of-sequence, or errored packets. In essence, whatever IP screws up, TCP can fix! TCP uses a system of sequence and acknowledgment numbers to account for all the traffic it sends. So if a user's router or a network service suffers a buffer overflow and a packet is dropped, TCP recovers it automatically. When a data packet is dropped and TCP is being used, the packet is not irretrievably lost. The packet is simply delayed until TCP discovers it is missing and orders a retransmission.

Layered protocols are implemented using a system of headers. As the PC or server prepares a message for transmission, software for each layer appends a header. The application layer may or may not append a header; that depends on which application protocol is being used and the particular stage of the transmission process. When the data is passed to the TCP layer, the TCP software appends its header that includes a port address, an error-checking field, and sequence and acknowledgment numbers. The sequence numbering is used to ensure that all of the pieces of the message arrive and that they are processed in the correct order. When a TCP message or segment is received, it is checked for errors and sequence, and the sequence number is then returned as an acknowledgment number. When using the unreliable UDP transport, there is a port address and an optional error checking field.

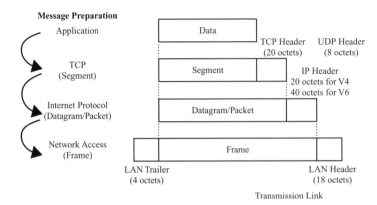

Figure 7-3: TCP/IP Protocol Implementation

7.5.2 IP Message Preparation

The TCP or UDP message is then passed to the IP software, which appends the IP header that includes the source and destination IP addresses. That IP address is a unique 32-bit number that guides the message through the network to its final destination. The IP message is called a packet or a datagram. To send the IP packet, the IP software passes it to a network interface (e.g., an Ethernet or Wi-Fi NIC), which wraps it up in a network access protocol for transmission. If the user device is connected to an Ethernet LAN, the IP packet is wrapped in an Ethernet frame. If the user is on a Wi-Fi network, the packet is wrapped in the Wi-Fi protocol. Let's take a closer look at how the process works. See Figure 7-3.

7.6 Dissecting TCP/IP

The major protocols in the TCP/IP model are shown in Figure 7-4. When we describe TCP/IP, we typically start with IP.

7.6.1 Internet Protocol (IP; OSI Layer 3)

All traffic, voice or data, is formatted as IP packets called datagrams (Note: While datagram is the technically correct term, the terms packet and datagram are used interchangeably when describing IP traffic). Routers are the packet switches that forward those IP datagrams. Each router analyzes the address in the IP packet, and refers to a forwarding table it has stored in its memory. Routers use

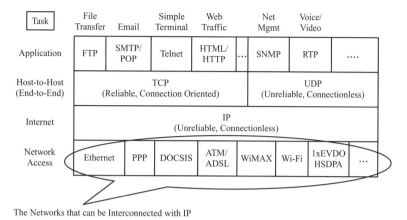

Figure 7-4: Major TCP/IP Protocols

a set of forwarding rules called a routing protocol to decide where to send the packet next. The packet moves router-to-router until it reaches its destination. Those packets can then be forwarded through a variety of network services using a different access protocol for each.

7.6.2 Network Access (OSI Layers 1 and 2)

One of the key features of IP is the range of networks IP packets can be forwarded through. A Wi-Fi user at home could generate a packet containing part of an email message, a Web page request, or a sample of voice bits. That IP packet would be forwarded over a WLAN radio link to the user's home router. From there it could be forwarded in a DOCSIS frame over a cable modem service. At the cable head end, the packet might be sent in a PPP frame through the Internet backbone. When it reaches the corporate network, the IP packet could be delivered to the server through a LAN switch using a 100 Mbps Ethernet connection. While the Layer 2 frame is stripped off at each router, the message that maintains its integrity and the address that guides it end-to-end is all part of IP. See Figure 7-5.

7.6.3 Host-to-Host Protocols: TCP and UDP (OSI Layer 4)

IP and the underlying networks provide an unreliable, best effort packet forwarding service. IP does not provide for error correction, sequencing, or guaranteed delivery. IP can be used with one of two transport protocols that are designed to provide different delivery guarantees.

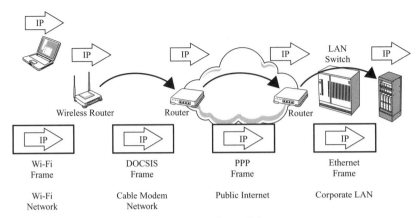

Figure 7-5: Layered Model Concept

1. **Transmission Control Protocol (TCP):** TCP is a connection-oriented protocol that provides reliable, error-free delivery. Connection oriented means that TCP employs a connection setup and tear down procedure. The TCP header includes an error checking field, a sequence number, and an acknowledgment number. The transmitter generates the error checking field, and sequentially numbers each element of the total transmission; the sequence number actually counts the number of octets rather than the number of messages. If the receiver gets the message error-free, it takes the sequence number from the received message and returns it as the acknowledgment number in the next message it sends. TCP also defines routines to fix anything that may go wrong. That would include reestablishing the sequence of messages if necessary and requesting retransmissions for lost or errored messages.

2. **User Datagram Protocol (UDP):** Not all of the traffic carried in an IP network uses TCP transport. The TCP/IP protocol set also includes an unreliable, low overhead, connectionless transport option called the User Datagram Protocol (UDP). UDP operates on a send and pray philosophy, which effectively says: send it out and pray that it gets to the destination. UDP is used for applications that are loss-tolerant, or for tasks where a recovery mechanism would be ineffectual; that latter category would include voice and video transmission. The retransmission system used in TCP would be useless for voice transport, as the retransmitted packet would arrive too late to have any relevance to the conversation. In short, you get one shot at delivering a voice packet.

To ensure more reliable delivery in conjunction with UDP, use an IP service like Multi-Protocol Label Switching (MPLS) that includes QoS. QoS is the generic term for techniques used in packet networks to give priority for some classes of traffic over others with the goal of providing faster, more consistent, and more reliable delivery. Different networks use different QoS mechanisms. Wireless LANs can use a modified access protocol called 802.11e that gives priority to voice and video frames over data frames. When applied to voice traffic, the result of that preferred treatment is shorter transit delay, less jitter, and fewer dropped packets. A user does not have to use QoS-capable services with UDP, but QoS does provide an additional level of protection. We will take a look at how those QoS protocols work in a moment.

7.6.4 TCP/IP Application Protocols (OSI Layers 5, 6, and 7)

The term application is used the same way in TCP/IP as it is in OSI. An application is a utility that will allow a program or a computer process to access particular network services. TCP/IP defines a number of application utilities, some of which use TCP for transport while others use UDP.

7.7 TCP and UDP Supported Applications

7.7.1 TCP Supported Applications

The following are several common applications supported by TCP.

- **File Transfer Protocol (FTP):** FTP is a batch transmission utility that operates through TCP and allows systems to send non-real time-sensitive bulk data. Downloading files from a Web site or downloading MP3s on a peer-to-peer service would be typical FTP tasks.

- **Simple Mail Transfer Protocol (SMTP)/Post Office Protocol (POP):** These are utilities that work with TCP to send and receive email messages.

- **Telnet:** This is another TCP-oriented application that supports remote terminal access. Telnet is designed for character-oriented terminals like the DEC VT-100 series.

- **Hyper Text Mark-up Language (HTML)/Hyper Text Transfer Protocol (HTTP):** HTML defines the formatting used in Web pages, and HTTP describes the rules for downloading a Web page from a server to a browser. HTTP also operates through TCP.

- **Secure Hyper Text Transfer Protocol (SHTTP):** This is an adjunct to HTTP that defines a mechanism for generating keys and encrypting information in a Web session. In Web commerce, a browser is typically transitioned into SHTTP when a user must enter a credit card number or other sensitive information.

7.7.2 UDP Supported Applications

The following are some UDP supported applications.

- **Real-Time Transport Protocol (RTP):** This is the protocol used to transport for real-time voice and video. The main function of RTP is to identify the payload encoding (e.g., G.711 or G.729A voice) and append sequence numbers and timestamps so that the receiver can process the packets in the correct order and reestablish the timing continuity. We will have more to say about RTP in a moment.

- **Real-Time Control Protocol (RTCP):** RTCP is a support protocol used in conjunction with RTP to provide information on the performance of the RTP connection. Each station in the RTP connection typically sends an RTCP sender and/or receiver report every five seconds. Those reports include statistics like number of packets lost, round trip delay, and jitter performance. RTCP uses a different port address than RTP, so a management system can intercept those reports and use them to drive a performance monitoring system.

- **Simple Network Management Protocol (SNMP):** SNMP is a protocol for forwarding management information between network elements (e.g., switches, routers, multiplexers, PBXs, etc.) and a network management platform like HP's OpenView. The standard Internet joke about SNMP is that it's not simple, and it's not a network management protocol.

7.8 TCP/IP Protocols for Voice: RTP and RTCP

While TCP and IP were originally developed to support the requirements of data applications, there are now protocols to support real-time voice and video services in an IP network. As we noted earlier, there are actually two sets of protocols that come into play: voice transport and signaling. First let's look at the protocols for transporting voice through an IP network.

To support voice over an IP network, two protocols are required:

1. **Real-Time Transport Protocol (RTP):** RTP transports real-time voice and video packets.

2. **Real-Time Control Protocol (RTCP):** RTCP reports on the quality of the network service and generates feedback messages that can be used to measure the performance of the network service.

7.8.1 Isochronous Service

Voice calls have traditionally been carried over circuit switched facilities that provide consistent delay. The term we use to identify a service that maintains timing continuity is *isochronous* (the term is derived from the Greek words *iso* for "same" and *chronos* for "time"). As they do not allocate capacity dynamically, circuit switched networks inherently provide isochronous transmission capability.

7.8.2 Real-Time Transport Protocol (RTP)

The dynamic allocation mechanisms used in packet networks result in *jitter*, or a variation in transit delay from packet to packet. Voice is typically carried in packets that contain 20 to 40 msec of human speech; each speech burst can be several seconds in length and so it will be carried in dozens of packets. Voice services require that the isochronous nature of the voicestream be reconstituded so the timing relationship between the speech packets must be reestablished to reproduce into understandable speech. RTP is used to ensure the packet sequencing and to restore that timing continuity. RTP was originally defined by the IETF in RFC 1989, but that standard was replaced by RFC 3550.

The RTP sender (e.g., an IP/Ethernet or Voice over Wi-Fi handset) places a timestamp and a sequence number in the header of each voice packet before it is sent into the network; the RTP header also includes a payload type field that identifies the type of voice or video encoding that is being used. It is expected that the buffering encountered in the routers the packet passes through on its journey through the network will disrupt the timing continuity of that packet stream. The process of sending frames over a contention-based WLAN also will disrupt the timing continuity. At the receiving end (e.g., the Wi-Fi or wired IP handset), the packets are buffered. The receiver resequences

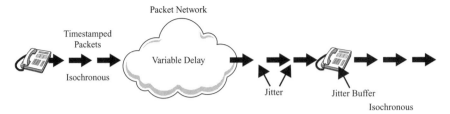

Figure 7-6: Real-Time Transport Protocol

the packets (if necessary) and then plays them out of the buffer according to the timestamps. That receive process effectively masks the jitter effect, and reestablishes the continuity in timing. The downside of this process is that the receive buffering increases the end-to-end transit delay. See Figure 7-6.

7.8.3 RTP Receive Buffer

One of the critical elements in the design of an IP voice system is sizing the receive jitter buffers. If the buffer is too short, the receiver automatically drops packets that are received outside of that range. On the other hand, the user should not set the jitter buffer too long, because that buffering adds to the overall transit delay. The problem is that different calls can result in different levels of jitter. Where the jitter introduced on calls between two wired stations on an IP PBX might be fairly consistent, the level of jitter encountered over a wide area network or a WLAN connection could be considerably greater.

7.8.4 Adaptive Jitter Buffers

With products that utilize fixed-length jitter buffers, the size must be set to accommodate the worst-case expectation (i.e., the call with the greatest amount of jitter). Typical settings range from 20 to 60 msec. Some devices now use adaptive jitter buffers that adjust the size automatically based on the amount of jitter that is found on each call. The ideal solution would be a network with QoS that minimizes the jitter, and an adaptive buffering system that would automatically reduce the size of the jitter buffer in response to better jitter performance.

IP voice uses the simplest implementation of RTP. In a multimedia conference, separate RTP streams are created for voice and video, and the two would have to be synchronized in the receiver. Separate packets are created for the voice and video elements, and the Contributing Source (CSRC) Identifier function is used to identify them. The voice and video packets will have the same Synchronizing Source (SSRC) identifier.

7.8.5 RTP Header Fields

The following are the parts of the header field and definitions.

- **V:** RTP Version.

- **P:** Padding bit indicates padding octets have been added to the payload (used with some encryption algorithms).

- **X:** Extension bit is used to indicate a header extension is included.

- **CSRC Count:** The number of contributing source identifiers included. A video connection will typically have two contributing sources, one for audio and one for video. A voice call will have only one contributing source.

- **M:** Indicates significant events (e.g., frame boundaries) to be marked.

- **Payload Type:** Identifies the format of the content (e.g., G.711 voice, G.729 voice, H.261 video, etc.) See Table 7-1.

- **Sequence Number:** Allows the receiver to restore packet sequencing.

- **Timestamp:** Indicates the time when the payload was sampled. There is no master clock source, so each sending device essentially generates its own timing reference (i.e., time elapsed since the last packet was generated).

- **SSRC Identifier:** The synchronizing source is a randomly chosen identifier for the RTP host. All packets from that host have the same identifier, which allows the receiver to group packets for playback.

- **CSRC Identifiers:** This is a list of the sources for the payload, and it is used when a mixer combines different streams of packets.

RTP voice and video packets are forwarded in the UDP. As we have noted, IP can lose packets or deliver them out of sequence. UDP provides no mechanism to ensure sequencing, but RTP does provide that function. However, there is no mechanism in either RTP or UDP to recover lost packets. As we noted earlier, even if RTP could request retransmissions, the retransmitted voice packets would arrive too late to have

Table 7-1: RTP Payload Type Values (RFC 3551)

Payload Type	Description	Audio/Video	Sample Rate (Hz)	Notes
0	PMCU	A	8,000	RFC 3551
1	Reserved	—	—	—
2	Reserved	—	—	—
3	GSM	A	8,000	RFC 3551
4	G723	A	8,000	[Kumar]
5	DVI4	A	8,000	RFC 3551
6	DVI4	A	16,000	RFC 3551
7	LPC	A	8,000	RFC 3551
8	PCMA	A	8,000	RFC 3551
9	G.722	A	8,000	RFC 3551
10	L16	A	44,100	RFC 3551
11	L16 (Dual Channel)	A	44,100	RFC 3551
12	QCELP	A	8,000	—
13	CN	A	8,000	RFC 3389
14	MPA	A	90,000	RFC 3551, RFC 2250
15	G.728	A	11,025	RFC 3551
16	DVI4	A	11,025	[DiPol]
17	DVI4	A	22,050	[DiPol]
18	G.729	A	8,000	—
19	Reserved	A	—	—
20	Unassigned	A	—	—
21	Unassigned	A	—	—
22	Unassigned	A	—	—
23	Unassigned	A	—	—
24	Unassigned	V	—	—
25	CelB	V	90,000	RFC 2029
26	JPEG	V	90,000	RFC 2435
27	Unassigned	V	—	—
28	nv	V	90,000	RFC 3551
29	Unassigned	V	—	—

Table 7-1: RTP Payload Type Values (RFC 3551)—cont'd

Payload Type	Description	Audio/Video	Sample Rate (Hz)	Notes
30	Unassigned	V	—	—
31	H.261	V	90,000	RFC 2032
32	MPV	V	90,000	RFC 2250
33	MP2T	AV	90,000	RFC 2250
34	H.263	V	90,000	[Zhu]
35–71	Unassigned	?	—	—
72–76	Reserved for RTCP Conflict Avoidance	—	—	RFC 3550
77–95	Unassigned	?	—	—
96–127	Dynamic	?	—	RFC 3551

any relevance to the conversation. If too many packets are lost, the user will notice degradation in the voice quality.

7.9 Real-Time Control Protocol (RTCP)

RTP uses RTCP to monitor the service quality. RTCP provides service monitoring reports to help manage the RTP process. These RTCP reports are typically generated by each device in the connection every five seconds, and they are sent with a different UDP port address than the voice packets. RTCP messages can also be monitored by a performance monitoring tool or a third party (e.g., a carrier) to track the performance of the network service. RTCP generates periodic reports, which include information like packet and octet counts, lost packet count, and inter-arrival jitter. See Table 7-2.

7.9.1 RTCP Reports

- **Sender Report:** Sent by an RTP source and provides transmit and receive statistics. Those statistics would include the sender's packet and octet counts as well as the receive statistics identifying packets lost, percent lost, and inter-arrival jitter experienced. The lost packet statistics would be based on the sender reports from the other party in the connection.

- **Receiver Report:** Includes the same information as a sender report but is generated if the device has not had any packets to send since the previous report.

Table 7-2: RTCP Extended Reports Statistics

IP Problems	Delay Problems	Signal/Noise Problems	Echo Problems	Configuration Problems
Packet Loss Rate, Packet Discard Rate, Burst Length, Burst Density, Gap Length	Packet Path Delay, End System Delay	Signal Level, Noise Level	Echo Level, Round Trip Delay	Jitter Buffer Configuration, PLC Type, End System Delay

- **Source Description Report:** Identifies the sender's capabilities and other identifying information (e.g., name, email address, location, etc.)

- **APP:** Application packet intended for experimental use.

7.9.2 Real-Time Control Protocol: Extended Reports

The IETF has recognized the value of the RTCP feedback, but they also see the limitations of the information provided. As a result, the IETF is now working on an enhanced version called Real-Time Control Protocol-Extended Reports (RFC 3611).

7.10 Quality of Service

While we have a protocol structure to support voice, basic IP operates on a best effort philosophy. That means the network provides no guarantees regarding delay, jitter, or loss. A traditional, best effort IP network:

- Can lose packets.

- Can deliver packets with errors.

- Can deliver packets out of sequence.

- Has no provision to ensure timing consistency.

- Can expose traffic to eavesdropping and other security vulnerabilities.

As we noted, RTP has the ability to resequence voice packets and reestablish timing continuity, but it does not recover lost packets. So the better and more consistent the underlying service, the better the overall voice quality will be. That is where QoS comes in.

The term QoS refers to a variety of techniques used in packet networks, the goal of which is to add predictability to the network service. The characteristics of network performance typically include guaranteed capacity or "bandwidth", latency (delay), jitter, and possibly, network availability. To minimize the transit delay, jitter, and packet loss, QoS allows us to recognize voice and video packets and give them a higher priority in the forwarding process.

There are different QoS mechanisms that can operate at either Layer 2 or Layer 3. As the QoS mechanisms span two protocol layers, one of the key design issues in an IP voice network is ensuring that the priority levels are mapped correctly as packets move through the network.

7.11 Layer 2 QoS Options

Layer 2 defines the functions of LAN protocols like Ethernet and Wi-Fi, and there are different QoS mechanisms defined for those two environments (see Figure 7-7):

1. **IEEE 802.1p:** The IEEE has developed a QoS mechanism for wired LANs in its 802.1p standard. Used in switched LAN networks, 802.1p inserts an additional field in the Ethernet header that includes a 12-bit virtual LAN identifier (802.1q), and a 3-bit priority indicator. With a 3-bit priority indicator, 802.1p can specify eight different priority levels. The idea is that a frame may

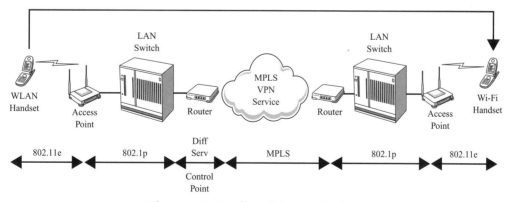

Figure 7-7: Quality of Service Options

be passed through several LAN switch modules on the way to its destination and each module can recognize and provide priority handling for voice frames.

2. **IEEE 802.11e:** While 802.1p provides priority handling through LAN switches, in a wireless LAN, users contend for access to a shared radio channel. The original 802.11 MAC's Distributed Control Function (DCF) described a wait-before-sending mechanism where all stations were provided equal access to the shared channel. The 802.11e standard (called Wi-Fi Multimedia or WMM Certified by the Wi-Fi Alliance) describes an Enhanced Distributed Control Function (EDCF) in which different wait-before-send intervals are used. High priority voice transmitters are assigned shorter waiting intervals and so get preferred access to the shared WLAN channel. We will look at 802.11e in detail in Chapter 11.

7.12 Layer 3 QoS

The IETF has developed two different strategies for delivering QoS that operate with IP:

1. **Differentiated Services (DiffServ):** In DiffServ, individual datagrams carry a priority indicator, called a DiffServ Control Point, in the Type of Service field of the IP header. DiffServ-capable routers are designed to recognize those priority indications and maintain separate queues for the different traffic categories. When packets arrive at each router, they are sorted into different priority queues based on the Diff Serv Control Points. The router gives priority to packets in the higher priority queues by serving them on a more frequent interval than lower-priority queues. DiffServ is typically implemented in enterprise router networks, though the Diff Serv Control Points are also used to define priority when packets are forwarded through a carrier's MPLS-based Internet service.

2. **Multi-Protocol Label Switching (MPLS):** MPLS is a more comprehensive QoS plan that is implemented primarily in ISP backbone networks; MPLS capability is now being provided in some enterprise routers as well. The ISPs use MPLS as a means to provide a premium priced, virtual circuit-based service they offer to enterprise customers called an MPLS Virtual Private Network (MPLS-VPN). The key feature of the MPLS-VPN is that it provides performance guarantees for delay, jitter, and packet loss. With MPLS-based services the carriers offer multiple priority levels to support voice, video, and

data with different performance guarantees for each. An MPLS service can guarantee the performance, because the capacity for each user is reserved as part of the service initiation. So an ISP can now offer two different Internet-based network services: the traditional best effort service, and a higher-priced MPLS-VPN service whose performance parameters can be guaranteed. It is important to note that those performance guarantees can only cover traffic that stays within that ISP's backbone network. As they will be a key element in wide area VoIP implementations, let's take a closer look at how MPLS networks work.

7.13 MPLS-Based VPN Service: RFC 2547bis

With MPLS, ISPs offer a new and different type of wide area service in their networks. These services are designed to address the performance and security requirements of enterprise customers, particularly VoIP users. Unlike traditional best effort Internet service, MPLS provides a structure whereby an ISP can provide a packet service with performance guarantees for jitter, delay, and packet loss.

7.13.1 MPLS Concepts

MPLS adds two important elements to traditional IP:

1. **Virtual Circuit/Label Switched Path (LSP):** Unlike traditional IP that is connectionless, in MPLS all of the packets for a particular session will be routed over a virtual circuit. The MPLS specifications do not call it a virtual circuit (that would make things too obvious), it is called a Label Switched Path (LSP). That LSP provides two basic advantages over traditional IP:
   a. Security: Transmissions cannot jump between virtual circuits within the network. As a result, the user should not need to encrypt transmissions. Users with particularly sensitive transmissions like financial information may still choose to encrypt MPLS traffic, though the security features offered in MPLS should be adequate for most enterprise customers.
   b. Ordered Delivery: A virtual circuit also ensures that all parts of the message arrive in order. As higher-level protocols (e.g., TCP, RTP) can reorder mis-sequenced packets, this feature has less user impact.

2. **Capacity Reservation/QoS:** Before MPLS will allow an LSP to be established over a link, it ensures there is sufficient capacity available to meet the

requirements of the connection. A carrier cannot ensure performance by simply assigning priorities; all a priority system does is treat some transmissions better than others. Priority does not mean the system treats anyone very well! Ensuring performance requires a capacity reservation mechanism that is one of the key features of MPLS, and it supports multiple service classes and defines different delay and loss parameters for each.

The other basic attribute of MPLS-VPN services is that they provide full mesh connectivity. Unlike earlier frame relay services that require a virtual circuit between any pair of points that will communicate directly, in an MPLS network, any end point can communicate with any other. When the user's network is initiated, a full mesh of LSPs is created among all end points. The user pays for access at each network location, not for virtual circuits, so a mesh network and a hub-and-spoke configuration have the same cost. Finally, as the MPLS capability is provided within the carrier's network, it is essentially transparent to the user's router configuration. All the user does is set the Diff Serv Control Points in each packet that will assign each packet to a particular service class (e.g., voice, video, data). See Figure 7-8.

7.14 MPLS Service Classes

As MPLS defines a guaranteed path over which each user's packets will be sent, the ISPs can provide services that offer worst case parameters for packet delay, jitter, and packet loss. Most carriers offer at least four classes of service designated gold, silver, bronze, and best effort; each of the carriers use a different set of terms for their service classes, but gold/silver/bronze/best effort provides the easiest vocabulary to track.

7.14.1 MPLS Pricing

There are three primary elements that go into the price of an MPLS service.

1. **Access Connections:** A private line connection is required from each user location to the nearest MPLS serving point. Most offer access rates up to at least DS-3 (i.e., 44.7 Mbps).

2. **Port Charges:** A flat rate monthly charge for connecting each site to the network. The port charge is based on the speed of the connection and the Class of Service (CoS) profile.

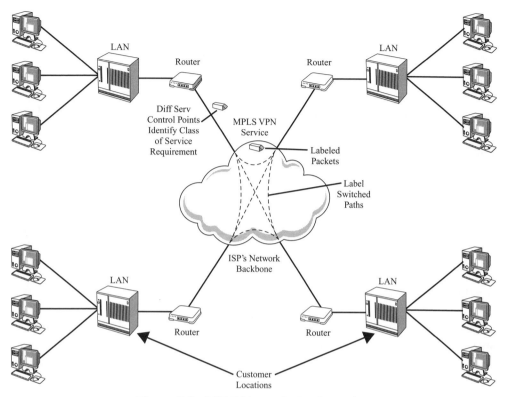

Figure 7-8: MPLS Network Configuration

3. **Class of Service (CoS) Profile:** In an MPLS service, the CoS profile defines the allowable percentage of gold, silver, and bronze traffic the user may send on that port. A profile with a higher percentage of gold traffic will be priced at a higher rate. The user specifies the handling of each packet by setting a Diff Serv Control Point in the type of service field in the IP header. If the user sends more than the allowable percentage in a particular service class, the excess will not be covered by the same performance guarantees as allowable traffic. The treatment for excess traffic varies by traffic category, and we will return to that issue in a moment.

7.14.2 MPLS CoS Guarantees

The carriers offer a different range of guarantees for each service class. Jitter performance is only defined for gold traffic. The delay guarantees are the same for all service classes, and vary based on the destination of the traffic. Domestic traffic will have the shortest delay guarantee, and delay to international points will depend on the amount of capacity the carrier has going into each country. The primary differentiator is the probability of packet loss, though even bronze traffic is typically guaranteed a delivery probability in excess of 99 percent. The performance guarantees are not defined for each packet, but rather for the average performance of the network over a 30-day period.

7.14.3 Performance Guarantee Limit: This Backbone Only!

One major limitation with regard to MPLS-VPN service guarantees is that the guarantees only cover traffic that stays within that provider's backbone. An ISP cannot guarantee traffic that is forwarded through an NAP or through another carrier's backbone. Some of the ISPs are discussing plans for a consortium of sorts that would ensure performance across all member backbones, but much work remains to be done.

7.15 Excess Traffic Handling

One little understood element in MPLS service is how traffic sent in excess of the CoS profile is handled. With the earlier frame relay service, when a customer bursted above the guaranteed capacity (i.e., sent traffic in excess of the committed information rate or CIR for that virtual circuit), the excess traffic was marked "discard eligible" and forwarded into the network if there was capacity available. Of course the discard eligible marking meant that the excess traffic had a higher probability of discard if the network experienced congestion.

MPLS actually has three different methods of handling excess traffic, depending on the class of service involved. See Table 7-4.

1. **Gold Traffic:** If any excess gold traffic is sent, it is discarded at the entry point. Excess gold traffic is not reclassified to a lower CoS level, it is discarded. As the gold category is typically used for voice traffic, it is very important that the customer determine how many voice trunks are to be provided, the number of bits required for each, and then ensure there is adequate gold capacity to support the load. Administrators can use Table 7-3 to determine the bit rate requirement for a voice trunk.

2. **Silver/Bronze Traffic:** Silver and bronze traffic have guarantees regarding delay and packet loss (not jitter). If excess silver or bronze traffic is sent, it is

Table 7-3: Packet Voice Transmission Requirements (Bits Per Second Per Voice Channel)

Codec	Voice Bit Rate (Kbps)	Voice Bandwidth (KHz)	Sample Time (msec)	Voice Payload (bytes)	Packets Per Second	Ethernet (Kbps)	PPP (Kbps)
G.711 (PCM)	64	3	20	160	50	87.2	82.4
G.711 (PCM)	64	3	30	240	33.3	79.4	76.2
G.711 (PCM)	64	3	40	320	25	75.6	73.2
G.722	64	7	20	160	50	87.2	82.4
G.722	64	7	30	240	33.3	79.4	76.2
G.722	64	7	40	320	25	75.6	73.2
G.726 (ADPCM)	32	3	20	80	50	55.2	50.4
G.726 (ADPCM)	32	3	30	120	33.3	47.4	44.2
G.726 (ADPCM)	32	3	40	160	25	43.6	41.2
G.729A	8	3	20	20	50	31.2	26.4
G.729A	8	3	30	30	33.3	23.4	20.2
G.729A	8	3	40	40	25	19.6	17.2

Note: RTP assumes 40 octets RTP/UDP/IP overhead per packet. Ethernet overhead adds 18 octets per packet. PPP overhead adds 6 octets per packet.

Table 7-4: MPLS CoS Description

Service Class	Performance Guarantee			Excess Traffic Handling
	Jitter	Transit Delay	Packet Loss	
Gold	Yes	Yes	Yes	Excess Traffic Discarded at Entry Point
Silver	No	Yes	Yes	Excess Traffic Marked "Out of Contract" and Forwarded
Bronze	No	Yes	Yes	
Best Effort	No	No	No	Not Applicable

marked "out of contract," but it is still forwarded into the network if there is capacity available. This is essentially the same treatment as frame relay's discard eligible marking. Out of contract traffic will have the same delay guarantee, but a higher probability of discard.

3. **Best Effort:** As there are no performance guarantees regarding best effort traffic, the user can send as much as they like.

7.16 VoIP Signaling Protocols

As we noted earlier, an IP voice service involves two protocols, one to transport the voice packets, and a second to establish the connection and activate features like hold, conference, and transfer. The voice path is typically referred to as the *media path* while the call setup and feature activation are covered by a signaling protocol. Digital voice is packetized and transported through an IP network using RTP, UDP, and IP, but there are a number of different signaling protocols that can be used.

7.16.1 Traditional Voice Signaling Protocols

The original signaling protocols for voice were developed for circuit switched networks and often assumed the end device (i.e., the telephone) had no processing power whatsoever. Traditional voice signaling involved closing a set of contacts when the receiver was lifted, opening and closing the circuit when the rotary dial was operated, and having the central office send a 90 V, 20 Hz ringing signal to alert a station to an incoming call. ISDN, the telephone industry's plan to convert the network to an end-to-end digital environment, introduced the idea of exchanging protocol messages sent over a separate signaling channel or D Channel between the central office and an intelligent subscriber handset.

7.16.2 VoIP Signaling Protocols

Like ISDN signaling, VoIP systems also assume an intelligent end device, either a VoIP handset, a softphone client in a PC, or a gateway. A telephony server or soft switch exchanges signaling messages with those end devices and coordinates connections between them. There are also peer-to-peer IP telephone systems that establish connections by exchanging signaling messages directly between the called and calling stations without the use of a telephony server.

There are four major signaling protocols that have been used in end user VoIP systems.

1. **H.323:** Originally developed to support multimedia videoconferencing over packet networks, the ITU's H.323 standard was later extended to include VoIP telephony requirements. Virtually all IP PBX systems use some version of H.323 signaling today. There are four versions of H.323 available, and all work over TCP. Each IP PBX vendor has developed its own proprietary signaling protocol based on H.323, so while all of these implementations are based on

H.323, IP handsets from one vendor cannot be used with a telephony server from another vendor.

2. **Session Initiation Protocol (SIP):** SIP was developed by the Internet Engineering Task Force (IETF) as an IP standard (RFC 3261-3264). Although SIP is gaining considerable attention, it will not become the dominant protocol in IP PBX systems for a few years. The attractions of SIP are potential interoperability among vendors, easier application development, operation similar to other existing IP protocols, and easier operation through firewalls. SIP operates over either TCP or UDP.

3. **Media Gateway Control Protocol (MGCP):** This is a VoIP signaling protocol used primarily with gateways. MGCP is defined in RFC 3435, an informational (i.e., non-standard) document by the IETF.

4. **MEGACO/H.248:** Another standard protocol, MEGACO/H.248 is a combined effort of the ITU and IETF (RFC 3525). It can be used with gateways and server-to-server communications.

7.16.3 H.323 Versus SIP

The major difference between H.323 and SIP is where the intelligence resides. H.323 depends on a centralized server that controls the call process and exchanges messages with the parties involved in the call. SIP assumes an intelligent end point, and the server, called a SIP Proxy, is primarily responsible for relaying signaling messages between them. As SIP is the developing standard for call control we will focus on it primarily. See Table 7-5.

Table 7-5: H323 Versus SIP

	H.323	SIP
Architecture	Intelligent Central Server	Intelligent End Point
Defining Body	ITU	IETF
Emphasis	Telephony	Multimedia, Multicast
Encoding	Binary, ASN.1	Text-based
Transport	Typically TCP	Typically UDP
Media Transport	RTP in UDP	RTP in UDP

7.17 Session Initiation Protocol (SIP)

The vast majority of IP telephone systems are migrating to SIP as their standard for signaling. SIP supports five major capabilities for establishing and terminating multimedia communication sessions.

1. **User Location:** Determination of the end system on which the user is located.

2. **User Availability:** Determination of the willingness of the called party to engage in different types of communications (e.g., voice, video, or IM); this capability is generally referred to as *presence* in IP PBX systems.

3. **User Capabilities:** Determination of the media options available for the connection (e.g., voice encoding to be used).

4. **Session Setup:** Establishment of session parameters for both called and calling parties using the Session Definition Protocol (SDP). That would include such issues as the voice or video encoding to be used on that connection.

5. **Session Management:** Features for call transfer and termination, modifying session parameters, and invoking services.

7.17.1 SIP Entities

The SIP standards identify four major entities.

1. **User Agent (UA):** The UA is the end point in a SIP connection. IP/Ethernet or IP/WLAN handsets or soft phones would be typical UAs, but a gateway that connects between a packet telephony and a circuit telephony network would also be a UA.

2. **Registrar:** A server that keeps track of where a user is located within the network. This is the element that allows for presence tracking or the ability to recognize where a party can be reached and with which modes of communication.

4. **Redirect Server:** A server that informs devices when they must contact different locations to complete a function.

5. **Proxy Server:** The primary entity involved in relaying signaling messages between end points in a call.

In reality, the last three functions are typically incorporated in one device generically called a SIP Proxy Server. So in essence they are two elements, User Agents and SIP Proxy Servers.

7.17.2 SIP Signaling Messages

SIP signaling depends on the use of text-based messages, which makes troubleshooting problems a far easier activity than with H.323s binary codes. There are about a dozen message types, but the main ones are:

- **REGISTER:** The message used to register a User Agent with a SIP Proxy Server.

- **INVITE:** The message used to initiate a SIP session or call.

- **BYE:** The command used to terminate a SIP session or call.

- **CANCEL:** The command used to abandon a connection attempt that has not yet been completed.

- **ACK:** A message used to acknowledge important notifications regarding the call.

7.17.3 SIP Response Codes

SIP also provides a series of Response Codes that are used to identify the progress of the call or to identify particular failure conditions. The major types of response codes are:

- **Provisional (100-series):** Trying, Ringing, Call Being Forwarded, and so on. A provisional code means that additional signaling messages will follow.

- **Success (200-series):** Currently limited to the 200 OK code, which means that a request has been satisfied.

- **Redirection (300-series):** Used to indicate problems establishing a connection (e.g., 300 Multiple Choices) or status indications (e.g., 301 Moved Permanently or 302 Moved Temporarily).

- **Failure Codes (400-, 500-, 600-series):** The 400-series identify specific call failures (e.g., 400 Busy Everywhere, 404 Not Found, 415 Unsupported Media Type), 500-series Server Failures, and 600-series Global failures.

The full list of SIP Response Codes is included in Table 7-6.

Table 7-6: RFC 3261: SIP Response Codes

Provisional 1xx

100	Trying	406	Not Acceptable
180	Ringing	407	Proxy Authentication Required
181	Call Is Being Forwarded	408	Request Timeout
182	Queued	410	Gone
183	Session Progress	413	Request Entity Too Large
		414	Request-URI Too Long
Successful 2xx		415	Unsupported Media Type
200	OK	416	Unsupported URI Scheme
		420	Bad Extension
Redirection 3xx		421	Extension Required
300	Multiple Choices	423	Interval Too Brief
301	Moved Permanently	480	Temporarily Unavailable
302	Moved Temporarily	481	Call/Transaction Does Not Exist
305	Use Proxy	482	Loop Detected
380	Alternative Service	483	Too Many Hops
		484	Address Incomplete
Request Failure 4xx		485	Ambiguous
400	Bad Request	486	Busy Here
401	Unauthorized	487	Request Terminated
402	Payment Required	488	Not Acceptable Here
403	Forbidden	491	Request Pending
404	Not Found	493	Undecipherable
405	Method Not Allowed		

7.17.4 Session Description Protocol (SDP)

SIP is designed to support connections for voice, video, and multimedia connections as well as Instant Message-type services. As a result, the descriptions of the call setup processes are generic in nature. To define the type of connection that is being made, SIP uses the Session Description Protocol (SDP). SIP essentially coordinates the connection between/among the parties in the call and SDP identifies the means by which the parties

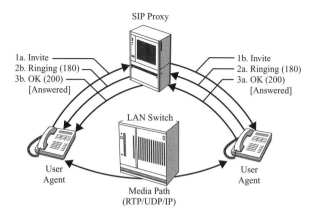

Figure 7-9: Simple SIP Call Setup

will communicate (e.g., voice, video, IM, and the type of information encoding that will be used, e.g., G.711 versus G.729a voice).

7.17.5 Call Setup Process

A call setup involves an exchange of signaling messages as described in the diagram in Figure 7-9.

1. To initiate the connection the calling party issues an INVITE addressed to the called party. The SIP address is called a Uniform Resource Identifier (URI), and looks like an email address preceded by "sip:" (e.g., sip:mfinneran@dbrnassoc. com). The SIP Proxy will use that address to route the INVITE to the appropriate end point or User Agent; the INVITE will also include the SDP with the specifications for the call being attempted. The INVITE may be forwarded through one or more SIP proxies.

2. The called station will typically return a 180 RINGING Response Code, which is relayed by the SIP Proxy to the caller.

3. When the called party answers, the called station generates a 200 OK Response code, which is relayed by the SIP Proxy to the caller. Now the media exchange (i.e., the voice or video connection) can commence. The media packets are sent directly between the called and calling parties and are not forwarded through the SIP Proxy.

To terminate the call, either party can generate a BYE message and the other party responds with a 200 OK.

7.17.6 Multi-Vendor Compatibility?

While everything else in the network has adopted multi-vendor standards, VoIP signaling clings to the idea of proprietary implementations. This was certainly the case with H.323-based systems. SIP seems to hold the promise of multi-vendor interoperability, however most vendors intend to support basic telephony features in an interoperable fashion but will still market more functional, proprietary SIP-based handsets.

7.18 Conclusion

The first thing to recognize about IP routing is that it was not designed to address the requirements of voice. Traditional IP is a best effort packet forwarding mechanism that does not include features to ensure delivery, sequence, or timing consistency. The addition of RTP provided a mechanism for the receiver to reestablish sequence and timing consistency. The problem is that RTP's receive buffering adds to the overall delay that the voice user will experience. Further, the more unpredictable the delay performance (i.e., the more jitter), the more buffering RTP will require in order to compensate for it.

However RTP does not have a means of recovering lost packets. In essence, we have only one chance to deliver a voice packet, and if we drop too many packets the voice quality will degrade. Hence, network services that include performance guarantees can be a major boon in providing high quality voice services.

In order to achieve the best overall service in an IP voice network, we will need an underlying network that provides more consistent performance. Fortunately, QoS mechanisms like DiffServ, MPLS, 802.1p, and 802.11e can help improve the underlying network performance. However, none of these mechanisms can deliver consistent delay by themselves. The best we can hope for with QoS in a packet network is the ability to minimize delay, jitter, and loss; we will still require RTP to reestablish true timing continuity.

In the next chapter we will look more closely at the major quality issues in IP telephony and how they relate to the service the user will receive.

CHAPTER 8

Quality Issues in IP Telephony

In the last chapter we described the overall operation of an IP routing network, and the TCP/IP protocols that are used to carry real-time voice and video traffic. In that discussion we introduced the major performance attributes of traditional and QoS-capable IP networks and how protocols like RTP can reestablish the packet sequencing and timing recovery that are essential for good quality voice.

In this chapter we will look at these issues from the standpoint of voice quality. We will describe the major quality issues that have been identified for packet voice, and the impact these various parameters will have on the user experience. Issues like delay, jitter, and packet loss are inherent in any packet voice system, and we will look at the additional challenge the will be introduced with the use of a wireless LAN as part of the transmissions path.

It is important to recognize that the wireless LAN is only *part* of the overall transmission path the voice packets will travel between the source and destination. The quality the user will experience is based on the overall performance of all of the elements within the path. Whether we are discussing packet loss or the delay factors, the performance of all of the elements in the path is additive. Whether the receive buffer, the local network, or the wide area network drops the packet, the result is the same: that packet does not arrive at the destination. Similarly, the delay introduced by the WLAN is added to the delay introduced to the wired LAN, the WAN, and the wired LAN at the far end.

In this chapter we will describe the major elements that impact the quality of packet voice. In particular we will focus on the known requirements for delay, jitter, and loss and identify the performance characteristics of the different network services the voice packet may encounter on the way to its destination. Fortunately, we now have a good base of experience gained from both local and wide area packet voice networks so we can say with a high degree of confidence how bad things can get before voice users will start to complain.

8.1 Quality Issues in Packet Telephony

The use of packet technology introduces three major quality issues into voice service: voice quality, transit delay, and echo control.

1. **Voice Quality:** The basic signal quality a user detects on a voice connection is a product of the voice coding technique that is used and the percentage of packets that fail to arrive at the receiver to be decoded. Packets can be lost two ways in a VoIP network:
   a. Packet networks can drop packets due to errors or buffer overflows.
   b. The RTP receive jitter buffer can drop packets if they arrive with a delay greater than the buffer can accommodate. So delivering a packet late is equivalent to not delivering it at all.

The impact of lost packets depends on the technique we use to encode the voice. We will look at loss tolerance and the other voice coding issues in a moment.

2. **Transit Delay:** Transit delay is the total delay the voice signal experiences as it travels through the network; this is also referred to as *mouth-to-ear delay*. A number of factors in the local and wide area network contribute to transit delay. These include voice coding/compression, packet generation, channel contention (in a WLAN), network transport/buffering, and jitter removal. The important thing to know is that once the one-way delay exceeds 150 msec, it will begin to affect the cadence of the conversation. Distance, router buffering, and WLAN contention are all contributing factors in end-to-end transit delay. Transit delay has been one of the major performance complaints we have seen in packet telephony.

The other timing issue is jitter or the variation in delay from packet to packet that is introduced by the dynamic buffering used in a packet network. Left untreated, jitter will render the voice unintelligible. As we noted in the last chapter, the RTP addresses jitter removal, but the downside of the RTP process is that it adds to the overall transit delay.

3. **Echo Control:** All telephone circuits introduce echo. However, when the one-way transit delay exceeds 35 to 40 msec, the echo becomes noticeable and annoying. When the delay exceeds that parameter, equipment must be used to remove the echo. Virtually all packet voice networks will exceed 40 msec one-way delay, so echo control will be one element that must be incorporated in the system design.

8.1.1 Voice Quality

The human voice produces a signal that is analog by nature. When you speak, your voice creates a pattern of pressure vibrations in the air, which is essentially an analog signal (i.e., a signal that is continuously varying). Before that analog signal can be transported over a packet network that carries digital information, we need a *codec* to convert that analog voice signal into a digital representation (Note: *Codec* is a contraction of the two terms *coder* and *decoder*).

There are three primary issues to consider in the selection of a voice coding system.

1. The digital transmission rate required.

2. The delay introduced by the coding process.

3. The loss tolerance, or the percentage of packets that can be lost before the voice quality degrades below the allowable quality threshold.

Any technique that converts voice into a digital representation introduces some degradation in the sound quality; in general, those signal degradations are indistinguishable to the human ear. However, if some of those bits are changed due to transmission errors, or if some of the bits are lost due to packet dropping, the quality of the recovered signal can be impacted. There are a number of voice coding techniques, and they vary with regard to efficiency, robustness, and encoding delay. The parameters for the major voice coding options are summarized in Table 8-1.

An efficient coding system reduces the number of bits we need for each channel and thereby increases the number of simultaneous voice calls we can support on a given amount of network capacity. As a general rule, the more efficient the voice coding system, the longer the encoding delay and the greater the impact of packet loss.

8.1.2 G.711 Pulse Code Modulation: A-Law and μ-Law

The wired telephone network uses a standard technique for encoding voice called Pulse Code Modulation (PCM), and it is described in ITU Recommendation G.711. In G.711 encoding, a voice-grade channel is represented in a digital bit stream of 64 Kbps. While PCM produces the highest bit rate, it is also the most tolerant to loss. A 64 Kbps voice channel can tolerate roughly 10 percent packet loss and still provide acceptable sound quality; an 8 Kbps voice signal will degrade significantly if the network drops even one percent of the packets.

Table 8-1: Voice Coding Options

Coding Technique	Bit Rate (Kbps)	Approximate Encoding Delay (msec)	Loss Tolerance (%)	Applications
G.711 Pulse Code Modulation (PCM)	64	0.13	7–10	Public Telephone Network, PBXs, and Most IP PBXs
G.726 Adaptive Differential PCM (ADPCM)	24, 32, or 40	0.4	5	T-1 Multiplexer Networks, and DECT Cordless Phones
G.729a	8	25	<2	WLAN and Wide Area Packet Voice Networks
G.723.1	5.3 or 6.4	67	<1	Limited Due to Coding Delay
G.722 Wideband (50 Hz to 7 KHz)	64	0.4	5	Radio Broadcasting and Conferencing Systems

The encoding process in PCM involves two primary steps.

1. **Sampling:** The PCM codec samples that analog signal 8,000 times per second. Each time it samples the signal, the codec measures the amplitude (i.e., height) of the analog signal.

2. **Encoding:** When the sample is measured, the codec rounds off the value and encodes it as an 8-bit data character.

What the codec sends is 8,000 8-bit measurements of the signal per second for a total of 64 Kbps. There are two different PCM standards in use: the North American standard and the international standard. In either case, the voice signal is sampled at 8,000 times per second, but each version uses a different technique to generate the encoding table; that encoding adjustment is called *companding*. The North American standard uses a companding adjustment identified as μ-Law, while the international version uses A-Law companding. No human being can hear the difference between μ-Law and A-Law companding, but the equipment must be optioned to recognize the difference. See Figure 8-1.

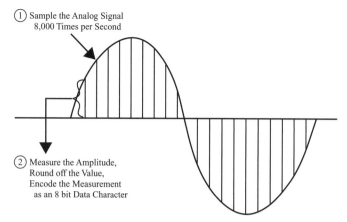

Figure 8-1: PCM Voice Coding

8.2 Voice Compression Techniques

While PCM has been the standard encoding for the public telephone network, packet voice systems can then use a variety of techniques to compress the digital voice signal to a lower bit rate. With a more efficient voice coding (i.e., fewer bits per second per channel) the result is that more voice channels can be carried on the same network capacity.

In analyzing the transmission requirements for voice it is important to factor in the packet overhead. Each voice sample will be carried in a packet that includes an RTP header, a UDP header, an IP header, and a Layer 2 header and trailer. For example, a PCM encoded voice signal using a 20 msec sample interval takes 82.4 Kbps when sent using PPP as the Layer 2 protocol. An 8 Kbps signal takes 26.4 Kbps! Table 7-3 (in the previous chapter) summarizes the total bit rate produced by each coding system using different voice codings, sample sizes, and transport protocols.

There are four alternative coding systems that could be used with IP voice.

1. **G.726 (Formerly G. 723) Adaptive Differential Pulse Code Modulation (ADPCM):** ADPCM was the first widely used compressed voice solution and it featured high quality and minimal delay (typically under 1 msec). ADPCM describes voice encoding in 24, 32, and 40 Kbps. In each case the voice is sampled at 8,000 samples per second and each sample is encoded in 3, 4, or 5

bits. ADPCM was used in T-1 multiplexer networks that were common in the 1980s, and the 32 Kbps version is also used with the DECT standard for cordless phones.

2. **G.729A:** G.729A is an updated version of an ITU standard designated G.729, and it defines a mechanism for compressing PCM to a bit rate of 8 Kbps. G.729A is widely supported in WLAN handsets and it is also used in wide area IP telephony. G.729A compression typically takes around 25 msec, and it produces a signal with minimal loss tolerance. The voice quality will typically drop below the acceptable threshold if just one percent of the packets are lost.

3. **G.723.1:** G.723.1 is an even more efficient compression system, yielding bit rates of 5.3 or 6.4 Kbps. The downside is that compressing in this format typically takes about 70 msec, and has a loss tolerance similar to G.729A. Given the inordinately long delay introduced, G.723.1 is rarely used in IP telephone systems.

4. **G.722:** G.722 provides greater voice bandwidth rather than improved efficiency. A typical voice codec (e.g., G.711, G.729A, etc.) captures frequencies up to about 3.1 KHz; according to Nyquist's theorem, with a coding system like PCM the highest frequency that can be encoded is one-half of the sampling rate or 4 KHz. G.726 uses an enhanced coding mechanism that can capture frequencies up to 7 KHz, and produces a bit stream at 64 Kbps. That means a G.722 encoding could deliver twice the analog bandwidth (i.e., high fidelity voice) in the same amount of digital transmission capacity as G.711. G.726 has been used in remote radio broadcast and video teleconferencing systems, and it is now being included in some high-end IP PBX handsets.

Besides these, there are also a number of vendor proprietary voice compression systems.

8.3 Other Voice Quality Issues

As we noted earlier, when voice packets are lost, there is no effective way to recover them in a timely fashion. The question then becomes: what does the receiver do to fill the empty time? IP voice systems will often incorporate tricks to mask the effect of lost packets. The techniques vary but they often include duplicating some of the packets that are received or inserting white noise to cover the gaps. The end result is that the content cannot be reconstructed, but at least there is no unwanted noise or dropouts.

8.3.1 Voice Activity Detection/Comfort Noise

If the IP voice network is using voice activity detection (i.e., allocating network capacity only when there is voice traffic to send), the channel may become perfectly quiet when no one is speaking. It can be so quiet the users think the call has been disconnected. As a result users keep saying: "Are you still there?" To address that difficulty, packet voice systems can inject white noise (i.e., low-level background noise) at the receiving end when there are no received packets to deliver; that is also referred to as *comfort noise*. Few VoIP systems actually use voice activity detection today, so voice packets are produced at a constant rate regardless of whether the party is speaking.

8.3.2 Foreign Language Speakers

One last note on voice quality deals with challenging quality environments. When the two parties in a conversation speak different native languages, they will typically require a better quality path on which to communicate. If the two parties are already struggling to understand one another, any distortion in the transmission path will only make matters worse. So while we generally look at voice quality as an objective standard, in reality, parties who are familiar with one another and speak regularly by phone can often get by with a poorer quality channel.

8.4 Voice Compression in WLAN Voice Networks

Virtually all WLAN voice handsets will support both G.711 (64 Kbps) and G.729A (8 Kbps) voice encoding. Given the limited capacity available on a WLAN you would think that it would be an ideal environment for compressed voice. In actuality, that is not necessarily the case. Along with the RTP/UDP/IP overhead that will accompany a voice packet, the WLAN access protocol adds additional overhead of its own. The end result is that the actual voice bits represent a rather minor portion of the overall transmission time.

The impact of voice compression is seen in the number of simultaneous voice calls an access point can support. While G.729A reduces the voice bit rate by a factor of eight, the increase in the number of simultaneous calls may increase by as little as 15 percent. Given that the vast majority of IP PBX systems use G.711 on calls between wired handsets, that same encoding will typically be used on WLAN calls as well.

8.5 Delay Tolerance

There are two time-related issues that can adversely impact the quality of IP voice services: transit delay and jitter. To the user, transit delay is the most obvious and annoying. However, it is important to recognize that the two issues are inextricably intertwined.

8.5.1 Transit Delay Tolerance: Maximum 150 msec One-Way Delay

The impact of transit delay in a voice transmission is essentially psychological. Human beings have a surprisingly accurate internal clock that governs the flow of a conversation. When a person asks a question, their internal clock starts. He or she is waiting a given span of time to hear a reply. If the answer is not heard within that time span, invariably he or she will repeat the question.

When the two parties can see each other (i.e., an in-person meeting), visual cues play a significant role. If you can see that someone is thinking, you will let them think. However, in a telephone call, there are no visual cues. In that case, the questioner must rely solely on his of her internal clock. If the questioner does not get a reply within the anticipated time frame, he or she will either repeat the question or ask, "Did you hear me?"

This human reaction is quite inconvenient in packet telephony. If the transmission path introduces an inordinately lengthy delay, that delay will trigger this follow-up reaction (e.g., "Did you hear me?"). That follow-up will typically collide with the response coming from the other end. Research has shown that if the network service introduces a one-way delay in excess of 150 msec, human speakers will begin encountering these conversational difficulties.

With the advent of IP telephony, a number of testing organizations set out to determine the range of delays that users would find acceptable. What they found was that one-way delays up to 70 or 100 msec were essentially unnoticeable. Once that one-way delay went past 100 msec, some people began to complain. When the delays reached 150 msec, virtually everyone was complaining. To get an idea of what that delay is like, a cell phone-to-cell phone call has a one-way delay of around 140 msec. We have found that users are willing to trade off voice quality of the convenience of mobility, so we should not try to equate that directly with user expectations on wired telephone systems.

8.6 Jitter/Delay Sources

As mentioned before, transit delay and jitter are the two sources of delay in packet telephony.

1. **Transit Delay:** The amount of time it takes for the signal to travel from the speaker, through all of the network elements, to the recipient.

2. **Jitter:** The variation in delay that is seen from packet to packet.

Transit delay is the total delay a voice signal will encounter when traveling through the network; this is called *mouth-to-ear delay*. Circuit switched voice systems introduce negligible delays, typically under 30 msec. IP PBXs can introduce delays on the order of 50 to 70 msec between wired stations. The addition of a WLAN connection increases the delay. The transit delays introduced in a wide area IP telephony application can be far greater. Given the distance and the number of routers involved, wide area systems can introduce delays in excess of 100 msec.

Transit delay is a combination of several factors:

- **Voice Encoding:** As we noted above, each voice coding system introduces some amount of delay. Where the 64 Kbps PCM encoding delay is under 1 msec, voice compression systems can introduce delays ranging up to 70 msec.

- **Packet Generation:** The packet voice system will collect some amount of voice information to insert in a packet; that delay will be a factor of the duration of the speech sample carried in each packet. To minimize the delay, we try to keep the voice packets short. The size of the voice sample carried in a packet will typically range from 20 to 40 msec, though we recommend a 20 msec sample size to minimize the packeting delay. The downside of a short packet size is that the amount of header information remains the same regardless of the size of the speech sample, so shorter packets increase the ratio of overhead to content.

- **WLAN Contention Delay:** If the voice is forwarded over a wireless LAN, the frame must be sent according to the WLAN access protocol (i.e., CSMA/CA). That means there is a short waiting interval before the frame can be sent, and if there is a collision or other transmission failure, the frame must be resent. Further, the entire frame must be received at the access point and tested for errors before it is forwarded through the wired network. The goal of WLAN

QoS (i.e., IEEE 802.11e) is to minimize the delay for voice packets. Even with the QoS feature, transmitting voice packets over a WLAN typically adds about 20 to 30 msec to the overall delay.

- **Serialization Delay:** The amount of time it takes to forward the packet onto the serial (i.e., bit-at-a-time) transmission link. The serialization delay is a factor of the packet size and the transmission rate of the link.

- **Propagation Delay:** The amount of time it takes the signal to travel over the physical transmission facility. In local networks, links are short and propagation delays are minimal; worst-case propagation delay for 100 m of EIA Category 5e LAN cable is 548 nsec. However, propagation delays do become significant when we introduce wide area facilities. Assuming the packet routing does not change (i.e., a virtual circuit service), this value should stay constant during the connection.

- **Switch/Router Delay:** Typically the biggest delay variable in IP voice systems will be the delay introduced by switches and routers in the path. Switch/router delays are composed of two elements.
  - Switching Delay: Each switch or router the packet passes through will take some amount of time to process the packet. That processing involves reading the address, doing a table lookup, and determining the best path if there are multiple entries in the table. The processing delay is typically constant and a factor of the processor speed and efficiency of the software.
  - Buffering Delay: Buffering is the major variable in the transit delay. Routers process one packet at a time, and the packet must wait in a buffer until its turn comes up. The waiting interval is determined by the volume of traffic in each router's buffer when the voice packet arrives. If QoS is used, the router will maintain separate buffers for each priority level, so the delay is impacted primarily by traffic in the same priority class. However, the highest priority queue is typically not given exclusive access to the facility until it is emptied. As a result, traffic in lower priority queues will also have an impact on the buffering delay.

- **Jitter Removal:** Once the packet arrives at its destination, the RTP receiver process must reestablish the timing consistency (i.e., remove the jitter). To do that, the packet is placed in a buffer and then played out according to the RTP timestamp. That receive buffer is the last element in the delay chain.

8.7 Real-Time Transport Protocol (RTP)

As we described in the last chapter, in an IP network the RTP is the application used to transport real-time voice and video. The basic trick in RTP is to place a timestamp and a sequence number in each packet before it is sent into the network. The packets will experience differing amounts of delay as they travel through the network; in a connectionless network they might also be delivered out of sequence. The RTP processing at the receive end reestablishes the sequence, buffers the packets, and then plays them out according to the timestamps. In effect, the RTP receive processing reconstitutes the isochronous nature of the information.

8.8 Echo Control

One last quality issue in packet telephony is echo control. Echo is the phenomenon of hearing one's own voice echoed back through the network. All telephone calls have echo. The issue is whether that echo is noticeable.

8.8.1 The Nature of Telephone Echo

Echo is generated in any voice transmission path. The culprit is a device called a hybrid or two-to-four wire converter. The long distance path is inherently four wire; that means there are separate paths for each direction of transmission. A telephone set, on the other hand, operates on a two wire connection. A hybrid circuit is used at each end of the four wire path where it connects to the two wire termination. When a party is speaking in one direction, the hybrid conducts most of the energy of the voice down the two wire termination. However, unless the impedance of the balancing network in the hybrid is perfectly matched to the impedance of the termination (something that never happens in the real world), some of the voice energy is conducted back onto the reverse path and is heard as echo.

8.8.2 Side-Tone Echo Is Always Present

The issue with echo is how quickly it comes back. The telephone set feeds back some of the transmit signal to the earpiece in a side-tone path, so the speaker can hear himself or herself while talking. Without the side tone, the user would think the call has been disconnected. Remote echo is created by the hybrid circuit at the far end of the line, and if the round-trip delay is short enough, the speaker cannot distinguish the return echo from the side tone. However, when a long transmission path delays the return echo, it becomes noticeable and annoying.

8.8.3 Echo Control Requirement

Research has found that when the one-way transmission delay is 35 msec or less (i.e., round-trip delay of 70 msec or less), the speaker cannot distinguish the echo from the side tone. However, when the one-way delay exceeds 35 msec, the echo becomes annoying and must be eliminated. The longer the delay, the more annoying the echo becomes. Virtually all packet voice systems will introduce one-way delays in excess of 35 msec, so some mechanism for controlling echo must be included. Typically the echo control is provided in a router, an IP telephony gateway, or an IP/Ethernet handset.

8.8.4 Echo Control Options

Two techniques have been developed to eliminate remote transmission echo.

1. **Echo Suppression:** This is an analog technique that effectively turns down the volume on the reverse path when someone is speaking. The echo is generated, but the power of the echo signal is reduced to a level at which the speaker cannot hear it.

2. **Echo Cancellation:** Most packet voice systems use a digital technique called echo cancellation. In echo cancellation, a circuit analyzes the digital signal in one direction, and cancels out that same pattern when it appears on the reverse path.

Echo cancellation systems can be tricky. When certain combinations of devices are involved in the path, the echo canceler can become confused and fail to cancel out the echo signal. Cell phones and cordless phones are often involved in these difficulties and they are very difficult problems to address.

8.9 Measuring Voice Quality

The public telephone network used 64 Kbps voice coding almost universally, and those voice calls were carried on circuit switched connections that introduced minimal delay, no jitter, and relatively few transmission errors. The telephone companies were essentially providing voice channels that were better than they needed to be. It became clear in IP telephone systems that designers could sacrifice the quality to some degree and the difference would not be noticeable to the user. The question is now, how bad can the quality get before users start complaining?

The growing interest in IP telephony has given rise to a number of techniques for measuring voice quality. The factors addressed are typically sound quality, which is affected by the voice coding and packet loss, and conversational quality, which is impacted by delay. There are two major approaches to measuring voice quality.

1. **Subjective Testing:** Using a panel of listeners who score the ability of different encoding systems to replicate challenging phrases. The Absolute Category Rating (ACR) or Mean Opinion Score (MOS) is the primary example.

2. **Machine-Based Testing:** These are systems that take a digital voice file, pass it through the encoding/compression system, and then compare the output to the input mathematically. The ITU's P.861: Perceptual Speech Quality Measurement (PSQM) and G.107 E Mode or R Factor are the primary examples.

8.10 Absolute Category Rating (ACR)/Mean Opinion Score (MOS)

The most widely referenced voice quality measurement, ACR/MOS, is a subjective test that uses a panel of listeners to measure the quality of different voice coding systems and rate them on a scale of 1 to 5. A score of 5 is considered "perfect," and 64 Kbps PCM produces a score of 4.4. A minimum score of 4.0 is recommended for business telephony, and 3.5 is considered marginally acceptable. The typical MOS value for a G.729A encoded connection is 4.0, and drops into the marginal range with one percent packet loss. The measurement can be computed for listening quality (LQ) and conversational quality (CQ).

To establish the listening quality, a set of phonetically balanced phrases like the Harvard Sentences (e.g., "The birch canoe slid on the smooth planks") is played to a panel of 16 or more listeners who score them. The subjective perception can vary widely with a small group of listeners, but as the number of panel members increases, the scores become more stable and a meaningful average can be computed. The University of Miami has an online MOS test that interested readers can try at: http://umsis.miami.edu/~gliu/.

To measure conversational quality, pairs of users are assigned a task to be conducted over the phone. Conversational quality is a more complex assessment as it addresses the overall call quality including listening quality, echo, and delay. The delay impact will vary based on the nature of the task (e.g., a business negotiation versus an informal chat).

8.11 Sample-Based Objective Testing: ITU P.861, P.862, and P.563

Unsatisfied with the subjective nature of ACR testing, engineers at the International Telecommunications Union (ITU) set out to develop a quantitative measure of voice system performance. In recommendations P.861 and P.862 they developed a technique for lab testing different voice codecs. A pure sample of a digital voice file is run through the codec or packet voice system and compared to a distorted version of that same file using Fourier Transform coefficients to compute a Perceptual Evaluation of Speech Quality (PESQ) score; the PESQ score has a range similar to MOS. As the PESQ score does not involve a conversational interaction, it is more akin to the MOS-LQ score. In 2004, the ITU developed P.563, a measurement that required only the received or distorted version of the sample file.

8.11.1 E Model-Based R Factors: ETSI TS 101 325-5 and ITU G.107

The E Model was originally developed by the European Telecommunications Standards Institute (ETSI) as a sample-based quality measurement that requires less processing than the PESQ. The E Model also takes conversational characteristics into account. Their E Model was developed as a planning tool for transmission systems, but it is now widely used as a quality measurement tool in VoIP networks. The E Model was adopted by the ITU as G.107 in 1998 and it has been updated and revised annually since then.

The E Model defines a transmission quality rating called the R Factor, with a range of values of 0–120. PCM produces an R Factor of 93, and a score above 80 is considered acceptable for business voice services. See Table 8-2.

The E Model is based on the premise that the effects of impairments are additive. The basic E Model equation is

$$R = R_\mathrm{o} - I_\mathrm{s} - I_\mathrm{d} - I_\mathrm{e} + A$$

where R_o is a base factor determined from noise levels, loudness, and so on; I_s represents signal impairments occurring simultaneously with speech, including: loudness, coding distortion (i.e., quantizing noise), and non-optimum side tone level; I_d represents impairments that are delayed with respect to speech, including echo and

Table 8-2: Typical Representation of Call Quality Levels

User Opinion	R Factor	MOS (ITU Scale)
Scale Maximum	120	5.0
Attainable Maximum (G.711)	93	4.4
Business Quality	≥80	≥4.0
Marginal	70–80	3.6–4.0
Poor	50–70	2.6–3.6
Not Recommended	≤50	≤2.5

conversational difficulty related to delay; I_e is the equipment impairment factor and represents the effects of VoIP systems on transmission signals; and A is the advantage factor and represents the user's expectation of quality when making a phone call. For example, people prize the convenience of a mobile phone, and are more forgiving in terms of quality problems. For typical R Factors and MOS levels of phone calls, see Table 8-2.

8.12 Other Factors in Voice Quality

There is more to the experience of a voice call than simply the quality of the received voice signal. In evaluating a packet voice solution it is important to address all of the factors in the calling experience. That overall assessment would include such factors as:

- Dial tone delay.

- Call setup time (end of dialing to ring initiation).

- Disconnect time.

- Recovery time to next dial tone.

- Percentage of blocked calls (i.e., all circuits busy).

- Feature activation time.

- Failure rate/mean time to repair.

- Percent of dropped calls.

- Percent of wrong numbers (not caused by dialing errors).

- Percent of calls dialed but not completed.

8.13 Conclusion

Users have high expectations regarding the quality of phone calls and there is a lot that can go wrong with a packet voice connection. The users' expectations have been built up over decades of experience using the public telephone network, and we now know that that network provided voice quality that was better than it had to be. The question is, how far can we let the voice quality slip before users will start to complain? Hopefully we will discover those difficulties before we have made a non-reversible commitment.

However, there is a perceptional factor in voice scoring, so MOS or R Factor scores should not be considered as hard and fast rules. We have found that users will trade off convenience for quality with cell phones. Further, challenging environments like calls between different native language speakers require better sound quality to be effective. If we are asking users to put up with poorer voice quality in an IP voice system, hopefully we will be able to deliver some offsetting benefit. The ability to support mobile voice calls over a wireless LAN would certainly fall into that category.

In the final analysis, the only true test for acceptable quality in a business organization will be a small-scale pilot test of the system in a real world environment. That pilot test should be conducted by the most demanding users to determine if the solution will be acceptable to them.

CHAPTER 9

Voice Network Design and Traffic Engineering

To deal with voice communications requirements, the network designer will require a basic understanding of telephone system components and design principles. The most ill considered phrase a data network designer may utter with regard to voice support is:

"Voice is just another application on my data network;"

or even worse

"There's nothing to know about voice."

Rest assured, there is plenty to know, and if that is the attitude that engineers are taking into the design process, they are in for a rude awakening. One of the fundamental difficulties in learning about the design and implementation of enterprise voice communication networks is that there are few if any decent texts written on the subject. Those that do exist focus primarily on the public telephone network rather than business telephone systems. Further, much of what is written deals with the history of telephone technology. How much background does the reader need on a technology that will soon be abandoned?

In this chapter we will provide some of that necessary enterprise voice background. First we will look at the service configurations used for business telephone services, and where VoIP-based telephone systems and services will fit. We will also look at some of the features of a PBX system that can be used to assist in the design and operation of an IP telephone solution. Finally, we will describe the basics of telephone traffic engineering. This is the process we use to determine how many channels or trunks will be needed to support a certain volume of telephone traffic. This design process is common to all telephone systems, and it will be absolutely essential in scaling of WLAN voice networks, particularly those that will support both voice and data requirements.

9.1 Business Telephone Systems: Electronic Key Telephone and PBX Systems

A WLAN voice network will typically be installed as an adjunct to a business telephone system. Business telephone users have traditionally invested in one of two types of telephone systems to support their voice networking requirements: Key Telephone Systems and PBXs. Key Telephone Systems or Electronic Key Telephone Systems (KTSs or EKTSs) are used in smaller installations with up to 50 lines. Larger businesses (i.e., >50 telephones) will typically buy and install a specialized telephone switch called a PBX (Private Branch Exchange).

As we noted in Chapter 1, both of these product categories are now being supplanted by IP PBX systems where telephone handsets supporting an IP/Ethernet interface are connected through LAN switch infrastructure. At the moment, however, the vast majority of business telephone systems are still traditional EKTS or PBX systems.

These legacy business telephone systems used the same type of circuit switching technology employed in the public telephone network, but users are now migrating to IP-based solutions with IP/Ethernet handsets and a telephony server to provide telephone service over their LAN switch infrastructure. Whether employing IP or traditional circuit switching technology, business telephone systems are used to provide four basic functions.

1. **Cost Savings:** In a PBX system, the connections to the public network are called trunks. The telephone company charges a monthly fee for each trunk, so the basic job of a PBX is to allow user stations to share those trunk connections. Different groups of trunks are used to support local and long distance calls. Large organizations might also lease *tie-lines*, which are dedicated voice channels connecting to other company facilities.

2. **Support Station-to-Station Calling:** PBXs also provide station-to-station connections without passing those calls through the public telephone network, further enhancing the cost savings.

3. **Provide User Features:** The telephone system will also support a set of features to improve convenience and efficiency (e.g., call waiting, call forwarding, conference calling, etc.)

4. **Provide Network Management Features:** Beyond the user features, the telephone system will also provide a set of capabilities to monitor and manage the system. Two of the more important features of the PBX are:
    *   **Automatic Route Selection (ARS):** ARS is a PBX feature that allows the network administrator to set up tables defining how each class of calls (e.g., local, long distance, intra-company) should be routed. From a management standpoint, we would like to ensure that every call is routed on the least expensive facility. On the other hand, administrators do not want to burden users with detailed instruction describing how calls to different locations should be dialed. With ARS, the user simply dials the telephone number they desire, and the PBX automatically routes the call on the least expensive facility that is available at that moment. If that first-choice facility is busy, the call can then be overflowed to a more expensive facility. The administrator can also define which users will have their calls overflowed and which will be given a busy signal. Besides selecting the most cost-effective facility, the ARS can also translate the dial digits so the correct dial sequence is sent on the trunk to complete the call (e.g., on a tie-line, only the last four digits are sent, but if that call is overflowed to a public long distance service, the ARS will send the full 10 digit number).
    *   **Call Detail Recording/Call Accounting System:** Call Detail Recording (CDR) is a feature that allows the PBX to output a record for each call that is placed. That record includes the date/time of the call, duration, the number dialed, the station placing the call, and the facility on which it was carried. The customer can buy an optional computer system that takes those CDR records, sorts them by extension, and computes a cost for each call so it can be charged back to the responsible departments. The call accounting system can also take that raw CDR data, sort it by date/time and number called, and produce the traffic information administrators need to size the various trunk groups.

9.2 User Stations

In a PBX or EKTS system, user telephone sets are referred to as *stations*. A business telephone system will typically support several different types of station sets to meet the requirements of different sets of users. The station equipment can represent a significant investment, as high-end digital sets can cost several hundred dollars apiece. The user

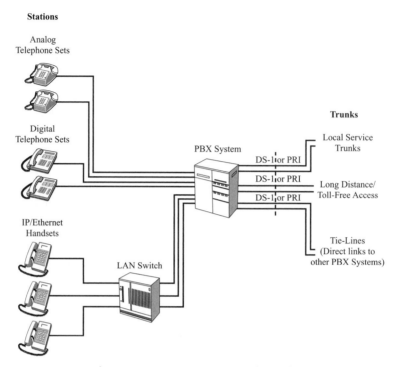

Figure 9-1: PBX System Configuration

purchases cards that are installed in the PBX to support the mix of station sets they require. See Figure 9-1.

- **Analog Sets:** The simplest station set is a single line analog telephone with a touch-tone (i.e., Dual-Tone Multi-Frequency) dial; this is the same type of telephone customers have at home. The industry designation for these is a *2500-set*. They operate on a single pair of wires and generally cost under $20.

- **Digital Sets:** Each PBX vendor also offers a line of proprietary digital sets that feature a hold button, multiple line appearances, one button access to features, and possibly a display. Depending on the features, these sets can cost up to several hundred dollars; they will operate on one to three pairs of wires. It is important to note that these sets use a proprietary interface and will only work on one vendor's PBX system.

- **IP/Ethernet Handsets:** Recognizing the migration to IP PBX systems, the PBX vendors are now offering IP-based handsets that use an Ethernet interface.

While these sets are designed primarily to be used on an IP PBX, the user can install a hardware upgrade in his or her existing PBX to support these IP/Ethernet handsets; that upgraded configuration is called a *Hybrid IP PBX*. Like the digital handsets, these units can cost several hundred dollars apiece, and will only work on one vendor's PBX. The connection between the PBX and the IP/Ethernet handset is made through the LAN switch infrastructure, and since it uses a 10 or 100 Mbps Ethernet interface, the connection to the desktop will typically be on two pairs of EIA/TIA Category 5, 5e, or 6 cabling. The PBX vendors are supporting this option to allow purchasers to cap their investment in traditional digital handsets and begin buying models that will be supported on the IP PBX system that will eventually replace the existing model. Of course those IP/Ethernet handsets will only work on that vendor's IP PBX solution, so the vendor is effectively hooking the customer into their product line for the long term.

9.3 Telephone Network Services

While the PBX or IP PBX is the visible part of the telephone system, the larger part of the voice communications budget goes to telephone services. Simply stated, you buy a PBX that lasts for several years, but you get a telephone bill every month! Given the volume of calls they generate, business customers will often be eligible for significant discounts on their telephone usage. Further, flat rate, dedicated channels can be used to interconnect company offices to further reduce the cost of calling between those locations. All of these network services will be connected to interfaces on the PBX or KTS.

9.3.1 Telephone Trunk Interfaces

As we noted above, trunks are the shared connections from the PBX. Trunks that handle outgoing calls fall into three primary categories: local network trunks used for placing calls in the local service area (also called "Dial 9 trunks"), trunks to access long distance services, and tie-lines.

Three different types of interfaces are used to connect to these trunk facilities:

1. **Analog:** An analog trunk supports one voice connection at a time, and may use different types of signaling to indicate whether it is idle or busy (e.g., loop start, ground start, or E & M). Analog trunks are typically used only with smaller systems like KTS today.

2. **DS-1/T-1:** These are higher capacity digital interfaces based on the T-Carrier systems used in the telephone network. Operating at a digital transmission rate of 1.544 Mbps, a DS-1 interface can carry 24 simultaneous conversations, each with a nominal bit rate of 64 Kbps. Some of the bits on each 64 Kbps channel (i.e., 1,333 bps) are reserved for signaling and are used to indicate whether the channel is idle or busy; that technique is called Bit Robbed Signaling. The 1.544 Mbps signals can be carried on two pairs of copper wire or a number of these interfaces can be combined and carried on a fiber optic link.

3. **ISDN Primary Rate Interface (PRI):** ISDN was a 1980s vintage plan to upgrade the public telephone network to end-to-end digital technology. While the overall plan never took off, the PBX interface that was defined has become the standard way of connecting telephone systems to the public network. Like a DS-1 interface, PRI operates on a 1.544 Mbps transmission channel, but uses a format called "23B + D". In that interface, 23 simultaneous 64 Kbps voice connections are supported along with a 64 Kbps D Channel that is used to carry the signaling information for all 23 channels—a plan called common channel signaling. Along with the basic idle/busy status, the D Channel also carries dialing, ringing, and feature activation messages like calling line identification. If multiple PRI interfaces are used, it is not necessary to dedicate a signaling channel on each one. A single 64 Kbps signaling channel can typically support 5 to 7 PRI connections, and those others can support 24 rather than 23 simultaneous connections (i.e., 24 B).

While those are the physical interfaces used, the user will lease a number of different services from local, long distance, and possibly IP telephony carriers to support voice requirements in the most cost-effective fashion. The telephone system allows users to access any of these facilities.

- **Local Telephone Service:** Connections to the local telephone network that typically use ISDN PRI connections.

- **Long Distance Service:** Long distance calls can be sent over local access trunks and passed on to the selected long distance carrier. However, the local telephone company collects an access charge for carrying those calls through its network. To reduce those access charges, business users typically install dedicated 1.544 Mbps links from their PBX or IP PBX directly to their selected long distance carrier's Point of Presence (POP). Those 1.544 Mbps connections can use either the DS-1 (24 channel) or PRI (23 or 24 channel) formats. The carriers offer discounted rates for high-volume users, and have special billing plans called virtual private networks (VPNs) or software defined networks (SDNs) for calls to other company locations.

- **Toll-Free (800) Service:** Toll-free services are used for inbound calls, particularly calls from customers. Toll-free calls are priced at a higher rate than outbound calls. These were formerly called 800 Services based on the special area code 800 that was used for all toll-free numbers, but now that the area codes 866 and 877 are also used for toll-free services, the generic moniker toll-free service is used.

- **Tie-Lines:** If a user has a high volume of calls between company locations, it may be possible to save money by installing dedicated lines to carry voice calls between them. While those connections are routinely called tie-lines, technically they are "tie trunks." Traditionally, 1.544 Mbps dedicated lines were used to support those tie-lines, and each link could support 24 tie-lines. Customers could also purchase equipment that provided ADPCM voice compression allowing 44 or 48 tie-lines to be carried on a single 1.544 Mbps channel.

These tie-line facilities could be replaced by a wide area voice over IP solution. In this configuration, the tie-lines are carried over the organization's router network rather than on separate leased channels. Tie-lines can also be configured to support other long distance calls using a capability called "tail end hop-off". In that arrangement, a call to a non-company location is carried over a tie-line to a remote PBX and is then extended to that outside organization with a local telephone connection. That call is then billed as a local call in that city as it originates from the remote PBX. See Figure 9-2.

9.4 Telephone Traffic Engineering

There are a range of network services offered for enterprise users, and the voice network manager is responsible for ensuring that there are an adequate number of trunks to support each type of service and that the most cost-effective mix of services is used.

In buying network services, the most basic question will be: "How many trunks do I need?" Telephone facilities are shared resources, and there is a monthly fee for each

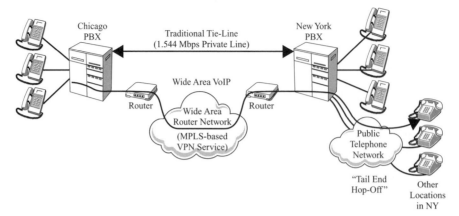

Figure 9-2: Voice Network Services

trunk connection. Administrators need a method to determine how many trunks will be required to support the user population without buying too many or too few. If too few trunks are provided, users will get busy signals or the ARS system will route calls onto more expensive routes. If too many trunks are provided, money is wasted on facilities that are never used.

9.4.1 Queuing Theory

Queuing theory is the branch of statistics that deals with the issue of allocating shared resources. Fortunately, for the practical task of sizing trunk groups, administrators do not need to deal with formulas or mathematical computations. The design process involves gathering the basic traffic data, sorting it into a useful form, and looking at a table (or entering information into a computer program) to determine the optimal number of trunks. Administrators do still have to be aware of the overall concepts and the limitations of the tools that are used.

9.4.2 Basic Concepts and Definitions

The three basic variables in telephone traffic engineering are summarized in the traffic engineering triangle. See Figure 9-3.

1. **Busy Hour Traffic Load:** The total volume of calling that must be supported during the busiest contiguous 60-minute period of the day.

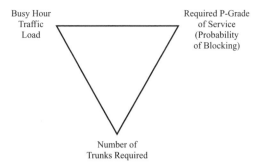

Figure 9-3: Traffic Engineering Triangle

2. **Probability of Blocking a Call ("P-Grade of Service"):** The probability that a call will be blocked during the busy hour because all of the trunks are in use.

3. **Trunks:** The number of channels required to support the traffic.

9.4.3 Handling Blocked Calls

The key concept in voice traffic engineering is that there will always be some probability of blocking a call. The only way to ensure that every call will find an available trunk is to provide as many trunks as there are potential callers; that would not be a cost-effective solution. Defining the acceptable percentage of blocked calls is a management issue, but the key is: what happens if the call is blocked?

In the worst case, the system provides a busy signal and the user has to hang up and try the call again. That is frustrating to the user and is not a productive use of his or her time. The other choice is to program the automatic route selection system to take those overflow calls and send them on a different, but more expensive, facility. The cost analysis becomes more complicated at this point because the administrator has to determine whether it is cheaper to add another voice trunk or just pay a higher cost for the calls that overflow.

9.5 Traffic Engineering Terms

Telephone traffic engineering involves a number of terms and concepts. The first concept is the period over which the calling activity will be studied. The volume of telephone traffic in a business varies widely during the day. A normal business that operates during daytime hours will typically have its busy calling periods around 10:30

AM and 3:00 PM. A pizza parlor will have its busiest period around mealtimes and will have a major calling spike at halftime during the Super Bowl.

9.5.1 Busy Hour

As calling activity is highly unpredictable from minute to minute, the convention in telephone traffic engineering is to size trunk groups based on the busiest 60-minute period during the day; that period is called the busy hour. The busy hour does not have to correspond to a clock hour, as it might run from 2:15 to 3:15 in the afternoon. The call detail records of the PBX can provide the raw information needed to analyze the calling activity, though a call accounting system will have to sort them by the called number, date/time, and duration to determine the busy hour.

9.5.2 Measuring Traffic Load

There are also standard measures that are used to define the volume of telephone traffic:

- **CCS or Hundred Call Seconds:** The Roman numeral C represents 100, and CS stands for Calling Seconds. A clock minute has 60 seconds, but 60 is an inconvenient number to deal with. To address this, telephone engineers came up with the idea of 100-second minutes or CCS. To convert regular clock minutes into CCS units, multiply by 0.6.

- **Erlangs:** Erlangs are the standard measure used to define traffic load today. An Erlang is equivalent to 60 minutes of calling. That could be one one-hour call or two 30-minute calls. An Erlang is the equivalent of 36 CCS or 3,600 seconds of usage.

9.5.3 Billable Time Versus Actual Use

In determining the actual volume of traffic, it is important to determine the actual amount of time the facility will be in use. If a network engineer is using billing records to compute usage, those records do not capture the entire use of the facility, only the amount of time the parties are actually connected. The trunk actually becomes busy as soon as the user finishes dialing. The system seizes the trunk, forwards the dialed digits, the call is set up, and the remote phone may ring a few times before the call is answered (each ring cycle is six seconds, two seconds of ringing followed by four seconds of silence). The billing time begins only when the call is answered. Further, there are no billing records for calls to busy numbers or for calls that are not answered.

After understanding what is or is not counted, the engineer may have to make adjustments to the raw data to recognize the actual use on the facilities.

9.5.4 Trunks

Typically, traffic engineering is used to determine the optimal number of trunks needed to support a particular type of calling. For example, requirements for local calling, long distance access, toll-free service, and tie-lines are determined separately. Traffic engineering is used to determine how many access points are needed in each area to support the expected volume of WLAN calls. In Chapter 11 we will look at how to determine the number of simultaneous calls and the total number of users an access point will be able to support. The number of trunks is always a whole number; it is not possible to buy a "half-trunk".

9.6 P-Grade of Service/Probability of Blocking

The probability a call will be blocked (i.e., not find a channel available in the assigned trunk group) is called the P-Grade of Service; management must decide what probability they find acceptable for blocking different types of calls. They might determine that local calls should get P.01 (i.e., a one percent probability of blocking during the busy hour), while long distance calls can get P.05 (i.e., a five percent probability of blocking during the busy hour), and tie-line calls can get P.10 (i.e., a 10 percent probability of blocking during the busy hour). Management usually accepts a higher probability of blocking long distance and tie-line calls because those facilities are more expensive.

9.6.11 The Erlang Loss Formula: "Erlang B"

Once an administrator has identified the busy hour, determined the volume of calling that is expected, and chosen the P-Grade of service, the number of trunks can be selected using the Erlang Loss Formula. The formula was developed by Agner Erlang (1878–1929), a Danish mathematician, statistician, and engineer who is credited with inventing both telephone traffic engineering and queuing theory. The Erlang Loss Formula is used to compute the probability of whether a call will be blocked; the formula is also called "Erlang B."

The Erlang B formula is written

$$E_{1,n}(A) = \frac{\dfrac{A^n}{n!}}{1 + A + \dfrac{A^2}{2!} + \ldots + \dfrac{A^n}{n!}}$$

where A is the traffic load in Erlangs, E is the probability of blocking, and N is the number of trunks.

Rather than struggling with the math, a copy of the table can be downloaded from the International Telecommunications Union (ITU) at http://www.itu.int/itudoc/itu-d/dept/psp/ssb/planitu/plandoc/erlangt.html.

Interested readers can also use a Web site like http://www.erlang.com/calculator/erlb/. The basic inputs are the traffic load measured in Erlangs and the P-Grade of Service to be provided.

9.6.2 Other Traffic Engineering Formulas: Extended Erlang B

One fundamental shortcoming of the Erlang B formula is that it assumes all call attempts are independent. In reality this is not the case, as users who get a busy signal will typically retry the call immediately. As a result, the assumption of independence is invalid. There is a version of the formula called Extended Erlang B that makes more realistic assumptions about users retrying blocked calls during the study period. A Web-based calculator for Extended Erlang B can be accessed at http://www.erlang.com/calculator/exeb/. If calls that are not serviced on the trunk group are overflowed to an alternative facility, Erlang B will provide the probability the call will be overflowed. If the caller is given a busy signal when all trunks are busy, Extended Erlang B will provide a more accurate estimate of the number of trunks required.

9.7 Traffic Engineering Process

Now that we have covered the basic concepts and vocabulary, we can look at the overall process of traffic engineering.

1. **Traffic Study:** After identifying the particular type of service to analyze (e.g., local, long distance, toll free, tie-line, WLAN, etc.) a traffic study is conducted to identify the busy hour and the total volume of calling that occurs during that particular 60-minute period. However, it is not possible to just take a one-time snapshot, because it might be an atypical day. Usage over a period of at least one week should be sampled, avoiding weeks that include holidays. If limited to looking at only one day, do not use Monday or Friday. It is very important to understand the organization's business and take seasonal fluctuations and changing business conditions into account. For example, it would be a mistake to upgrade the network facilities just when management is getting ready to downsize the organization.

2. **Adjust to Recognize Actual Usage:** Once the raw traffic data is availably, it is important to determine if it reflects the actual usage those facilities will see, and make adjustments as necessary. Remember, no billing record is generated for calls to busy lines or for calls that are not answered. It may be necessary to consult the PBX supplier to confirm how the call detail system actually times the call. One important note: If the facilities will be supporting calls in both directions (i.e., inbound and outbound), make sure to count the traffic in both directions! While this may sound comical, it is actually one of the most frequent mistakes that is made in sizing voice systems.

3. **Confirm the Objective for P-Grade of Service:** Most large organizations will have a policy that specifies the grade of service to be provided. If no such policy exists, it is important to define the issues for the responsible managers and ensure they understand the cost and operational impact of the decision.

4. **Size the Trunk Bundle:** With the raw traffic data in hand, and a definite knowledge of the grade of service to be provided, it is time to determine the number of trunks required by looking it up in a traffic engineering table or using a Web-based calculator (e.g., http://www.erlang.com).

5. **Sensitivity Analysis:** Many administrators will simply install the number of trunks identified by the Erlang computation and call it a day. However, a thorough analysis should consider the cost of calls that are overflowed to other facilities to determine the total cost. Remember, the busy hour on which the Erlang computation is based is only one hour out of the day. If an inordinately large percentage of calls occurs during that busy hour, the administrator will be

installing a large number of trunks to support those calls, but most of those trunks will be sitting idle for most of the day. A thorough analysis requires that an administrator look at several different scenarios and compute the expected cost for each.

6. **Confirm the Results:** Once the required facilities have been ordered and installed, it is also important to go back and monitor the trunk group to determine 1) if the traffic volume predicted is actually occurring, and 2) if the service is performing as planned.

One fundamental weakness of traffic engineering is that it is based on activity in the past while planning for the future. Even when making changes to an existing network, several months could elapse between doing the traffic study, getting management approval for the additional expenditure, placing the necessary service orders with the carriers, getting the services installed, making the required changes in our telephone system, and actually putting the changes in place.

9.7.1 Periodic Reviews

While that process will take care of the initial installation, telephone traffic patterns and calling rates change over time. Each part of the network must be reexamined periodically to determine if the configuration is still optimal. In short, traffic engineering is an ongoing management process, not a one-time event. In budgeting for a solution, be sure to consider the ongoing personnel cost required to maintain that optimal configuration.

9.8 Voice Traffic Engineering for IP Voice Systems

While that is the basic story of voice traffic engineering, the use of IP technology introduces another complication. We now need to know how many voice channels the IP voice system can support, and ensure that we do not overtax its capabilities. If we do, the system will start dropping packets, which in turn degrades the voice quality. If an administrator allows 20 voice calls to be routed over a packet service that can only support 10, there will not be 10 good calls and 10 bad ones; there will be 20 bad calls!

To determine the number of voice calls the packet network can support, it is imperitive to know the amount of capacity the network guarantees to provide on each route, and the number of bits per second required for each voice channel. After making that determination, steps must be taken to prevent too many voice calls from being sent. This computation is done differently for WLAN and wide area packet services.

In traditional IP services, there is no guaranteed capacity (that is what "best effort" means). Enterprise oriented packet services like frame relay and MPLS-VPN will define how much traffic can be sent before the network starts dropping packets. As we described in Chapter 7, where traditional IP provides best effort service, an MPLS offers several Class of Service (CoS) categories with different guarantees for delay and loss defined for each. We use the terms gold, silver, and bronze for the service categories, where gold service is used to support voice. The CoS profile for the service configuration will define the maximum allowable percentage of each service class that can be sent. The administrator marks each packet going into the network to assign it to one of the available categories.

9.8.1 Excess MPLS Gold Traffic Is Discarded

It is very important to understand how the service treats excess traffic in each category. The standard operation today is that any excess gold traffic is automatically discarded; excess traffic in the other categories may be delivered, but it has a higher priority for discard. Recognizing the treatment for excess gold traffic, it is critically important that only the supportable volume of voice traffic is sent over the MPLS gold category. To determine the gold capacity that should be ordered, the administrator should first determine the number of voice trunks required and then use Table 9-1 to identify the transmission requirement for each trunk.

The variables that will affect the bit rate are:

- **Voice Codec:** G.711 (64 Kbps), G.729A (8 Kbps), G.723.1 (5.3 Kbps)

- **Voice Sample Size:** Longer voice samples are carried in larger packets. Those larger packets are more efficient with regard to the ratio of payload to header, however they also increase the transit delay (i.e., it takes longer to collect the voice needed fill a larger packet). A voice sample size of 20 msec is recommended.

You multiply the bit rate per trunk by the number of trunks to determine the gold capacity that will be required.

Table 9-1: Packet Voice Transmission Requirements (Bits Per Second Per Voice Channel)

Codec	Voice Bit Rate (Kbps)	Voice Bandwidth (KHz)	Sample Time (msec)	Voice Payload (bytes)	Packets Per Second	Ethernet (Kbps)	PPP (Kbps)
G.711 (PCM)	64	3	20	160	50	87.2	82.4
G.711 (PCM)	64	3	30	240	33.3	79.4	76.2
G.711 (PCM)	64	3	40	320	25	75.6	73.2
G.722	64	7	20	160	50	87.2	82.4
G.722	64	7	30	240	33.3	79.4	76.2
G.722	64	7	40	320	25	75.6	73.2
G.726 (ADPCM)	32	3	20	80	50	55.2	50.4
G.726 (ADPCM)	32	3	30	120	33.3	47.4	44.2
G.726 (ADPCM)	32	3	40	160	25	43.6	41.2
G.729A	8	3	20	20	50	31.2	26.4
G.729A	8	3	30	30	33.3	23.4	20.2
G.729A	8	3	40	40	25	19.6	17.2

Note: RTP assumes 40 octets RTP/UDP/IP overhead per packet. Ethernet overhead adds 18 octets per packet. PPP overhead adds 6 octets per packet.

9.9 Voice Traffic Engineering in WLAN Networks

WLANs pose a different problem because WLAN devices do not drop voice packets, they delay them. If a WLAN network becomes congested, the time it takes to get access to the channel increases, but eventually the packet will be sent. That philosophy works fine with data traffic, as TCP will automatically adjust its window size, and all that the user will notice is an increase in response time. That philosophy can wreak havoc on a voice network, however, as a burst of activity will have a negative effect on jitter. If the RTP receive jitter buffer is set to a fixed size and voice frames arrive outside of that boundary, they will be dropped. You will recall that with RTP, delivering frames late has the same result as not delivering them at all.

WLAN voice design is further complicated by the fact that the network engineer must guess where a user is going to be when he or she makes or receives a call. The engineer must look at the call capacity of each access point and the specific area it covers rather than the total capacity of the network. Available capacity on the second floor does the user no good if he or she is trying to place a call on the 30th floor!

Allowing too many voice calls to be set up on any access point will cause increased delay, jitter, and possibly packet loss for all of those calls. One of the critical elements in the WLAN voice design will be call access control, where we define the maximum number of simultaneous calls an access point should accept. Potentially, a busy access point might be able to redirect additional voice service requests to another, less utilized access point (i.e., load balancing). We will discuss the overall design process in Chapter 11, and identify the monitoring and management functions in Chapter 14.

9.10 Conclusion

Voice and data networks have traditionally been served by different networks using different technologies. That architecture is clearly coming to an end in business services. That technology shift brings with it a greater degree of interaction between the voice and data communications groups within the IT organization. Some of the biggest problems we have seen relate to the management issues that convergence brings about.

9.10.1 There's Nothing to Know About Voice

The first step in addressing that convergence is to recognize the basic requirements of a business telephone system. Everyone has used a telephone, and that familiarity has led many in the data networking community to assume, "There's nothing to know about voice." Nothing could be further from the truth! Most business telephone systems deliver a high degree of performance and reliability, but that did not happen by chance. The performance the user receives from that voice network is the end result of a hundred years of progressive improvements.

CHAPTER 10

WLAN Voice Configuration

Now that we have laid the groundwork regarding wireless LANs and voice over IP technology, we can look at how these come together in a voice over wireless LAN configuration. This is the first of several chapters that will deal specifically with WLAN voice. In this first chapter we will look at the major hardware elements that go into a WLAN voice network, how they are configured, and how they interface with the wired telephone system. In the following chapters we will describe the technical requirements, design, planning, and network management functions to be addressed. We will also investigate the potential of Fixed Mobile Convergence, the idea of integrating WLAN voice with cellular technology to provide an all-encompassing mobile voice capability.

WLAN voice began in earnest in 2003, when companies like SpectraLink (now part of Polycom), Symbol Technologies (now part of Motorola), Vocera, and Cisco introduced WLAN voice handsets and the equipment to support them. A number of enterprises, particularly in health care and materials handling, deployed small-scale VoWLAN networks. Those installations were small-scale by choice as customers found that the initial crop of Wi-Fi products lacked many of the important features that would allow these networks to provide the scale, security, and other features required for a large-scale deployment. Handsets with those required features arrived on the market in 2007.

In this chapter, we will look at the architecture, infrastructure, and device requirements for a WLAN voice solution. We will begin with the overall architecture choices and then review the infrastructure requirements. Finally we will look at the range of handset choices that can be used to implement the solution while comparing the functionality that can be had with each combination. With regard to handsets we will look at the full range from simple Wi-Fi cordless phones to softphone clients that would operate over the WLAN interface.

10.1 Voice over WLAN Architectures

The first decision with regard to a WLAN voice deployment will be whether voice and data devices will be supported on the same or different WLANs. Closely related to that will be the choice of which channels to use for the voice versus the data users. There are a number of design and support issues that will be affected by each course of action.

10.1.1 Shared WLAN for Voice and Data

The primary advantage of using a shared network approach is that most of the WLAN infrastructure equipment that is currently installed can be reused. However, there is a good chance that the network capacity will have to be increased to support the additional traffic, and the coverage area may also have to be expanded. Increasing capacity in a WLAN involves adding additional access points to increase the number of available channels and to shorten the maximum range over which a client will need to transmit. Shortening the range has the beneficial effect of increasing the transmission rate the device will be able to achieve. Of course this entails buying additional equipment, and paying the cost of installing wired LAN drops for each of the new access points. Further, the radio coverage plan will have to be reworked as more access points are added to the environment. That redesign must be based on predictions of where WLAN voice users will access the network, and the number of calls that will have to be supported in each of those dense areas. See Figure 10-1.

With regard to the coverage, we will have to determine the areas where we want the voice service to be available, and ensure those areas are covered with services that have

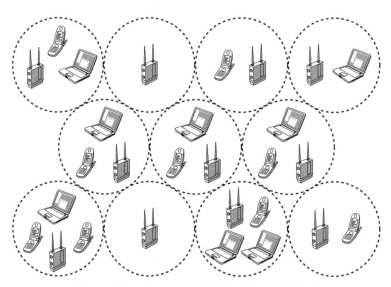

Figure 10-1: Integrated Voice/Data WLAN

the capacity to support the projected volume of voice usage. Many enterprise WLANs were designed to provide spot coverage or coverage only in areas where it was assumed laptop users would congregate. Those areas would typically include conference rooms, auditoriums, and other public areas. Very often it was possible to estimate the potential number of users by counting the number of chairs in the conference room! That approach works for data traffic, because the available network capacity will be spread over the users who are associated with the access point. As more users joined the network, the capacity available per user declined, and hence the network performance would slow.

Where data users will see their application performance degrade as more users jump onto the shared channel, voice is an all-or-nothing environment. To ensure the quality of the calls we support, we must limit the number of voice users that can associate with an access point; that function is referred to as call access control. If the maximum number of voice users is already associated with that access point, the next call attempt gets a busy signal. As we described earlier, allowing too many users to access the channel will result in increased jitter and delay for all of the users as they contend for the channel. The process for defining the maximum number of simultaneous calls will be described in Chapter 13.

The other major feature that will be required if voice and data users are sharing the same channel is the 802.11e QoS mechanism. We will be looking at how 802.11e works in the next chapter, but its basic purpose is to ensure that voice and video users get preferred access to the shared radio channel. The important thing to recognize about any QoS mechanism is that when treating one class of traffic better, by definition other classes of traffic are treated worse. To put it another way, QoS does not increase capacity, it merely manages scarcity! So in determining the number of simultaneous voice calls to allow on an access point, designers also have to reserve some capacity for data users.

The other issue to address in building a shared voice/data WLAN is the requirement to support different security policies. Legacy WLAN voice devices may not support the same range of security options as the data devices. Fortunately, most current WLAN access points do support the ability to provide multiple virtual WLANs on the same radio channel. In this configuration, the access point is supporting one physical WLAN channel, but it can broadcast beacon messages for a number of virtual WLANs, each with a different SSID (i.e., network name) and supporting different security options; on the wired side, those separate networks can be associated with different virtual LANs.

Data devices can associate with the data SSID and voice devices can associate with the voice SSID. The major issue to bear in mind with this configuration is that both virtual networks are sharing the same physical WLAN radio channel (i.e., one shared 11 or 54 Mbps channel); the separate networks are separate *virtual* networks. Both voice and data devices will be vying for access to the same channel, so if 802.11e QoS is used for the voice devices, they will get preferred access to the channel over data devices even if they are on a different virtual WLAN.

10.1.2 Separate WLANs for Voice and Data: Dual Overlay Network

A dual overlay network or the idea of implementing separate WLANs for voice and data is not as far-fetched as it once was. With early WLAN access points, a dual network implementation would essentially double the cost of the network. With modern access points and WLAN switches, much of the infrastructure can be shared between the two networks. Many WLAN access points today include dual radios allowing separate 2.4 GHz and 5 GHz networks to be run out of the same unit. Unlike a virtual WLAN configuration, these are truly separate physical networks, so the voice users would not be vying for capacity with data users. If the voice and data networks are implemented in different frequency bands (i.e., 2.4 and 5 GHz), there will be no radio frequency (RF) interference between them, even if there is 100 percent overlap in the coverage areas.

Even though there is no RF interference between the two networks, there are still a number of radio issues to address. The 2.4 and 5 GHz radio signals have different loss profiles, so you cannot simply add a 5 GHz channel to every 2.4 GHz access point and expect to have adequate 5 GHz radio coverage. Rather, a separate RF coverage plan is required for the 5 GHz network. See Figure 10-2.

The other question is, which frequency band should be used for voice? The initial idea was that voice users should be served on the 5 GHz network. The reason for this was that the data users were already occupying the 2.4 GHz band, and the voice users could take advantage of the greater number of channels available in the 5 GHz band. The problem with this idea is that many early VoWLAN handsets only worked in the 2.4 GHz band using the lower-speed (i.e., 11 Mbps) 802.11b radio link. The 802.11b radios were more power efficient than the early OFDM transmitters so the battery life was longer. Power consumption and battery life for OFDM devices has improved significantly with the new generation WLAN handsets, so the user is now free to choose.

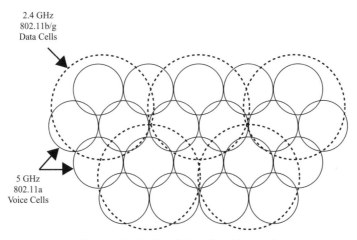

Figure 10-2: Dual Overlay Network

Users also have less experience deploying 5 GHz networks. Based on the greater signal loss at 5 GHz, we do know that more access points will be needed to cover the same area. Users will expect high reliability in a voice network, so we will have to learn how to deliver that on a 5 GHz network. The other factor to bear in mind with regard to a shared network that must support legacy 802.11b voice handsets is that the throughput for the 802.11g devices drops significantly when 802.11b and g devices share a network.

10.1.3 802.11n Impact

The choice of which radio band to use has become more complex with the advent of the 802.11n radio link. As we noted in Chapter 5, 802.11n supports data rates up to 289 Mbps (in a 20 MHz channel), and can operate in either the 2.4 or the 5 GHz band. However, the throughput for 802.11n devices will drop significantly if they use the mixed mode option where they share channels with legacy 802.11a, b, or g devices. One proposed solution is that we leave the legacy 802.11b/g devices in the 2.4 GHz band, and save the 5 GHz channels for 802.11n.

For the moment there are no voice devices that support the Draft 802.11n specification; the final specification is due in 2009. By then we may have 802.11n voice handsets to support. Fortunately there is considerable breathing room in the 5 GHz band. In the current 555 MHz allocation, there are 23 non-interfering 20 MHz channels available: 4 are designated for indoor use only, 4 are outdoor only, and 15 can be used indoors or outdoors. To avoid supporting voice and data users on the same channels, users could divide the available 5 GHz channels among 802.11a voice and 802.11n data users.

The other question that comes up when we look at the idea of a dual overlay network is the utility of 802.11e QoS. As all of the voice users are on the same network and have the same priority for accessing the channel, is there any value in implementing a QoS mechanism? The answer is "yes." There are other features included with 802.11e that will be beneficial to all voice users. The waiting intervals and back-off counters are shorter in the 802.11e protocol, which should lead to shorter transit delay and improved jitter performance. There is also an option to conserve battery power called WMM-Automatic Power Save Delivery. The 802.11e specification also includes a signaling protocol called the Traffic Specification (TSpec) for requesting capacity on the WLAN and implementing call access control. There is even an option called Direct Link Protocol that recognizes when two parties on a WLAN call are on the same access point and allows them to communicate directly rather than through the access point.

So even if we intend to support voice users on a separate WLAN, there are still a number of features in 802.11e that add to the functionality of that configuration.

10.2 Wireless LAN Switch

We introduced WLAN switches in Chapter 3, but their importance in voice over WLAN configurations should be reemphasized here. To support voice on a wireless LAN, there is a requirement for a WLAN infrastructure that is capable of providing the capacity, manageability, and functionality required to meet those voice requirements. A WLAN switch will be a critical element in any WLAN voice configuration of any significant scale.

Home users who are looking to support Wi-Fi cordless phones over their wireless LAN to access VoIP services like those from Vonage or Skype or WLAN/cellular services like T-Mobile's HotSpot @Home can still get by with a single access point. That type of solution can also serve the requirements of a small branch office, though support for enterprise features like 802.11e QoS will increase in importance as the traffic load on the network increases.

10.2.1 Handoff Capability Is Key

Any wireless LAN implementation that requires more than a few access points should be built on a centrally controlled solution, a WLAN switch. As described in Chapter 3, a WLAN switch network is composed of a number of access points whose functions are coordinated and controlled by a central controller that is connected to the wired LAN. A WLAN switch can provide sophisticated RF management, centralized security, and rogue AP identification/mitigation. For voice, however, the ability to support handoffs will be critical. WLAN switches support fast secure handoffs allowing voice users to roam freely within the coverage area. Typically those handoffs are accomplished within 150 msec so the user is unaware that his or her connection has been transferred to another access point. There is a pending standard for handoffs designated 802.11r, but its ratification is not expected before 2009. Until then, a centrally-controlled WLAN switching system will be necessary to support the handoff requirement.

The economics of a WLAN switch can be difficult in a small office environment. In a WLAN switch implementation, the cost of the central controller is distributed over the number of access points in the network. That investment is easier to justify in a large installation where there are hundreds of access points. To address those small office requirements, the WLAN switch manufacturers have pursued two different options. Some offer a small version of their controller that might support a maximum of six access points. The other approach is to provide the capability for the controller in a large office to support a number of small remote offices as well. The controller will communicate with the access points in those remote offices over a wide area connection.

10.3 Voice over WLAN Handsets/Clients

Once there is a secure and functional WLAN infrastructure that provides the coverage, capacity, and features to support voice users, then comes the need for handsets or other devices that can support WLAN voice calls. There are four major types of VoWLAN devices a customer might install.

1. Consumer Wi-Fi Cordless Phones.

2. Enterprise-Grade Wi-Fi Voice Systems.

3. Integrated WLAN/Cellular Phones.

4. VoWLAN Softphone Clients in Laptops, PDAs, and Mobile Computers.

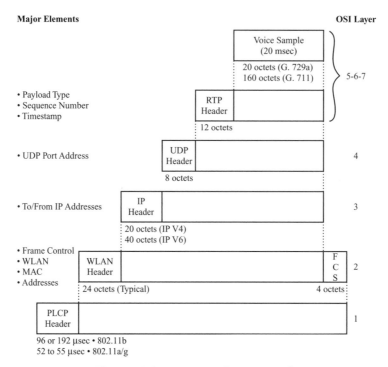

Figure 10-3: WLAN Voice Protocols

As we have described in the past few chapters, any VoIP implementation will require protocols to support both signaling and voice transmission. Signaling deals with the process of setting up/tearing down connections and activating features, while voice transmission addresses the fundamental task of carrying voice packets between the stations. Both of these tasks will be far more challenging in an enterprise grade Wi-Fi voice system than in a single access point home network.

As we described in Chapter 7, the transmission of the packetized voice between users universally employs the Real-time Transport Protocol (RTP). Voice samples with RTP headers are sent using the User Datagram Protocol (UDP), over the Internet Protocol (IP), which is then placed in a wireless LAN frame. The protocol stack for a WLAN voice frame is shown in Figure 10-3.

With regard to the voice path, there are two primary design issues:

1. **Voice Coding Technique:** Typically G.711 at 64 Kbps or G.729A at 8 Kbps.

2. **Voice Sample Size Per Packet:** Typical range of 20 to 40 msec, though 20 msec samples are recommended.

Given the limited capacity available in a WLAN, it would appear that using the lower bit rate G.729A voice coding would be the obvious choice. It turns out that the WLAN protocol adds so much overhead that the number of simultaneous voice calls supported is less than might be expected. Even though the voice stream is compressed by a factor of 8, the maximum number of simultaneous calls may increase by as little as 15 percent. The increase in the number of simultaneous voice calls ranges between 15 and 100 percent and is more pronounced as the WLAN data rate decreases. It is also important to remember that G.729A voice compression adds about 30 msec of delay.

10.3.1 WLAN Voice Signaling Options

As virtually all Wi-Fi voice devices support the same range of media options, the major difference is the signaling protocol they employ; the signaling must be compatible with the telephony server. There are three different architectural models being used:

1. **SIP-based:** The Session Initiation Protocol (SIP) is becoming the accepted signaling standard for IP telephony and multimedia communications. In its current implementation, there is a standard implementation of SIP that can support basic call setup/tear down and support about a dozen standard features like hold, conference, and transfer. If the Wi-Fi handsets use SIP signaling, they should be able to operate with any compatible SIP Proxy (i.e., the name for a SIP telephony server). Most consumer handsets support SIP signaling.

2. **Wi-Fi Vendor Proprietary:** Many enterprise-grade Wi-Fi voice handsets incorporate their own proprietary signaling systems; Polycom/SpectraLink and Vocera would be two widely used examples. As the signaling used over the Wi-Fi link is proprietary, these systems are sold with a controller that attaches to the wired LAN and interfaces with the wired telephone system. That interface can be either packet-based (i.e., IP over Ethernet) or circuit-based (e.g., DS-1 or PRI). One distinguishing characteristic with this solution is the level of integration with the user features on the wired telephone system. The level of functionality will vary with the Wi-Fi voice product being used and the model

of the wired PBX it is connected to. The obvious shortcoming of a vendor-proprietary solution is that you will be limited to the range of handsets offered by that vendor. On the other hand, a mix of SIP-based and proprietary handsets could be used as long as there are servers available to support each group.

3. **IP PBX Proprietary:** Some IP PBX suppliers have developed their own Wi-Fi handsets that use the same signaling protocols as their wired IP handsets. Two notable examples in this category are Cisco, whose Wi-Fi handsets use their Skinny Call Control Protocol (SCCP), and Siemens, whose handsets support their proprietary H.323-based signaling (Note: Siemens has also promised a SIP version of their OptiPoint Wi-Fi handsets). While this again ties the user to a vendor proprietary solution, using the same protocol for wired and wireless devices generally ensures a higher level of feature integration than can be had with either a SIP-based or proprietary Wi-Fi solution. As all of the IP PBX vendors have indicated their intention to migrate to SIP-based signaling at some point, this architecture will merge with the SIP-based option introduced above. It is important to note that the IP PBX vendors are promising to support both basic SIP features as well as an enhanced, proprietary set that will be unique to their handsets.

10.3.2 Consumer Wi-Fi Cordless Phones

Consumer Wi-Fi phones are wireless handsets designed to operate over an 802.11-compatible wireless LAN and are targeted primarily at residential and small office applications. These devices are typically marketed in conjunction with IP telephony services like Vonage, Skype, Packet8, or BroadVoice. As the handset simply associates on the wireless LAN and passes signaling messages and voice packets through the access point to the IP telephony service, they can be used with virtually any home-based wireless LAN.

There can be difficulties in using these handsets with fee-based public Wi-Fi Hot Spot services. Fee-based services like those from T-Mobile and Wayport require some type of authentication process before the device is allowed to communicate on the wireless network. In that process, the user must input a user name and password. Even free Wi-Fi services often require that the user accept their standard service agreement before the device is allowed to connect to the network. As the handset does not have a full keyboard and display, there is no convenient way to execute those basic functions. In effect, a small software client in the device must be programmed to perform the required initiation functions.

Some of the consumer VoWLAN handsets are supplied by companies that manufacture consumer WLAN routers like Cisco/Linksys, D-Link, and Netgear, while others are made by companies like UTStarcom. The physical design is often superior to enterprise models, and many support up to four hours of talk time.

The primary attraction of consumer-oriented Wi-Fi handsets is the price. Typically these units cost around $100 while enterprise models can be five times that. Also, they typically use SIP signaling so they do not require a separate controller. However, they also lack many of the features customers look for in enterprise applications:

- Many support only the 802.11b radio link.

- Few have WPA or WPA2 encryption.

- Most do not support 802.11e QoS.

- Few support WMM-APSD for battery conservation.

Given their limited capabilities, the consumer-oriented devices will not likely receive widespread acceptance in enterprise environments in the near term. However, there is strong market pressure for open SIP-based VoIP solutions, paving the way for multi-vendor VoIP networks. If that trend continues, a market for multi-vendor, SIP-based WLAN handsets could develop. Someone will recognize the requirements of enterprise buyers and jump in to deliver a product that fits that category.

10.4 Enterprise Voice over Wireless LAN Systems

Most enterprise customers who implement VoWLAN capabilities opt for handsets designed specifically for their requirements. The major manufacturers are:

- Polycom/SpectraLink.

- Cisco.

- Vocera.

- Siemens.

- Samsung.

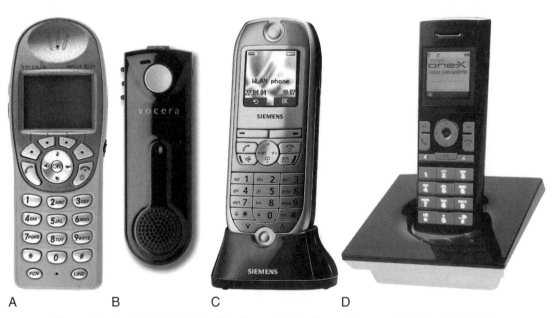

**Figure 10-4: VoWLAN Voice Handsets: (a) Polycom/SpectraLink NetLink 8020,
(b) Vocera Communications Badge, (c) Siemens optiPoint WL2,
(d) Samsung's Avaya 3631**

A selection of the most popular WLAN voice handsets is pictured in Figure 10-4.

Polycom/SpectraLink and Vocera employ proprietary signaling systems and require a dedicated controller/telephony server while the Cisco and Siemens solutions use the same signaling and the same telephony server as their wired IP handsets. SpectraLink has a long history in wireless PBX systems, having sold a wireless PBX adjunct called the Link WTS for over a decade; Nortel marketed the SpectraLink system under the name *Companion*. The Link WTS uses a proprietary 900 MHz radio system rather than an 802.11 WLAN, but the expertise they gained with that product has helped them in integrating PBX features in their NetLink WLAN phones.

The first enterprise Wi-Fi handsets models appeared in 2003–2004, but offered a limited set of features. Those first generation handsets supported only the 802.11b radio link and WEP-based encryption. Some eventually got a firmware upgrade that allowed them to support WPA-based encryption. Most of these vendors introduced a second generation product in 2006–2007; these second generation products would include Polycom/

SpectraLink's 8020/8030, Cisco's 7921, and Siemens' OptiPoint WL2. Most of these newer models feature 802.11a/b/g radio link support, WPA2 encryption, and 802.11e/WMM QoS. They also feature significantly improved battery life with talk times up to eight hours.

The health care segment is recognized as the single largest market for WLAN voice equipment, and so there are models specifically targeted for that market. Vocera's Communication Badge is a primary example. The 1.9 ounce unit is 4.2 in by 1.4 in and does not have a dial pad; all dialing and other functions are voice-activated. The feeling is that nurses and other health care workers may need to have their hands free to attend to patients while placing calls.

Shop floor and materials handling are also prime markets for WLAN voice. Devices that are designed for that environment will typically feature ruggedized designs that can survive drops and general rough handling. Those ruggedized models will also support push-to-talk (PTT) capabilities as they often replace walkie-talkies as well as providing PBX access.

10.5 Wi-Fi Cellular Handsets

Wi-Fi/cellular handsets combine a traditional cellular capability along with Wi-Fi; the cellular capability can be either GSM or CDMA. Most of these devices use SIP-based signaling for Wi-Fi calls, though Nokia does have models that support Cisco's proprietary SCCP signaling. Having an integrated device will allow a user to carry one handset rather than two, and make or receive calls on either network as long as there is a compatible Wi-Fi voice network available. The obvious benefit of such a device is that a user could make calls over the wireless LAN while at his or her office, thereby reducing the cellular charges, and the WLAN should provide better indoor signal coverage than the cellular network.

One note of caution with regard to integrated Wi-Fi/cellular handsets: make sure the Wi-Fi interface includes all of the features needed for an enterprise-grade Wi-Fi voice implementation. The cellular handset manufacturers obviously have extensive experience with cellular technology, but that does not ensure competence with new wireless technologies. Companies like Polycom/SpectraLink and Cisco have a lot more experience with Wi-Fi than Nokia, Motorola, or Samsung. Many of the early Wi-Fi/cellular handsets lacked critical features like 802.11e QoS and WPA2 encryption.

The integration of Wi-Fi and cellular in the same handset will be a major step toward moving Wi-Fi voice into the management ranks. Most initial Wi-Fi voice deployments addressed nurses, retail clerks, and warehouse personnel; those are employees who are not given company-provided cell phones. Management level employees are typically provided with cell phones, and an integrated handset would be an obvious benefit. Those cost and convenience benefits are based on the assumption that the WLAN infrastructure is truly voice capable. Finally, as a WLAN voice network will be an IT-controlled installation, employees will no longer be able to buy their own cell phones and recover the cost through expense vouchers.

The long-term vision for these devices is described as Fixed-Mobile Convergence (FMC) or the ability to fully integrate the public cellular and private Wi-Fi networks, transparently handing off calls between them. The ability to provide with the ability to full integration lies entirely in the hands of the cellular carriers. Japan has become a leader in this area, but T-Mobile has now introduced a consumer-oriented FMC service called HotSpot @Home. In the US., Sprint/Nextel is also testing an enterprise FMC service called Wireless Integration. The dual mode Nokia 6086 GSM/Wi-Fi handset used with the Hot Spot@Home service is pictured in Figure 10-5.

Even without the cooperation of the cellular carriers, there are a number of products that can be used to simulate Wi-Fi to cellular handoffs. There is also one school of thought that says that a handoff capability is not worth the bother. They note that the incidence of these handoffs will be minimal, and how big an inconvenience will it be to say, "I'm going out of the building, let me call you back on my cell"? We will look more closely at the options for Fixed-Mobile Convergence in Chapter 12.

10.6 Wi-Fi PDAs and Softphone Clients

When we think of voice, the obvious user device is a handset or something that looks like a phone. In mobile devices today we are seeing a merger of telephony and various levels of computing functionality in the same device. In effect those products form a continuum from smart phones, to voice-enabled PDAs, to laptops and mobile computers that include softphone clients allowing a user to make and receive phone calls. As all of those devices can now be equipped with Wi-Fi interfaces, they all become potential Wi-Fi voice devices. Two of the more popular WLAN voice-data devices, the RIM Blackberry 7270 and the Motorola/Symbol Technologies MC-70, are pictured in Figure 10-6.

Figure 10-5: Nokia 6086 Dual Mode GSM/Wi-Fi Handset

Research In Motion (RIM), the manufacturer of the enormously successful Blackberry handheld, now sells the 7270 model that includes a Wi-Fi interface. Like the Blackberry models that operate over cellular data service, the 7270 supports the full range of email and data functions as well as Wi-Fi voice. Currently the network selection is an either/ or proposition; there is only one network interface, so a device will operate over either a Wi-Fi network or a cellular network. Clearly, the 7270 is targeted at users who do not leave the office (i.e., the coverage area of the wireless LAN), but they are planning models that will operate on both Wi-Fi and cellular networks.

There are a number of softphone clients that work in Windows Mobile-based PDAs like the Hewlett Packard iPAQ series. HP also makes the h6315 model that includes a GSM cell phone, GPRS cellular data, and Wi-Fi in the same device. Motorola/Symbol Technologies manufactures ruggedized mobile computers that can also be used for specialized applications like bar code scanning. Those devices can be equipped with radio interfaces to operate over either Wi-Fi or cellular voice/data services, and also

A   B

Figure 10-6: WLAN Voice-Data Devices: (a) RIM Blackberry 7270 (b) Motorola/Symbol Technologies MC-70

support push-to-talk (PTT) or walkie-talkie type voice over the Wi-Fi network. Mobile computer suppliers like Motorola/Symbol Technologies typically work with a network of software developers and integrators who can develop and deliver a complete wireless network solution.

If the user typically has access to a Wi-Fi equipped laptop, a softphone client in the laptop can allow them to make calls over the Wi-Fi network. There are a number of SIP-based softphone clients available from companies like PingTel, NCH Swift Sound, and eStara. The IP PBX vendors also provide softphones that are compatible with their systems. Clearly a laptop is about the least convenient voice device imaginable, and it will typically not be able to receive calls when it is in sleep mode. However, if the user's job requires him or her to use a laptop continuously, integrating voice capability in it might be a reasonable solution.

A few notes of caution with regard to using these non-handset devices. First, support and integration requirements increase dramatically with the support for multiple device types. This is particularly true if you require softphone clients for multiple platforms (laptops, PDAs, scanners) and operating systems (Windows, Mac OS, Windows Mobile). Further, there can be issues with regard to the Wi-Fi drivers in those devices. The driver incorporates a number of features that determine how the Wi-Fi interface will operate. If that driver was developed for a data device, the design might not work ideally for voice applications. For example, voice devices need to roam quickly in order to maintain the voice quality of the call. Data drivers are typically sticky, which means they will try to hold on to the access point as long as possible even if the data rate drops.

10.7 What to Look For in Wi-Fi Handsets

As we noted earlier, the first generation of WLAN handsets lacked many of the features enterprise customers would need to deliver a stable, large-scale voice service. The following are the major issues to consider in selecting any voice device for use on a wireless LAN.

- **Wi-Fi Certification:** The best way to ensure interoperability with other devices on a network will be ones where all of the required features are certified by the Wi-Fi Alliance. Certifications are issued for each major feature including radio links, Quality of Service (i.e., Wi-Fi Multimedia), and encryption (i.e., Wi-Fi Protected Access/2). You can check for certifications on all products at: http://www.wi-fi.org

- **Basic Functionality:** Will the device be used solely for Wi-Fi network access or will it also require a cellular capability? Does the user need a handset, a PDA, a specialized mobile computer, or a laptop? If a user requires a device that includes mobile computing capabilities, he or she must also identify the operating system, processor/memory capacity, and interfaces required.

- **Form Factor:** Will appearance and cell phone type features be a factor in the selection?

- **Accessories:** Will the device require a wired or Bluetooth headset, integrated camera, carrying case, travel and car chargers, spare batteries, external battery chargers, and so on?

- **Environmental Factors:** Will the device be used in a safe, clean office environment, or in an area that will expose it to shock, drops, or environmental or explosive hazards?

10.7.1 Technical Requirements

The technical requirements will vary based on the capabilities of the WLAN infrastructure, but it is important to select devices that will continue to operate with the longer term infrastructure improvements. Among the major features to be considered are:

- Support for 802.11a, b, and g radio links (802.11n capability in the future).

- Security based on 802.11i/WPA2 or WPA as a minimum.

- Support for WMM/802.11e QoS using the Enhanced Distributed Control Access (EDCA). It does not appear that 802.11e Hybrid Controlled Channel Access (HCCA) will be a factor in the near future.

- Wi-Fi Multimedia (WMM) Power Save/802.11e Auto Power Save Delivery (APSD) for improved battery life.

10.8 DECT (Digital Enhanced Cordless Telecommunication) Standard

One technology that can serve as an alternative to smaller-scale WLAN voice implementations is DECT (Digital Enhanced Cordless Telecommunication Standard). DECT is a radio standard developed by the European Telecommunications Standards Institute (ETSI) for cordless telephones. While it is not a wireless LAN standard, DECT-based phones could be an alternative to WLAN voice, and DECT actually offers a number of important advantages. In particular, DECT is a circuit-based rather than a contention-based interface, so there are no problems with delay or jitter found in WLAN voice. Further, DECT devices are far better at power conservation with the result that they can typically provide 10 hours of talk time and 90 to 120 hours of standby time per battery charge. Finally, DECT devices operate in a different frequency band so there is no interference with WLAN devices. See Table 10-1.

Table 10-1: DECT Frequencies

Region	Frequency Range (MHz)	Channels
North America	1,920–1,930	5
Europe	1,880–1,900	10
South America	1,910–1,930	10

The DECT radio interface uses a time division multiple access (TDMA) radio transmission technique with time division duplex (TDD) over a number of frequency division multiplexed channels. TDD means that the client and the base station take turns sending on the same channel.

DECT operates on 1.728 MHz radio channels; in Europe and South America there are ten channels available and in the North America there are five. Each transmission channel operates at a digital rate of 1.152 Mbps, and transmits 10 msec frames, each of which is divided into 24 time slots (12 inbound and 12 outbound); up to 12 simultaneous calls can be supported on each channel. Each time slot has a traffic capacity of 480 bits for a total of 48,000 bps per user. Voice coding is done using 32 Kbps ADPCM, and there is also a 4.8 Kbps control channel for each user. The signaling is based on a version of the ISDN D Channel called LAP C. With 10 channels and the ability to support 12 simultaneous calls per channel, a DECT network can support 120 simultaneous calls (60 simultaneous calls in a five channel North American system).

A DECT systems is composed of portable handsets (PHs) that communicate over the radio link to base stations called Radio Fixed Parts (RFPs). Multiple RFPs are connected to a Cordless Controller (CC). The handover process is requested autonomously by the portable terminal and the radio fixed parts, according to the carrier signal levels. A Generic Access Profile defines a minimum set of requirements for the support of speech telephony.

DECT has received far more support in Europe where it is used for four separate applications:

1. **Cordless Phones:** Residential DECT handsets are connected to a base (radio fixed parts), which is in turn connected to the public telephone network. A single base can support a number of DECT handsets. This configuration would be an alternative to residential Wi-Fi voice.

2. **Wireless PBX:** Business DECT handsets connected to a PBX; this configuration would be an alternative to enterprise WLAN voice. In this case, a number of radio fixed parts can be used to provide coverage throughout an extended area. In the US, Ericsson markets a DECT-based wireless solution with small PBX systems and Polycom added a DECT product line with the acquisition of Kirk Telecom in 2007.

3. **Public DECT:** This implementation would provide DECT-based access to the public telephone network, essentially an alternative to cellular telephone service.

4. **Wireless Local Loop:** A DECT radio link could also be used as an alternative to the normal wired connection between the local central office and a subscriber. These latter two implementations are rarely encountered.

10.9 Conclusion

The purpose of this chapter was to introduce the major issues and considerations in building a Wi-Fi voice network. Those issues include the architecture (shared or dedicated), the infrastructure, and the type of handsets or client devices that will be used. While the original Wi-Fi handsets all used proprietary signaling mechanisms, as the business matures, there is a trend toward standards, in particular, Wi-Fi capable phones that use SIP signaling. That type of development will be critical as we move toward handsets that include both a cellular and a Wi-Fi capability. No cell phone manufacturer is going to want to build 10 different models of their Wi-Fi/cellular handset to operate with 10 different Wi-Fi voice signaling systems.

Having a wide array of handsets that meet the requirements of different users will be a key element in Wi-Fi voice. Managers will need integrated Wi-Fi/cellular models, and those handsets must be available in the form factor and with the accessories those users have come to expect with cell phones. Users whose work confines them to the coverage area of the wireless LAN will not need cellular capability, but they may need PDA type functions for messaging or data retrieval, and they may need holsters or other accessories to keep their hands free for other tasks. Many wireless voice applications are appearing in the warehouse or on the factory floor, and for those environments a device must be rugged enough to withstand vibration and drops, and may also need a bar code scanner and PDA functions with applications geared for the specific workplace.

Handsets address the user side of the VoWLAN network, and we may have many different types of users. Simple Wi-Fi cordless phones for use with home-based wireless networks could turn out to be a mass-market item on their own. Those devices may be usable as a low cost option in enterprise networks as well, but they will certainly help to drive down the price of Wi-Fi handsets across the board. In the next chapter we will investigate the technical requirements of the Wi-Fi network that will be needed to support secure, reliable, high quality voice service.

Technical Requirements for WLAN Voice

In the last chapter, we introduced the overall configuration and hardware components for a voice over WLAN system. While the handset might be the visible element of a WLAN voice solution, the critical design element will be the wireless infrastructure we put in place to support those handsets. No one is going to be happy with their WLAN handset if they cannot use it to make a call.

Earlier we described the access points, controllers, antennas, protocols, and security mechanisms we use in WLANs, but now we have to examine how those elements must be modified to support voice. The original 802.11 wireless LAN protocols were designed to address the requirements of data users. Having all of the users vying or contending for the channel introduces transit delay and jitter, but for the most part, those issues are inconsequential for data users. The 802.11 Distributed Control Function is designed to treat all *data* users equally; when we support voice, we will need to treat some users better than others.

To address these and other voice requirements, the WLAN protocols are being modified to address real-time voice and video. The most important of those developments is the 802.11e standard for QoS. While QoS is the most obvious requirement, there are a number of steps that must be taken in the WLAN network design to address real-time voice and video. Those issues include security, load balancing, handoffs, and network management capabilities.

While some companies have deployed VoWLAN networks, most of those have been small-scale installations. There is a major step involved when we go from supporting dozens of handsets to hundreds or thousands. A small number of WLAN handsets will typically not produce enough traffic to impact the overall performance of the network.

In this chapter, we will look at the protocol options and other network features that will be needed to support large-scale, enterprise-grade voice services over a WLAN infrastructure. That will include:

- A general overview of the infrastructure.

- The QoS options including 802.11e/Wi-Fi Multimedia as well as the vendor-proposed solutions.

- The requirements and options for Call Access Control and Load Balancing.

- Voice security issues including E911 compliance.

- Handoff support.

11.1 Pervasive WLAN Infrastructure

In order to support VoWLAN, the first issue to address is the WLAN infrastructure. Most organizations that have deployed WLANs have implemented only "spot coverage" with service in conference rooms, public areas, and other locations where people are expected to congregate with laptops. Voice users are far more mobile than data users. That is particularly true of data users with laptops; a PDA or a mobile computer is a far more portable device than a laptop. Voice users can wander all over the facility, and the WLAN will have to be available wherever they go.

The basic design assumption is that voice service will require a pervasive WLAN infrastructure (i.e., WLAN coverage throughout the facility). If a company is not providing coverage throughout the facility, the administrator must ensure that the users know where they can expect service or they will be calling the help desk reporting problems. While a small office might be covered with a single access point and a single WLAN channel, the preferred way to build a large-scale, enterprise-grade wireless LAN is with a centrally controlled WLAN switch. Along with features like RF management, these systems provide the ability to hand off connections from one access point to another, a critical feature for mobile voice users. A centrally controlled solution can more readily provide features like load balancing to address unexpected demand spikes in certain parts of the coverage area.

11.2 VoIP Quality Issues in WLANs

To begin with, no IP PBX supplier recommends using shared media LANs for voice traffic. However, that is exactly what happens on a wireless LAN. All of the WLAN users in a given area vie for access to a single, shared radio channel on a contention basis. All of that traffic is carried on a relatively low-capacity, half duplex radio link, which will exacerbate the problems of jitter and delay. Even if the network provides QoS, all of the voice users will still be contending for the channel on an equal basis.

In Chapter 8, we introduced the three major quality issues that come into play with any packet voice service, and we should reintroduce them here:

1. **Voice Quality:** Voice quality is a factor of the voice encoding system that is used and the percentage of packets that are dropped by the network. Packet loss can result from transmission errors, buffer overflows, or frames received outside of the RTP receive jitter window.

2. **Transit Delay:** This is the overall delay the signal will experience as it passes from the speaker, through the network, to the recipient. Once the one-way delay exceeds 100 msec, it will begin to affect the cadence of the conversation. One-way delay in excess of 150 msec will almost certainly result in user complaints. Tests have shown that forwarding voice over a WLAN introduces an additional 20 to 30 msec of delay.

   Jitter is the variation in packet-to-packet delay that results from the dynamic allocation of network capacity (e.g., WLAN contention). The Real-Time Transport Protocol (RTP) addresses jitter removal, but the RTP buffering process adds to the overall transit delay. Further, if the VoIP devices use fixed-size jitter buffers, they will automatically drop packets that are received outside of their time window, which in turn degrades voice quality. Adaptive jitter buffers are the preferred solution with WLAN voice systems.

3. **Echo Control:** All telephone circuits introduce echo, and once the one-way transit delay exceeds 35 to 40 msec, the echo becomes noticeable and equipment must be used to remove it. Virtually all packet voice networks will exceed the delay parameter, so echo control must be incorporated into the system design.

11.3 IEEE 802.11e WLAN MAC Quality of Service Enhancements

The first key development in supporting high-quality voice over WLANs is the IEEE 802.11e protocol. Ratified in late 2005, 802.11e defines a new MAC option called the Hybrid Coordination Function (HCF), which can provide a QoS capability within the WLAN structure. The IEEE 802.11e standard defines two options:

1. **Enhanced Distributed Channel Access (EDCA)/Wi-Fi Multimedia (WMM):** A contention-based access mechanism with four priority levels or *access categories* that can provide voice users with preferred access to the shared channel. The Wi-Fi Alliance identifies products that have tested in compliance with the EDCA option as Wi-Fi Multimedia (WMM) certified.

2. **Hybrid Controlled Channel Access (HCCA):** A polled access mechanism that can provide consistent delay (i.e., isochronous service) for time sensitive traffic. The Wi-Fi Alliance identifies products that have tested in compliance with the HCCA option as Wi-Fi Multimedia-Scheduled Access (WMM-SA) certified. Currently, no devices have received this certification and the Alliance has put the development of the certification plan on hold.

11.3.1 IEEE 802.11e Implementation

While the IEEE standard defines two separate options, most vendors are planning to support only the EDCA/WMM implementation. It is important to understand the capabilities of both options, but buyers should focus on EDCA primarily. Importantly, the implementation of either 802.11e option should require only a software, rather than hardware, upgrade. This is particularly important, because any feature that requires a hardware upgrade typically involves buying new access points, Wi-Fi NICs, and handsets.

There may still be difficulties upgrading existing Wi-Fi handsets given the limited amount of processing power and memory built into the device. If older handsets are not upgradable they will still be usable, however, they will not be able to take advantage of higher priority channel access (i.e., they will have the same priority as data devices), and hence will suffer greater delay and jitter.

11.4 802.11e EDCA

EDCA/WMM builds on the basic operation of Wi-Fi's Distributed Control Function; the *Enhanced* part is that we can now define four different priority levels called Access Categories (ACs) for accessing the shared channel.

11.4.1 Arbitrated IFS (AIFS)/Multiple Back-off Counter Ranges

In EDCA/WMM, user priorities are implemented by enhancing two mechanisms from the original WLAN access protocol: the inter-frame spacing (IFS) and the back-off counter. In a traditional WLAN, a station wishing to send a frame must wait for an interval called the DCF Inter-Frame Spacing (DIFS). All users have the same access priority because they all use the same DIFS interval; they also use the same range of back-off values to address collisions and busy channel conditions (i.e., CW_{min} and CW_{max}). EDCA defines different waiting intervals and back-off ranges for each of four access categories, and shorter intervals are assigned to higher priority traffic.

The four EDCA/WMM defined priority levels or access categories are designated as follows:

Access Category (AC) 1: Voice.

Access Category (AC) 2: Video.

Access Category (AC) 3: Best effort data (identical to current DCF devices).

Access Category (AC) 4: Background data.

11.4.2 EDCA Implementation: Arbitrated IFS (AIFS)

The new inter-frame spacing interval is called an Arbitrated IFS (AIFS). Two of those intervals (i.e., Voice and Video) are shorter than the current DIFS value, one is the same (i.e., Best Effort), and one is longer (i.e., Background Data). As a result, DCF and EDCA stations can be mixed in a wireless LAN, and transmission from the legacy stations (i.e., pre-802.11e) will all be categorized as Best Effort. All pre-802.11e VoWLAN handsets use the legacy DIFS waiting interval, so if they were used in an 802.11e network, they would not have the same access priority as 802.11e-compliant devices. See Table 11-1 and Figure 11-1.

Table 11-1: IEEE 802.11e Default Parameters

Parameter	DSSS PHY (802.11b)				ODFM PHY (802.11a/g)			
	Access Category				Access Category			
	1	2	3	4	1	2	3	4
AIFS (SIFS + x Time Slots)	2	2	3	7	2	2	3	7
AIFS Time (µsec)	50	50	70	150	28	28	37	73
TxOP Limit (µsec)	3,264	6,016	0	0	1,504	3,008	0	0
CW_{min}	7	15	31	31	3	7	15	15
CW_{max}	15	31	1,023	1,023	7	15	1,023	1,023
Note: The SIFS interval is 10 µsec, and the time slot duration is 20 µsec for 802.11b and 9 µsec for 802.11a/g.								

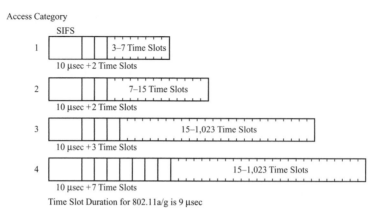

Figure 11-1: IEEE 802.11e Default Parameters (802.11a/g)

With EDCA/WMM, if a voice or video user is trying to access the channel at the same instant as a best effort user, the voice or video user will always send first. So it should be virtually impossible for a voice and a data user to collide; however, two voice users or a voice and a video user could still have a collision. Stations could also collide if they are hidden nodes. The standard also specifies different ranges for the back-off counters, so lower-priority transmissions will always select a longer interval than higher-priority transmissions in the event of a collision. If a voice and a video user collide, the video user will almost certainly get a longer back-off interval.

The shorter waiting interval will give the voice frame preferred access to the channel, but the access point needs to be able to recognize voice frames so they can be forwarded into the wired LAN with the correct priority (i.e., 802.1p-defined priority). To accomplish that, an additional 2 octet field is added to the frame header following the last address. The elements in the QoS field are (see Figure 11-2):

- **User Priority (3 bits):** Filled with the same values used with the 802.1p priority mechanism that would be used over the wired LAN. The correlation is shown in Table 11-2.

- **End of Service Period-EOSP (1 bit):** Set to 1 at the end of a triggered service period.

- **ACK Policy (2 bits):** ACK or Do Not ACK for this frame.

- **Unused (10 bits)**

Table 11-2: 802.11e/WMM to 802.1P Mapping

WMM Access Category	WMM Description	802.1P Priority	802.1P Designation
1 (AC_VO)	Voice	7 (111)	NC
		6 (110)	VO
2 (AC_VI)	Video	5 (101)	VI
		4 (100)	CL
3 (AC_BE)	Best Effort Data	3 (011)	EE
		0 (000)	BE
4 (AC_BK)	Background Data	1 (001)	BK
		2 (010)	–

When 802.11e is used, every beacon must include a WMM Information Element (IE) or Parameter Element (PE) that identifies options to be used on the network. The IE defines 802.11e-specific network features that are supported, such as the maximum number of frames the access point can deliver on one access and flags for the APSD function. The PE defines the $CW_{MIN/MAX}$ and the TxOP limit. The TxOP is the maximum amount of time a station can hold the channel while streaming a series of frames and is measured in 32 μsec increments; the TxOP covers the frame and acknowledgment transmission times.

11.4.3 Other Features of EDCA

While the AIFS and variable length back-off provide the primary functions of EDCA, there are a few other features designed to improve voice performance.

11.4.3.1 Streaming

In traditional DCF access, the device had to gain access to the channel to send each packet, and a voice packet typically contains 20 to 40 msec of speech. A speech burst will typically last several seconds and require dozens of frames, and a number of packets might be stored in a device before it can gain access to the channel. EDCA defines the ability for a voice device to send all of the packets it has stored before relinquishing the channel. Once the device gains access to the channel, subsequent voice frames are sent after an SIFS interval. The maximum duration a station can hold the channel is referred to as a TxOP limit; if the TxOP limit is greater than 1, multiple frames can be sent before relinquishing the channel. Some care must be used in defining the TxOP limit, as other stations will be frozen out for the duration of that multi-frame transmission and so experience greater jitter. The maximum amount of time defined is for video (access category 2) on an 802.11b network and that is 6.016 msec.

While allowing one station to hold the channel for an extended period of time can impact the jitter experienced by other users, it is important to recognize that a five seconds burst of speech will not keep the network busy for five seconds. Voice is being generated at 64 Kbps (plus the associated overhead), but bits are traveling over the air at up to 54 Mbps. So there is a considerable time compression factor.

11.4.5.2 Block Acknowledgment

With streaming, each frame may still require an acknowledgment from the receiver, increasing overhead and decreasing throughput. The EDCA also supports a Block Acknowledgment to use in conjunction with streaming. In this case, the receiver can return a single acknowledgment for an entire stream of frames.

11.4.3.3 No Acknowledgment

The utility of acknowledgments is questionable in a WLAN voice network. ACKs are used in wireless LANs because stations cannot detect collisions, so the failure to receive an ACK alerts the sender that the frame must be resent. That procedure works fine for non-time-sensitive data traffic, but it might be useless in WLAN voice. If the increase in delay causes the frame to arrive outside of RTP's receive jitter buffer, it would be discarded. As delivering the frame outside of the RTP buffer's time window would not benefit the voice quality, that retransmission would essentially be wasting transmission time. With 802.11b (which uses the longest back-off intervals) a CW_{MAX} value of 15 would result in a back-off interval of 0.35 msec (i.e., 300 μsec back-off + 50 μsec AIFS). Fixed duration RTP jitter buffers usually range from 20 to 60 msec, allowing for a number of WLAN retransmissions.

11.4.3.4 Direct Link Protocol

In a WLAN, clients send to the access point, and the access point sends to the clients. If two users on the same WLAN are engaged in a conversation with one another, every frame will be sent twice over the radio channel: once inbound to the access point, and once outbound to the other client. The 802.11e protocol defines an option called Direct Link Protocol where the access point can recognize when two of its stations are involved in a conversation, and allow them to send frames directly to one another in a

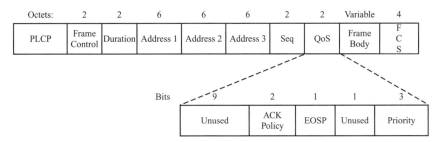

Figure 11-2: WLAN Frame Format with QoS

mode similar to an Ad Hoc network. While rarely implemented today, this feature would reduce the network capacity required by that call by 50 percent.

11.4.4 Not Consistent Delay Service

It is important to note that EDCA does not provide for true consistent delay or isochronous service. All devices in the same priority level vie for the channel on an equal basis, so collisions among stations in the same access category can still occur. The controlled channel access option, however, does provide for consistent delay service.

11.5 HCF Controlled Channel Access (HCCA)/Wi-Fi Multimedia-Scheduled Access (WMM-SA)

The 802.11e Hybrid Coordination Function also defines a polled or contention-free access called Hybrid Controlled Channel Access (HCCA). The Wi-Fi Alliance identifies devices that are tested to comply with the HCCA options as Wi-Fi Multimedia-Scheduled Access (WMM-SA) Certified. The mechanism defined for contention-free access is very similar to the rarely used Point Control Function (PCF) in the original 802.11 MAC. Initially, most vendors are not implementing the HCCA option.

In HCCA, the access point or *hybrid coordinator* takes control of the channel by sending a field in the beacon message that causes all stations to set their NAV timers; all stations will assume the channel to be busy for that period of time. The access point essentially declares a contention-free access period for time sensitive traffic. The access point determines the required duration of that contention-free interval based on the number of stations receiving HCCA access, the transmission requirements, and transmission rates for each. At other times, stations will vie for access to the channel using the EDCA mechanism.

During the contention-free periods, the access point will poll each station that has time sensitive traffic to send. This is essentially the same mechanism used with PCF, but there are some important enhancements.

1. Stations will use the TSpec signaling protocol to request a transmission profile that specifies requirements for bandwidth, latency, and jitter.

2. If the access point does not have sufficient capacity to meet the requested profile, it will return the equivalent of a busy signal.

3. If the connection can be supported, the station will be assigned a slot in the access point's polling table and the contention-free interval will be increased by the required amount. HCCA stations can transmit only when polled.

4. During each contention-free cycle, the station will be polled. When the connection is no longer required, the station sends a disconnect message to the access point, and its entry is removed from the polling table. See Figure 11-3.

Needless to say, the use of HCCA will have a major effect on the capacity and delay experienced by non-HCCA users. All of the contention-based traffic will now be squeezed into some percentage of the available transmission time. The longer the contention-free period, the less capacity will be available for other users. Users can make an estimate of the network capacity available for contention-based traffic by determining the maximum percentage of time allocated for HCCA traffic and multiplying by the expected throughput of the network.

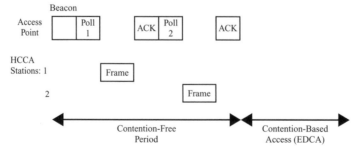

Figure 11-3: Hybrid Controlled Channel Access (HCCA)

11.6 Capacity and Quality of Service

The biggest question regarding WLAN voice is, "How many calls will the WLAN be able to support while providing adequate performance and delay?" Capacity and QoS are grouped together because the issues of providing adequate capacity while minimizing transit delay and jitter can be addressed by increasing the overall network capacity or by ensuring that voice packets are recognized and given priority over data packets (i.e., voice QoS).

Matthew Gast, author of the book, *802.11 Wireless Networks: The Definitive Guide,* did a mathematical analysis to determine the theoretical maximum call capacity of a wireless LAN and published it on his Web page (http://www.oreillynet.com/pub/a/ etel/2005/12/13/how-many-voice-callers-fit-on-the-head-of-an-access-point.html). His analysis looked at the amount of time it would take to send a voice frame and the number of frames required to support a two-way voice path. The calculation assumes 20 msec voice samples and no silence suppression (i.e., frames are sent at a constant rate, 50 frames per second, regardless of whether or not the parties are speaking). He also assumes perfect scheduling, which means

- The channel is available for each user as soon as he or she is ready to send.

- The users all transmit at the same data rate.

- No collisions occur.

- There is no network overhead.

In short, this is not a situation that would ever occur in the real world. It does provide a definitive maximum call capacity for WLAN voice. See Table 11-3.

The actual number of calls that could be carried would likely be about half of that theoretical maximum. There has been far more experience with 802.11b implementations as most of the current handsets use that radio link, and vendor recommendations range from 6 to 14 simultaneous calls on an 11 Mbps network. The actual number will vary based on the quality of the RF coverage (e.g., the percentage of callers operating at 11 Mbps versus 1 Mbps) and can be reduced by using call access control to reserve some portion of network capacity for data use.

Table 11-3: Theoretical Maximum Calls per WLAN (20 msec Voice Sampling)

Codec	802.11b Network				802.11a or g Network			
	11 Mbps	5.5 Mbps	2 Mbps	1 Mbps	54 Mbps	36 Mbps	18 Mbps	6 Mbps
G.711	23	16	8	4	78	69	51	24
G.729A	30	24	14	8	92	86	73	45
G.723.1	44	36	21	13	138	129	110	66
Skype iLBC	28	22	13	7	89	83	69	40

Source: www.oreilly.com, Matthew Gast, "How Many Voice Callers Fit on the Head of an Access Point," December 13, 2005. Available at http://www.oreillynet.com/pub/a/etel/2005/12/13/how-many-voice-callers-fit-on-the-head-of-an-access-point.html.

11.6.1 Impact of 802.11e EDCA/WMM Quality of Service

EDCA/WMM gives voice and video packets priority access to the shared radio channel, however, that will not substantially increase the number of calls that can be supported. QoS should lessen the delay and jitter that voice packets will experience, so the impact will be better voice calls rather than more voice calls. Again, it is important to remember that giving voice users preferred access to the channel means giving non-preferred access to data users. If call access control parameters allow too many simultaneous calls on the WLAN, the data traffic could be squeezed out entirely.

11.6.2 VoWLAN Overhead

Added together, a considerable percentage of the transmission time on a WLAN is taken up with waiting intervals and overhead elements. Those overhead factors become the more dominant element when the data payload becomes smaller. In Table 11-4, we look at the percentage of time spent on overhead versus content in a 1,500 byte Ethernet frame versus a 160 byte voice frame (64 Kbps voice encoding, 20 msec sample size). With the 1,500 byte Ethernet frame, overhead accounts for 31 percent of the transmission time, but with the voice frame it is 86 percent. If you compress that 20 msec voice sample to 8 Kbps using the G.729A technique, the actual voice content of the frame drops to 20 bytes, and the overhead percentage jumps to 97 percent!

11.7 Access Control and Load Balancing

While QoS will give voice users higher priority than data users in accessing the shared WLAN channel, it does not prevent too many voice users from trying to make calls

Table 11-4: 802.11b Overhead Factors

1,500 Octet Data Field (Maximum Size Ethernet Frame)			200 Octet Data Field (64 Kbps Voice, 20 msec Sample, 40 Octets RTP/UDP/IP Headers)	
Bytes	Transmission Time		Bytes	Transmission Time
—	50 µsec	DIFS Interval	—	50 µsec
18	192 µsec	Preamble/PLCP Header	18	192 µsec
1,542 Header: 42 Content: 1,500	1,120 µsec Header: 30 µsec Content: 1090 µsec	Data Frame	242 Header: 82 Content: 160	176 µsec Header: 60 µsec Content: 116 µsec
—	10 µsec	SIFS Interval	—	10 µsec
18	192 µsec	Long Preamble/PLCP Header	18	192 µsec
14	10 µsec	ACK Message	14	10 µsec
—	1,574 µsec	Total Transmission Time	—	630 µsec
31%		Total DIFS/SIFS Preamble/PLCP Overhead	86%	

through the same access point. The great irony of a priority system is that priority only ensures you will treat some users better than others; it does not mean that you will treat any of them very well! If too many equal-priority voice users attempt to place calls on the same WLAN network, they will all experience greater jitter and lengthier transit delays.

11.7.1 Call Access Control: Protecting Voice from Voice

The contention problem among voice users on the same WLAN channel is referred to as *protecting voice from voice*. The key to doing that is to implement some form of call access control.

There are three primary steps in implementing a call access control system:

1. **Define the Number of Simultaneous Calls to Support on Each Access Point:** That can be universal definition, or you might define different maximums in each area. The major factors that will impact that are:

a.   The radio link employed (802.11a, b, g).
b.   The transmission rate required per call (voice encoding, sample size, silence suppression).
c.   Capacity reserved for data access.

Typically the WLAN switch vendor will make a recommendation based on the capabilities of their equipment, however, the network engineer will be well advised to pilot test different settings to determine how well they perform in their environment.

2. **Specify the Mechanism Used to Identify Voice Calls and Reject Excess Callers:** The access point will need some mechanism to recognize voice calls for what they are. The access point could do that by noting the number of MAC addresses that are generating Access Category 1 (i.e., voice) frames. If 802.11e is used, there is an optional Traffic Specification (TSpec) signaling mechanism that allows a station to request a specific amount of capacity from the access point and specify other features about the connection.

The other option would be to have the access point monitor for SIP signaling messages. SIP messages can be identified by the Port Address 5060, and the SIP Invite message includes a Session Definition Protocol that will identify the capacity requirements for the connection. Cisco has also implemented a call access control mechanism as part of their Cisco Compatibility Extension (CCX) program though they support TSpec as well. Some WLAN switches also have the ability to recognize non-voice traffic that is incorrectly marked for voice priority and change the priority setting accordingly.

3. **Identify the Mechanism for Handling Excess Calls—Deny or Divert:** If accepting an additional call will exceed the access point's maximum call capacity, that call can be blocked or diverted to an alternate access point. When a call is blocked, the user will receive a busy signal and have to attempt the call later. The process of diverting the call to an alternate access point is called *load balancing*. The access point can also invoke different policies for new calls versus calls that are roaming into the area. As a general rule, the policy should be more flexible for calls roaming into an area, as the alternative would be to disconnect a call in progress.

Call access control is critical in voice services, because the result of accepting too many calls is that the quality for all of them will decline. Voice users are highly mobile and calling patterns may be dynamic and unpredictable. Providing good service on an ongoing basis will require a monitoring system that tracks dropped or blocked calls by area so that the network capacity can be increased in those areas before user complaints escalate. We will look more closely at the type of performance statistics that should be collected in Chapter 14.

11.7.2 Load Balancing (802.11u)

A call access control mechanism that blocks excess call attempts is the equivalent of addressing the problem of network congestion with a blunt instrument. A more finessed approach would be to divert or load balance excess calls to alternate access points. While 802.11e/WMM does have mechanisms to implement call access control, the standard for load balancing, IEEE 802.11u, is still in development. In the meantime, many WLAN switches are incorporating load-balancing features by tricking the client into seeking out another access point. For example, the busy access point could instruct a neighboring access point to respond to the station's Association Request. As these techniques work on features that are standard in all Wi-Fi clients, they will work with virtually any handset.

11.7.3 Neighbor Reporting (802.11k)

Neighbor reporting is another developing area in Wi-Fi voice. Standards for this capability will be described in 802.11n. The idea of neighbor reporting is to give the access point the ability to instruct a client to measure the received signal strength from other access points in its immediate area and report that information to the central controller. With that information, the central controller can better decide when and where a client should be handed off. This is not a new idea. GSM cell phones use a similar technique called a *mobile assisted handoff* where the base station can instruct mobile devices to measure and report signal quality from other base stations, and improve the overall handoff process. In WLANs, neighbor reporting can also assist in maintaining the security association. Neighbor reports can also allow a client to cache security codes from adjacent access points to speed the secure handoff process.

11.8 Vendor Proprietary Voice QoS Techniques

Prior to the development of 802.11e, a number of VoWLAN vendors developed other techniques to improve the performance of voice. The challenge is to make these modifications in a way that does not violate the basic operating principles defined in the 802.11 MAC. The downside is that there may have to be different QoS capabilities for inbound and outbound transmissions.

11.9 SpectraLink Voice Priority (SVP)

The most widely deployed voice QoS mechanism came from VoWi-Fi handset maker SpectraLink (now part of Polycom) in what they called the SpectraLink Voice Priority (SVP) protocol. SVP was incorporated in SpectraLink's handsets and they have a program to certify access points from different vendors who are compliant with the protocol; compatible products are identified as *SpectraLink VIEW Certified*. Most WLAN switch products are certified VIEW-Compatible. To confirm if a particular access point is View-Certified, go to: http://spectralink.com/partners/view_certification. jsp. SVP provides a different set of capabilities for outbound and inbound transmissions.

11.9.1 SVP Outbound

The basic requirement for a VIEW-Compatible access point is that it must be able to differentiate voice and data packets and maintain separate transmission queues for each. That distinction can be met by noting 802.1P priority indicators or Diff Serv Control Points in the frame or packet headers. On the outbound side, SVP compatible access points will attempt to send voice packets before data packets and will employ *streaming* to reduce delay and jitter for voice. So once the first frame is sent, subsequent frames are sent after a SIFS interval and no other station has the opportunity to seize the channel. Unlike 802.11e that specifies shorter inter-frame spacing intervals for high priority traffic, in SVP, voice packets use the same DIFS interval as data traffic, but that wait only applies to the first frame in the stream. Further, in busy channel or collision conditions, a VIEW-compatible access point will set its back-off counter to zero so when the channel becomes idle, it will typically grab it quicker than a data device.

11.9.2 SVP Inbound

On the inbound side, SVP is more limited. In order to remain compatible with the 802.11 standards, SpectraLink handsets must use the same DIFS interval as data devices. However they also use the zero back-off trick, so they get a de facto advantage in busy channel and collision conditions. As a result, the retransmission of a voice packet is almost instantaneous. While their handsets can support either 64 or 8 Kbps (i.e., G.729a) voice encoding, SpectraLink recommends 64-kbps G.711 noting there is so much overhead accompanying each voice packet, there is little efficiency gain in compressing voice. Further, they only use silence suppression if it is also used on the wired LAN; as silence suppression is rarely used on wired LANs, it is rarely used on WLANs either.

11.10 Meru Networks: Air Traffic Control

While SpectraLink manufactures handsets and has a protocol solution that works with a number of different access points and WLAN switches, WLAN switch manufacturer Meru Networks has taken a different approach to VoWLAN QoS. While Meru does not disclose the full details of their patented Air Traffic Control (ATC) technology, it uses features inherent in any 802.11-compatible device to schedule voice and data client devices to transmit in specific time slots. In essence, Meru's proprietary ATC control system takes the randomness out of the 802.11 DCF access protocol. Meru does not poll stations as in HCCA; rather, the switch can manipulate the network so that different stations see the channel idle at different times. By reducing the incidence of collisions, Meru claims to be able to support 30 simultaneous 8 Kbps voice conversations on an 802.11b WLAN where other vendors might get 6 to 14.

The Meru approach also uses a different configuration for the radio layout. In a Meru network, the coverage areas of the access points overlap, and any device can be made to associate with any access point. Their switch allows up to 12 channels to be overlaid in an area (those channels can be a combination of 2.4 and 5 GHz channels). This feature supports instantaneous handoffs and effectively provides load balancing.

Given the efficiencies they have incorporated in their basic design, Meru claims to be ambivalent about 802.11e. They say they can already provide the same type of improved delay and jitter performance that 802.11e is designed to provide, and can do so with legacy 802.11 handsets. Certainly this is a unique solution to operating a large-scale WLAN, but customers must approach proprietary solutions with caution. On paper the Meru solution looks elegant, but what happens if they go out of business? The key to success in marketing a proprietary solution is to build a large enough customer base to ensure ongoing financial viability. As Meru is privately held, information on their finances is not publicly available.

As proprietary solutions go, Meru's approach has some strong features going for it. It can deliver an appreciable increase in call capacity and incorporates many of the other features required for WLAN voice like load balancing and handoffs. Also it does not violate the standards and can deliver its performance benefits even with legacy WLAN voice devices.

11.11 WLAN Handoffs

The other major technical requirement for WLAN voice is the support for handoffs, or the ability to transfer a WLAN connection from one access point to another without interrupting the call. As we know, voice users are more likely to be mobile than data users, and the security association must remain intact as the handoff is made. There is a rudimentary handoff mechanism included in the original 802.11 standards. However the operation is far too slow for voice and can take several seconds, particularly if the client must reauthenticate with the new access point. The goal for voice handoff is to complete the exchange within 50 msec so that the user notices no interruption in the call.

11.11.1 Handoffs: IEEE 802.11r

The IEEE is developing a standard for fast (i.e., <50 msec) and secure handoffs designated 802.11r. Providing the handoff function while maintaining the security association and QoS features is proving to be a daunting task. Further, the handoff function will have to work in conjunction with any load balancing capabilities, as a voice user may be moving into a cell that is already saturated. Under the current timetable, it appears that the standard will not be complete before late 2008, which would mean that products based on it would not be available before 2009.

11.11.2 WLAN Switch Solution

In the meantime, most WLAN switch vendors do have a workable handoff capability today. In a WLAN switch environment, the user associates with the central controller rather than the access point. Once associated, the user can move between access points freely and the handoff time is typically below 50 msec. Further, those handoffs operate in conjunction with the security and load balancing capabilities that particular system offers. The handoff function is controlled entirely by the WLAN switch and so it will inherently support any WLAN handset.

There are variables in how quickly a WLAN handoff can take. Roaming between IP subnets will typically increase the handoff time. Further, the security mechanism, and how it is implemented, can impact the handoff delay. The WLAN switch vendors recognize that handoff performance will be a critical element in the acceptance of their systems for voice applications and have developed a range of capabilities to address those requirements. In virtually all cases their worst-case handoff performance is under 150 msec, which should be barely perceptible to the user.

The downside of the WLAN switch solutions is that they are vendor proprietary; you must buy the access points and controller from the same vendor. A standards-based solution would allow the customer to buy the various elements from any number of vendors with a high degree of assurance they would interoperate. Further, the availability of multiple sources also protects the user in the event their particular supplier goes out of business or elects to no longer support that particular product line. In short, the decision comes down to choosing between the bird in the hand and the bird in the bush. We can have the handoff functionality immediately, but we have all the risks associated with implementing a proprietary solution. On the other hand, if we want a standards-based solution, we will have to defer WLAN voice installation until 2009.

11.12 WLAN Voice Security Exposure

WLAN voice security concerns can be organized into three major groups of threats:

1. **Privacy:** Protection for eavesdropping and message alteration.

2. **Authentication:** The ability of a hacker to access the network and make or receive calls (i.e., toll fraud).

3. **E911 Compliance:** While not essentially a security issue, some states are now mandating that PBX systems be able to report the location of stations placing calls to emergency services. If the system cannot report location on WLAN stations, these requirements may preclude their use in those states.

11.12.1 Privacy

Security will be a particular concern with WLAN voice, as users expect their telephone conversations to be private. Further, that secure association must be maintained as calls are handed off from one access point to another. Unfortunately, some handset manufacturers have fallen a generation behind in implementing the latest Wi-Fi security features.

11.12.2 Privacy Options

As we described in Chapter 6, there are three generations of Wi-Fi privacy solutions, and we expect that newer Wi-Fi handsets will incorporate the latest options.

- **Wired Equivalent Privacy (WEP):** This is the woefully inadequate static-key, RC4-based encryption mechanism that was part of the original 802.11

standards. Data users often employed a work-around solution using a VPN secure tunnel connection when operating over a WLAN, however that feature is not available in most Wi-Fi handsets. Secure tunnels provide security measures that are used in spite of the WLAN, and we would much prefer a solution that operates as an integral part of the WLAN network.

- **Wi-Fi Protected Access (WPA):** WPA has been an acceptable standards-based security alternative for Wi-Fi voice since it first appeared in Wi-Fi handsets in 2005. WPA uses the same RC4 encryption as WEP, but uses a 128-bit key and changes the key on every packet to thwart brute force attacks (i.e., Temporal Key Integrity Protocol). The security establishment has been cool toward WPA based on its dependence on RC4 encryption, but there have been no successful hacks against correctly implemented WPA. That correct implementation involves using WPA with 802.1x for key distribution (i.e., Enterprise WPA) or selecting a challenging key in pre-shared key implementations to thwart dictionary attacks.

- **802.11i/WPA2 Advanced Encryption Standard (AES):** The IEEE has developed a new encryption technique for WLANs in the 802.11i standard that uses the Advanced Encryption Standard (AES). The Wi-Fi Alliance describes full 802.11i compatibility as *WPA2 Certified*. Support for 802.11i/WPA2 is growing among VoWLAN handset vendors after it first appeared in early 2006. Starting in mid-2006, the Wi-Fi Alliance required 802.11i/WPA2 capability on all new Wi-Fi Certified products, so we can anticipate it on all new Wi-Fi handsets.

Clearly, 802.11i/WPA2 will be the preferred mechanism to ensure privacy on WLAN voice connections. However, there are a number of scenarios that might preclude its use. If a device was not designed to support 802.11i/WPA2, it can typically not be upgraded. Encryption is typically performed on a chip, and those hardware components are not upgradeable.

11.13 Authentication/Toll Fraud

Toll fraud through business telephone systems has been an ongoing concern over the years. While not as widely reported as security threats against data systems, toll fraud represents a significant financial exposure, particularly if the outside party has the ability to place long distance or international calls that would be billed to the PBX owner's account. Toll fraud in wired telephone systems has typically been perpetrated through a Direct Inward System Access (DISA) port, which is installed to allow legitimate traveling users the ability to make use of the telephone system when out of the office. DISA ports require the user to enter a password before being allowed to place calls, but if the hacker gets the password, then he or she can make free calls—with the cost picked up by the company.

For VoWLAN toll fraud, the user would require a Wi-Fi-compatible handset or softphone, and hack the system for authenticating phones. Thus far, the toll fraud threat has not been significant. Enterprise Wi-Fi handsets cost a few hundred dollars and are not widely available. Further, the authentication and signaling protocols are proprietary, so exploiting them would require considerable engineering expertise. Current handsets use a version of 802.1x authentication, and all telephone stations must be registered with the telephony server, so hacking the authentication process would be challenging. Finally, as there have been so few VoWLAN implementations, hackers would have a hard time locating one.

The picture could change dramatically with the support of SIP-based VoWLAN handsets. As SIP is an open protocol, most of the basic challenge in gaining network compatibility will disappear. SIP also includes an authentication mechanism, but its security capabilities have been questioned. Improving the security capabilities of SIP is a high priority in VoIP, and an improved SIP authentication mechanism in conjunction with a solid Wi-Fi authentication theme should keep user networks secure from toll fraud. However, as SIP authentication is still new and evolving, special attention should be paid when implementing SIP-based Wi-Fi systems.

11.14 E911 Compliance: WLAN Station Location

While not specifically a security threat, regulations regarding compatibility with emergency calling services could impact the use of WLAN voice. When a 911 call is placed to any one of the over 7000 Public Safety Answering Points (PSAPs) or 911 Centers, the phone number of the calling party is provided. With Enhanced 911, a database then uses the calling number to look up the address where the phone is installed. VoIP and wireless services have created a major challenge for caller location.

VoIP services like Vonage allow subscribers to receive calls to their registered number anywhere they have a broadband connection to the Internet. Those subscribers must now register their location with the carrier so that if a 911 call is placed from their line, the call is routed to the correct PSAP and the correct location information is provided by a service bureau.

11.14.1 *National Emergency Number Association (NENA)*

Some states are now requiring location information on 911 calls placed from PBX stations; there are requirements for different levels of E911 compatibility in Colorado, Illinois, Kentucky, Mississippi, Texas, Vermont, and Washington. The National Emergency Number Association (NENA) is the agency responsible for monitoring and promulgating information regarding 911 compatibility; they can be located at http://www.nena.org. Given the range of issues brought about by the increase in voice network technologies and options, regulators and lawmakers are having a difficult time keeping abreast of our rapidly changing technology options. As a result, there has been little guidance on their site regarding requirements for WLAN voice. NENA collects and disseminates information from the various state requirements, and none of those seem to address the challenge of WLAN voice location.

The elements required to provide WLAN E911 compliance are known. The wired IP PBX system will require a database called an E911 Server that contains the location of each wired handset. To extend that capability to VoWLAN phones, the wireless LAN must have a location appliance to determine the location each Wi-Fi handset. WLAN vendors have developed capability for WLANs to determine a device's location by the received signal strength measured at three or more access points. Given the imprecision of signal loss characteristics in an indoor environment along with the multipath characteristics, precise location is challenging. However, the 10- to 20-foot accuracy provided is typically within the acceptable parameters for the E911 regulations. The location database could be updated continuously as the VoWLAN phone moves around, but given the infrequency of 911 calls, it would likely be more practical to determine the WLAN handset's location when it actually places a 911 call.

The enterprise will also require a service bureau that can deliver that location information to the PSAP. Companies like Tele-Communications Systems Inc. (http://www.telecomsys.com) provide that service for both VoIP providers and PBX customers. The WLAN location appliance would have to deliver the location information to the E911 server, which in turn would forward it to the service bureau for delivery to the PSAP. The utility of all of this is questionable as the user could still move after they placed the 911 call! See Figure 11-4.

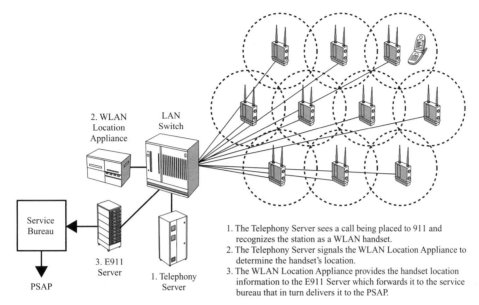

1. The Telephony Server sees a call being placed to 911 and recognizes the station as a WLAN handset.
2. The Telephony Server signals the WLAN Location Appliance to determine the handset's location.
3. The WLAN Location Appliance provides the handset location information to the E911 Server which forwards it to the service bureau that in turn delivers it to the PSAP.

Figure 11-4: WLAN E911 Support

WLAN location systems today are geared toward locating things rather than people. Wi-Fi tags are being used to locate equipment (e.g., Stat Carts in a hospital environment). Location tracking is also used to locate rogue or malicious access points. Most WLAN switches provide some location capability as part of their security features.

Predicting developments on the regulatory front is even more challenging than divining technical directions. What we do know is that state requirements for E911 compatibility with PBX systems are on the rise, but those mandates still do not address the full range of technologies including VoIP and WLAN systems. However, it will be important that some party in the organization be tasked with monitoring E911 compliance requirements in whatever states the enterprise operates.

11.15 Battery Life

Battery life continues to be a problem with Wi-Fi voice handsets. Unlike cell phones where the designers were focused from the outset on building small handheld devices that would yield long battery life, the initial design vision for Wi-Fi was an interface in a laptop computer. The handset designers are now coming to grips with the challenges of adapting the Wi-Fi radio interface to the demands of small devices.

In a Wi-Fi device, both the transmitter and the receiver consume power, but the transmitter is by far the greater power draw. However, the transmitter will typically be active only during a call while the receiver must remain on continuously to sense and receive incoming calls. So when looking at battery life we identify two components: talk time and standby time. Cell phones typically deliver talk/standby times of 10/100 (i.e., 10 hours talk time, 100 hours standby). Early Wi-Fi handsets, on the other hand, typically provide 2/20 to 3/30. For a heavy user, the battery might not last for an entire 8-hour shift.

As they generally do not use voice activity detection, Wi-Fi handsets will be sending a voice frame and acknowledging a received frame every 20 msec for the duration of the call; this is a total of 50 frame transmissions and 50 ACK transmissions per second. One advantage of the No ACK option in 802.11e will be to eliminate those 50 ACK transmissions per second.

The first step in improving battery performance will be to have a robust WLAN infrastructure. In very simple terms, if a device can send its frames at 11 Mbps rather than 1 Mbps, its transmitter will be on for a shorter period of time. Further, a more robust network will also require fewer retransmissions.

11.15.1 Wi-Fi Multimedia: Power Save/Unscheduled Automatic Power Save Delivery

The original 802.11 standards included a power save feature, but it required that the station wake up every 100 msec to hear the beacon message from the access point and determine if there were frames waiting to be delivered. The client would then send a PS Poll message to receive each frame. Polling for each incoming frame wastes power, and a wake up interval of 100 msec will add 100 msec of transmit delay for the voice packets. To help improve battery life, the 802.11e standards include an Unscheduled Automatic Power Save Delivery (U-APSD) function, which the Wi-Fi Alliance calls WMM-Power Save.

As the handset knows it will be receiving a frame every 20 msec for the duration of the call, it can now wake up on that schedule. While an APSD station wakes up more frequently, it does not have to transmit polling messages to retrieve each voice frame. ASPD uses a trigger frame to retrieve all stored traffic. Importantly, the trigger frame can be an indicator in the header of a traffic frame the station is sending. The station will be waking up to send a frame every 20 msec so it will not be waking up unnecessarily and will not be generating additional frames. There are also features to increase that sleep period if voice activity detection is added.

There are a number of other power saving enhancements that are being considered in the standards process, and many of these are already being incorporated in vendor proprietary solutions.

- **802.11h Transmit Power Control (TPC):** TPC is a mechanism whereby the access point can order a client device to reduce its transmit power. TPC was originally developed to meet the requirements for European 5 GHz installations, but the same feature can be used to limit battery drain. Normally, the handset will transmit at full power regardless of its proximity to the access point; Cisco's CCX does include a client power control function. By having the client transmit only at the power required to reach the access point, users can minimize the battery drain and also limit interference with other networks operating on the same channel.

- **Maintaining Association/Improving Handoffs:** Even if a client has not been involved in a call, it must still reinitialize its encryption key periodically, an exchange that normally involves four frames (i.e., a four-way handshake). The 802.11i standard includes a feature called Opportunistic Key Caching (OKC) that allows a station to store keys that it can use when it roams to another access point, eliminating the need to re-key. The developing 802.11k standard for neighbor reporting also includes a function where radio and authentication information about nearby access points can be provided to a client to guide and streamline the handoff process. The developing 802.11r roaming standard also includes a feature that will reduce the number of frames needed to re-key.

- **ARP/Broadcast Proxy:** Even when a Wi-Fi handset is idle, it must still process ARP requests and other broadcast traffic received over the radio link. Many access points now have the ability to intercept that traffic and determine if it really needs to be forwarded over the radio link; as it knows the MAC addresses of the stations associated with it, the access point can actually respond to ARP requests on the client's behalf. Disencumbering the handset of the task of processing these overhead frames can have a surprising impact on battery life.

- **Paging Server:** One option being considered by the 802.11v Task Group would allow the handset to register with a paging server and essentially turn off its Wi-Fi interface when not in use. The paging server would use a separate, more power-efficient, radio mechanism to alert the device that it had an incoming call, at which point its Wi-Fi interface would be activated. The idea of paging to announce incoming calls is borrowed from the paging process used in cellular networks.

- **Fixed-Mobile Convergence:** Battery life is a particular concern in fixed-mobile convergence implementations as the handset will have two radios, one Wi-Fi and one cellular, that need to be powered. One option in that environment is to turn off the Wi-Fi interface when the handset determines there are no Wi-Fi networks within range.

11.16 Conclusion

It is possible to provide high-quality voice services over a wireless LAN, but there are a number of issues that must be addressed. At the outset, the WLAN must provide pervasive coverage so that users can wander anywhere within the defined coverage area and maintain voice connectivity. For any large-scale VoWLAN deployment, a centrally controlled WLAN switch network will be the key infrastructure element. Not only do these systems ease the practical problems involved in ensuring RF coverage and detecting rogue access points, they also have addressed the critical requirement to support fast, secure handoffs.

Good radio coverage makes everything work better in WLAN voice. When devices can transmit at higher bit rates, their transmitters are on for a shorter period of time. Besides improving battery life, that also increases the efficiency and throughput of the WLAN. Better radio coverage also means that fewer frames will have to be retransmitted.

Support for 802.11e QoS will be at the top of the list of critical features for WLANs supporting both voice and data traffic. The ability to prioritize voice is less critical if we are building a separate WLAN network exclusively for voice users. However, even in a dual network strategy, some of the other features of 802.11e, like streaming and no acknowledgment operation, can still provide benefits for voice users.

Beyond the basic task of setting up and maintaining a mobile voice call over the wireless LAN infrastructure, there are also concerns regarding privacy, toll fraud, and remaining in compliance with regulations regarding E911 services. As WLAN voice migrates from proprietary to SIP-based solutions and increases in popularity, all of these issues will become more pressing. Success will depend initially on a sound network design, but ongoing success will depend on monitoring and maintenance systems that will ensure that the network can be maintained and continue to support the required services.

CHAPTER 12

Fixed-Mobile Convergence: WLAN/Cellular Integration

The ability to support voice services over a wireless LAN is important, but the functionality of that solution would increase exponentially if we could integrate that capability with a cellular service. Solutions that are geared toward merging those two environments are referred to as Fixed-Mobile Convergence (FMC). The ultimate solution would be one where the user could place a "free" call over the WLAN, and then have that call handed off to the cellular network when he or she moves out of range.

While there has been considerable interest in Wi-Fi/cellular integration, from the cellular carrier's standpoint there are challenges on two fronts: technology and marketing. From a technical standpoint, Wi-Fi and cellular networks use completely different technologies, and the task of handing off calls between the two is not trivial. The bigger challenge is on the marketing front, as the cellular carriers will have the primary say in whether or not these offerings come to market; thus far they have been decidedly lukewarm on the idea of converged offerings.

Faced with that reluctance, the industry has responded with a bevy of products that simulate integration but require no support from the cellular carriers. The simplest of these are handsets that include both Wi-Fi and cellular capability, but provide no ability to hand off calls between the networks. While these offer the benefit of carrying one device instead of two, this is a far cry from a truly integrated network offering.

In this chapter we will begin with a description of the various fixed-mobile convergence strategies that are being pursued. Then we will overview the two major cellular technologies, GSM and CDMA, and compare the technologies they use with voice over WLAN. The cellular carriers are also looking at strategies to improve the indoor performance of their networks as an alternative to WLAN voice, and we will examine the options they are deploying: microcells, femtocells, and distributed antenna systems. With regard to fixed-mobile convergence, the solutions fall into two major categories, network controlled and PBX controlled. We will take a detailed look at both sets of options, the degree of functionality they deliver, and the likely costs involved. Finally, we will look at the prospects for the other developing wide area wireless technology, WiMAX.

12.1 Customer Motivation

Before we look at the "how" in fixed-mobile convergence, we should first take a look at the "why." Specifically, what is the customer's motivation and justification for pursuing an integrated WLAN/cellular solution? There could be several different reasons.

- **Cost Savings:** The most typical justification is cost savings, in particular, reduced cellular charges. A significant percentage of cell phone calls are made or received while the user is inside a company facility where there is a more cost-effective communications network readily available. In reality, the cost savings potential may be overstated. First, we have to ensure that the calls are actually placed on the less costly service when it is available; any solution that depends on the user to "do the right thing" will see only a fraction of those savings. Also, there will be a charge for the fixed-mobile convergence service, so unless the reduced calling charges exceed that cost, a company is still not saving money. Finally, if the contract with the cellular carrier is based on a specific volume of usage (i.e., a bucket of minutes plan), there will be no reduction in calling charges until the next contract cycle.

- **Improved Accessibility of Improved Management Productivity:** Rather than focusing on direct cost savings, it might be easier to justify this solution by looking at the productivity improvements. One of the biggest drags on office productivity is the inability to reach critical people at critical times. Typically the person with the answer or the person with the authority to say "yes" is in a meeting or otherwise unreachable at a critical time. This typically sets off a frenzy of communications as staff member try office numbers, cell phone numbers, emails, text messages, and door-to-door searches to try to reach the critical party. The ability to reach people with one number regardless of where they are located can be a major productivity boon. Further, if that one number forwards calls to one voice mail, it will be easier for that person to catch up with contacts when he or she is again available.

- **Better Integration with IP PBX-Based Features and Services:** Cellular telephones are an anomaly in modern business communications. Cell phones represent a separate and independent communication service that is not integrated with anything else in the corporate network. Having users continuously check their PBX and their cellular voice mailboxes is an obvious waste of time. Any incoming call to a user's desk phone is free; incoming cellular calls are charged (unless they are coming from another cell phone on the same carrier's network). If users are placing outgoing phone calls on cell phones while in the office, they are avoiding the more cost-effective network services that are connected to the PBX. There is no way the PBX's automatic route selection system will be able to divert that cell phone call onto the more cost-effective alternative.

- **Better Indoor Coverage:** Lastly, indoor cellular coverage is inherently problematic, particularly in the core and basement of the building. We can address those coverage problems with distributed antenna systems, mini-cells, and other cellular network enhancements, but all we are doing is encouraging people to use the least cost-effective means to place their calls.

12.2 Fixed-Mobile Convergence Options

There has been considerable confusion regarding the range of WLAN/cellular convergence solutions and the functionality they provide. The image that has stuck in the minds of most people is a dual mode Wi-Fi/cellular handset and a mechanism to hand off calls transparently between the two environments. While that might be the standard image, there are lots of different technologies and configurations that are being proposed for fixed-mobile convergence.

The major dividing line among the solutions is whether the handoff function is controlled by the cellular network operator (network controlled) or by the customer (PBX controlled). In the network-controlled solutions, the indoor option could be Wi-Fi, but there are also solutions that employ Bluetooth and short-range cellular. In fact, there is one school of thought that argues that cellular to Wi-Fi handoffs will occur so infrequently, we shouldn't bother with the handoff at all. In that view we simply provide a dual mode handset and let the user choose the appropriate network.

First, let's take a quick overview of the options. See Table 12-1.

Table 12-1: Overview of FMC Options

	Manual Handoff	Automatic Handoff
Network Controlled	Sprint/Nextel *Wireless Integration*	• Unlicensed Mobile Access (UMA) • IP Multimedia Subsystem (IMS)
		Local Service Options
		• Wi-Fi (T-Mobile HotSpot @Home) • Bluetooth • Femtocell Cellular
PBX Controlled	• Non-Integrated Wi-Fi/Cellular Handsets (No Handoff) • Extension-to-Cellular/ Simultaneous Ring • User Initiated Handoff	• Seamless Convergence • DiVitas Mobility Controller

12.2.1 Network Controlled Solutions: Fixed-Mobile Convergence

To truly hand off calls between the carrier's network and a customer's Wi-Fi network, the customer's network must be recognized as an integral part of the cellular network; in that way calls can be passed back and forth in a manner similar to the handoff between two cellular base stations. This level of integration would require a full buy-in by the cellular operator, and the cellular operators have shown some reluctance in introducing this option.

The situation is starting to change, though we must distinguish between business and consumer services. In the US, the carriers have approached business services cautiously. Sprint/Nextel has tested a service called Wireless Integration that requires the user to input a command on their handset to transfer a call from the public to the private network (i.e., manual handoff). In 2007, US-based T-Mobile introduced a consumer fixed-mobile convergence service called HotSpot @Home that combined GSM cellular and Wi-Fi, and it could work through a home Wi-Fi network or through any of T-Mobile's public Wi-Fi Hot Spots.

12.2.2 PBX Controlled Solutions

Sensing the reluctance of the carriers to integrate Wi-Fi into their network offerings, the PBX vendors have developed a range of solutions that incorporate cellular services and emulate the handoff functionality.

12.2.2.1 Integrated Wi-Fi Cellular Handsets

The simplest solution integrates Wi-Fi and cellular capabilities in the same handset and allows the user to choose between them. This is nothing more than two separate wireless phones built into the same package. The packaging allows a user to carry one device instead of two, and with some of the other PBX features like Simultaneous Ring and User-Initiated Handoff, we can simulate the functionality of an integrated solution.

12.2.2.2 Simultaneous Ring/Extension to Cellular

The Simultaneous Ring and Extension-to-Cellular features supported in most IP PBX systems do provide a rudimentary mechanism to integrate cell phones. With the Simultaneous Ring feature, when a call is placed to the user's desk phone, the PBX system simultaneously rings the call on the desk phone and on the cell phone (i.e., the user's cell phone number is stored in the PBX, and the PBX places a call to that number). When either device answers, the ringing is stopped on the other. This solution does not require a dual mode Wi-Fi/cellular handset. There is no automatic handoff function, though the user may be able to transfer an in-progress call from one device to the other.

12.2.2.3 PBX-Controlled Automated Handoff

The PBX vendors have found that by adding a special server to the PBX that monitors the availability of devices on the WLAN network, they can automate that call transfer or handoff. The first system to deliver that functionality was announced in mid-2004 by Avaya, Motorola, and Proxim under the name Seamless Convergence. Motorola provided the server (called the Wireless Services Manager) that connected to the Avaya IP PBX and managed the handoff function using dual mode GSM/Wi-Fi handsets. In the end, the cellular partner, Cingular Wireless (now AT&T), pulled out of the program, refused to certify the Motorola dual mode handset for use on their network, and the product died. DiVitas Networks is now reintroducing the idea with a controller that will work on a variety of PBX systems and support a range of dual mode Wi-Fi/cellular handsets.

12.3 Cellular Technology

The first thing to recognize about cellular telephone technology is that it is very different from Wi-Fi. Table 12-2 compares the VoWLAN technology with the two most popular cellular standards the Global System for Mobile Telecommunications (GSM)

Table 12-2: Comparison of Wi-Fi and Cellular Technologies

	Wi-Fi	GSM	CDMA
Switching Technology	Packet	Circuit	Circuit
Radio Technology	CCK-DSSS (802.11b) OFDM (802.11a and g)	Time Division Multiple Access (TDMA with 8 Slots and a 4.6 msec Cycle Time)	Walsh-Coded Direct Sequence Spread Spectrum (DSSS)
Channel Size	22 MHz (b) 20 MHz (a and g)	200 KHz	1.25 MHz
Duplex Operation	Time Division (TDD)	Frequency Division (FDD)	Frequency Division (FDD)
Frequency Band	Unlicensed 2.4 GHz (b and g) 5 GHz (a)	Licensed 824–890 MHz (US) 1.85–1.99 GHz (US)	Licensed 824–890 MHz (US) 1.85–1.99 GHz (US)
Voice Coding	G.711 (64 Kbps) G.729A (8 Kbps)	RPE-LTP 13 Kbps (Full Rate) 6.5 Kbps (Half Rate)	QCELP 8 Kbps (Variable Rate)
Authentication	Shared Key, 802.1x with Various Authentication Protocols (TLS, TTLS, PEAP, etc.)	GSM Standard A3/A8	Cellular Authentication and Voice Encryption (CAVE)
Encryption	RC4 (WEP) RC4 TKIP (WPA) AES (WPA2)	GSM Standard A5/1: Europe A5/2: No America A5/3: 3G	P/N Long Code Scrambling (Cellular Message Encryption Algorithm)

and Code Division Multiple Access (CDMA). There are virtually no common technologies between them. The cellular radio technologies for voice are also far more advanced than Wi-Fi, and the carriers have over 25 years of experience in building and maintaining large-scale wide area wireless networks.

Cellular networks operate on licensed spectrum, which protects them from interference from other applications. Those radio licenses represent a significant investment for the carriers. The most recent US cellular auctions were for the 90 MHz of Advanced Wireless Services (AWS) spectrum; they were held in 2006 and raised close to $14

Table 12-3: Cellular Telephone Frequency Allocations

System	Inbound (MHz)	Outbound (MHz)	Bandwidth In or Out (MHz)
AMPS (US)	824–849	869–890	25
PCS (US)	1,850–1,910	1,930–1,990	60
AWS (US)	1,710–1,755	2,110–2,155	45
GSM 900 (Europe)	890–915	935–960	25
GSM 1,800 (Europe)	1,710–1,785	1,805–1,880	75

billion. The regulatory authorities in different countries allocate different frequency bands for cellular service. So while a customer may have a GSM handset, to use it internationally he or she will need a model that can operate on the frequencies that are allocated in the particular country. See Table 12-3.

Despite the elegance of the radio technology, the current Second Generation cellular technologies both depend on circuit switching technology. A dedicated radio channel is provided for the duration of the call. The cellular networks are in the midst of a major technology upgrade to a Third and eventually a Fourth Generation solution. Those new solutions will be built on packet radio technologies and will feature higher data transmission rates and greater bandwidth efficiency. Thus far only Third Generation technologies are being deployed, and those are being used to deliver high-capacity data services primarily. In the GSM environment, the 3G solutions are designated Universal Mobile Telecommunications System (UMTS) and High Speed Downlink Packet Access (HSDPA), while in the CDMA world, they are 1xRTT (Radio Transmission Technique) and 1xEV-DO (Evolution-Data Optimized).

12.3.1 What's 4G?

The International Telecommunications Union (ITU) and the 3G Partnership Projects (3GPP for GSM and 3GPP/2 for CDMA) are still considering the technical underpinnings for the next or Fourth Generation cellular technologies. In the GSM camp, the 4G solution is referred to as Long Term Evolution (LTE) while in CDMA it is called Ultra Mobile Broadband (UMB). While they have the names, customers are still waiting for a full description of the technology. The primary elements that are expected to be incorporated include:

- Packet-based service.

- Wideband radio channels (>1.25 MHz).

- Orthogonal Frequency Division Multiplexing (OFDM) modulation.

- Multiple input, multiple output antenna technology.

- Support for multi-megabit data services.

The technologies being considered are the same as those that have been incorporated in the 802.11n radio link standard and the WiMAX technology we will be looking at later. In the meantime, Sprint/Nextel has declared WiMAX to be *their* 4G solution, but that is a vendor announcement and should not be misconstrued as an endorsement from the standards community.

12.4 Cellular Network Configuration

Cellular telephone networks are built around a specially modified telephone central office called a Mobile Switching Center (MSC). The carriers typically have one MSC in each city where they provide service, though large metropolitan areas might require two or three. Each MSC controls a network of Base Transceiver Stations (BTSs) with electronics, antennas, and towers that provide access to mobile subscribers. The base station network is managed by an element in the MSC called a base station controller. The base transceiver stations connect back to the mobile switching center over private microwave radio channels or T-1 rate facilities (i.e., 1.544 Mbps channels) leased from the local telephone company. The MSCs are in turn connected to one another with a common channel signaling network or Signaling System 7 (SS7) network that is used to authenticate users and support inter-system roaming. See Figure 12-1.

The cellular carriers have developed considerable expertise in building and maintaining radio networks that use hundreds of base stations, cover large areas, and can be fine tuned to address difficult terrain or coverage environments. CTIA, The Wireless Association (formerly the Cellular Telecommunications and Internet Association, http://www.ctia.org) claims there are almost 200,000 base stations in the US. In rural areas, a BTS might cover dozens of square miles, while in a downtown area that distance would be a few blocks. Cells can also be sectorized using directional antennas with different channels configured in different directions. There will often be cellular antennas mounted on triangular structures that allow the carrier to divide the

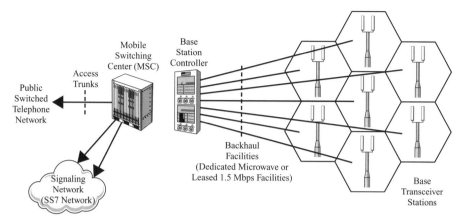

Figure 12-1: Cellular Network Configuration

cell into three 120° sectors. While indoor coverage presents a difficult challenge for the cellular carriers, they can deploy specially designed distributed antenna systems to improve reception in airports, sports arenas, train stations, and other locations where people congregate.

Recognizing that increased customer satisfaction (and retention) requires a sound network infrastructure, the carriers spend considerable resources monitoring the usage pattern on their networks and the signal strength throughout the coverage area to locate and address problem areas. Many of the techniques we use in building and maintaining WLAN networks were pioneered in the cellular industry.

12.5 Network Controlled Convergence

To be truly functional, a fixed-mobile convergence solution would require that the customer's private network be recognized as an integral part of the cellular network so that calls could be passed back and forth in a manner similar to the handoff between two cellular base stations. This level of integration would require a full buy-in by the cellular operator, and the ability to interface the cellular carrier's signaling network that controls the handoff process.

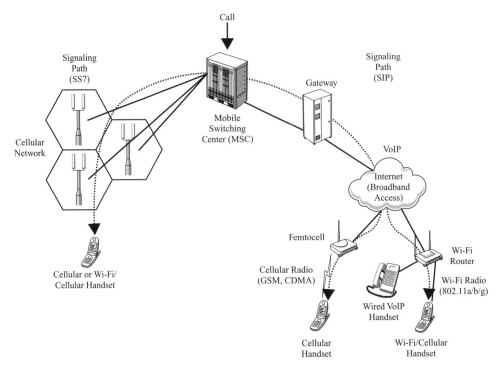

Figure 12-2: Network Controlled FMC Options

12.5.1 Defining the Private Network

We continue to use the term *fixed-mobile convergence* rather than *Wi-Fi/cellular convergence*. The reason is that from the cellular standpoint they are looking at handing off a call from their *mobile* network to a wired VoIP network. That VoIP call would be carried on the Internet and delivered over a broadband connection to either a wired or wireless subscriber device. If wireless, that VoIP call could be delivered using Wi-Fi, Bluetooth, or cellular technology. See Figure 12-2.

12.6 Cell Phone Signaling Interconnect

As we described earlier in this chapter, cell phone networks are interconnected by a common channel signaling network that is used to authenticate users and support inter-system roaming and handoffs. When a user moves between cells in a cellular carrier's network, the handoff is controlled by the base station controller. When a user moves out of the coverage area of their home cellular carrier, the handoff of their call to a roaming partner is coordinated through the common channel signaling network.

To support true Wi-Fi/cellular convergence, the user's private network would have to be connected to that signaling infrastructure. The cellular carriers control who gets access to the signaling network, and currently only cellular carriers have access. The basic fact to recognize is that the cellular carriers have the fundamental say about the degree to which WLAN/cellular convergence happens.

Should the carriers decide to endorse WLAN/cellular convergence, there are two technical solutions that could be employed.

1. **IP Multimedia Subsystem (IMS):** IMS is a next-generation common channel signaling protocol being developed by the ITU to support fixed-mobile convergence. The vision would include media gateways to pass calls between cellular networks and VoIP-based private networks using a feature called virtual call continuity (VCC). This type of connection would require both media (i.e., voice path) and signaling gateways. The signaling gateways would translate call commands from the SS7 environment used in the cellular network to the Session Initiation Protocol (SIP) used in the VoIP networks. The media connection would then be transferred from the cellular infrastructure over the wired IP network where it could be delivered to either a wired or wireless handset.

   The IMS view is to have one handset, one number, one voice mail, and a seamless user experience regardless of the network the call is carried over. There is still work to be done on the SIP side to address such issues as: QoS, fraud prevention, privacy/encryption, billing/chargeback, and voice mail integration. Carriers are now investing in IMS equipment, which is a good sign. It is important to remember that while IMS provides the *potential* for a carrier to implement fixed-mobile convergence, the carrier must still decide whether to introduce such a service. While the carriers do seem to be warming to the idea, the timetable for those offerings is still unknown.

2. **Unlicensed Mobile Access (UMA):** There are a number of gateway products from companies like Kineto Networks and Bridgeport Networks that provide fixed-mobile convergence today. Essentially, these devices provide trunk and signaling gateways between cellular and private networks, and allow calls to be handed off between them. The most widely endorsed of these strategies is a plan called Unlicensed Mobile Access (UMA). Promoted initially by Kineto

Wireless, UMA is now endorsed by the 3G Partnership Project (i.e., the coordinating body for 3G cellular migration), and it is included in Version 6 of the 3GPP standards. UMA will allow calls to be handed off between cellular networks and unlicensed radio networks including wireless LANs. There are other gateway options as well from companies like Bridgeport Networks, but for the moment, UMA appears to be the architecture of choice.

IMS, UMA, or some other gateway interface could provide the ability to hand off calls between cellular and either wired or wireless private networks, however, it is still the cellular carriers' choice to deploy those capabilities. Having a standard is not a guarantee that you will get a product.

12.6.1 Will the Cell Carriers Cooperate?

If the cellular carriers want to provide a converged WLAN/cellular offering, they certainly have the tools at their disposal to deliver that capability immediately. From the cellular carriers' standpoint, there are several factors arguing in favor of such an offering:

- **Improved Indoor Coverage:** Handing off calls to a WLAN would allow the cellular carrier to provide better indoor coverage, a well-known shortcoming of cellular services. The cellular operators could also address that problem by implementing distributed antenna systems, femtocells, and other indoor coverage solutions we will look at in a moment.

- **Load Shedding:** Passing calls to the WLAN or other private network also clears channels on the cellular carrier's network, allowing the carrier to support more traffic without investing in additional cellular infrastructure.

- **Respond to Market Demand:** Enterprise customers have seen their cellular costs skyrocket over the past several years, and they are clearly interested in a converged solution. The first carrier to deliver such a service could gain a significant competitive advantage in the enterprise market.

Weighed against those potential benefits, the cellular carriers have some valid concerns regarding the technical and business impact of a converged service offering.

- **Quality Control:** The cellular carriers have struggled for years to improve the quality of service they provide. Now they fear they could be held responsible if quality degrades or a call is dropped when it is handed off to a poorly designed WLAN. That would be particularly true if the user could not determine whether their call is being carried on the WLAN or the cellular service when the failure occurs. The cellular carriers do not want to take the rap (and the resulting customer service calls) for handing off calls to badly designed wireless LANs.

- **Security:** The cellular carriers were badly burned by theft of service with their First Generation AMPS networks. They addressed those problems with the move to Second Generation digital cellular technologies like GSM and CDMA, which essentially eliminated the problem of cloned cell phones. Given the widely-recognized security faults in the WLAN technology, the carriers fear a converged solution could potentially open security vulnerabilities in their networks. If a hacker thwarted the authentication mechanisms on the WLAN network and placed an unauthorized call, could that call be handed off to the cellular network?

- **Privacy:** WLAN and cellular networks use completely different encryption mechanisms, and while encryption is an integral function of the cellular service, WLAN encryption is optional. As we noted, the early WEP-based encryption was woefully inadequate. Will the cellular carrier be able to insist (and ensure) that adequate security mechanisms are used and maintained on any WLAN they hand off calls to?

- **Customer Control:** While the other issues are primarily technical in nature, the major business question gets down to who controls the customer. The cellular carriers have a unique and enormously successful franchise in providing mobile telephone service for which they can charge a premium price. If calls can move transparently between cellular and other networks, will that mean that the cellular service is no longer something special? The carriers are very cautious about diluting their franchise; a fixed-mobile convergence solution could leave them as just another element in the customer's integrated telephone system.

12.7 Fixed-Mobile Convergence in the US: T-Mobile's HotSpot @Home

The first true fixed-mobile convergence service in the US market was introduced on a trial basis by T-Mobile in early 2007; they announced a nationwide rollout in mid-2007. The service integrated WLAN and the carrier's GSM service, though it was aimed at residential customers rather than business users. The customer would get a dual mode Wi-Fi/cellular handset; only two handset models were offered initially: the Samsung t409 and the Nokia 6086. The customer would also require a wireless LAN and a broadband Internet connection (i.e., DSL or cable modem) at home.

When the customer was out, their handset would automatically go to the T-Mobile cellular network. Once the handset came within range of the home Wi-Fi network, it would associate with the wireless LAN and register with the T-Mobile service over the Internet indicating that it was now able to receive calls over the Internet/Wi-Fi path. If the handset was engaged in a call when it came into the Wi-Fi coverage area, the T-Mobile network would automatically drop the cellular connection (and stop the cellular billing) and forward the call in a VoIP format over the Internet and out to the Wi-Fi interface to the customer's phone. The Wi-Fi capability would work over the customer's home Wi-Fi network or through any of T-Mobile's 2000+ Hot Spot locations. See Figure 12-3.

T-Mobile charges $20 per month on top of the user's wireless service plan for the HotSpot @Home service; they also offered an introductory plan for $10 per month. To break even, the customer would have to be able to reduce their regular cellular charges by $20 per month. This service is intended primarily for users who do not have a traditional, wired telephone service and who rely entirely on cellular service.

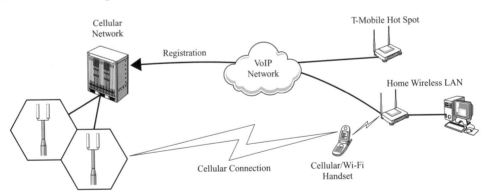

When the cellular/Wi-Fi handset comes within range of
a usable Wi-Fi network, it automatically registers with the
T-Mobile service over the Internet connection.

If there is a call in progress, it is automatically switched to a
VoIP call and delivered over the Internet to the customer's
broadband Internet connection and continued on the customer's
home WLAN.

Figure 12-3: T-Mobile Hot Spot @ Home

British Telecom's BT Fusion or BluePhone

British Telecom introduced the first truly integrated fixed-mobile convergence solution in 2006. Originally called BluePhone, the service is now marketed under the name BT Fusion, and like T-Mobile's HotSpot @Home service, it too is targeted at residential rather than enterprise users.

The original service used GSM for the cellular service and Bluetooth for the private network connection. Subscribers would get a Motorola V560 or RAZR V3B GSM/Bluetooth handset and a home base station/router called the BT Home Hub. That home hub connected to BT's DSL-based broadband Internet service for VoIP telephone service. The Home Hub could pair with up to six handsets and support up to three simultaneous calls. The handset used a Class 1 Bluetooth interface and claimed to support a range of 60 m through free space or an indoor range of 25 m in ideal conditions. Outside of the home, the handset operates on BT's GSM cellular service, but when the user comes within range of the Home Hub, the call would be seamlessly handed off to the Bluetooth connection and the call is transferred to the VoIP service.

In 2007, BT dropped the Bluetooth element and opted for Wi-Fi. Like T-Mobile's service, they still require a broadband Internet connection, only now the home wireless technology is Wi-Fi. The handsets also work over BT's Openzone Hotspots.

12.8 Cellular Service as an Alternative to VoWi-Fi

The other option for providing mobile voice service in the enterprise is to simply make use of the cellular network. Many cellular carriers now provide free, unlimited calling between subscribers on their networks. If most of a company's mobile calling is intra-company, the cellular network will meet those requirements for the basic subscription rate. Of course, a company would still be paying cellular rates for calls to and from other cellular networks and wired telephones. In short, a company could have free mobile calling among employees on the cellular network and simply use the Wi-Fi network for data applications.

There are a number of advantages to simply using the cellular carrier's service:

- The network is already in place and was designed by professionals.

- The security features are world-class and require absolutely no design or coordination by the user.

- The cellular carrier monitors the coverage and capacity on an ongoing basis, and the company's users will automatically benefit from any improvements made in the network.

- There is no concern with the impact of voice calls on the Wi-Fi network.

- There is no need to expand the Wi-Fi network to provide pervasive coverage or implement the other Wi-Fi features required for voice support.

- The user can still have a wired desk phone and use the PBX simultaneous ring feature to ring incoming calls to the desk phone and to the cell phone. With that feature, the user can remain accessible inside and outside the facility. It is important to note that calls that are extended from the PBX are charged at the regular cellular billing rate (i.e., only in-network cellular-to-cellular calls are free).

12.9 Improving Indoor Coverage: Distributed Antenna Systems, Microcells, and Femtocells

The cellular carriers are using two different options to address the perennial problem of indoor cellular operation. The options are different for commercial versus residential/ small business locations.

12.9.1 Commercial Installations: Distributed Antenna Systems

In commercial installations, the carrier must first determine if the facility simply requires better indoor radio coverage, or if it also requires additional calling capacity. To address the coverage problem, the carrier could deploy a radio repeater or a full Distributed Antenna System (DAS) inside the customer's facility. A DAS is a custom-designed network of amplifiers, cabling, and antennas that is used to distribute radio signals in an indoor environment. An outdoor antenna could be used to capture the radio signal from the nearest base station and the DAS could distribute it throughout the facility. See Figure 12-4.

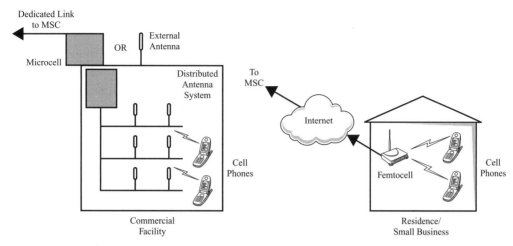

Figure 12-4: Options for Improving Indoor Cellular Coverage

12.9.2 Microcells

If the facility also requires additional call capacity, the carrier can deploy a small base station called a microcell as well as the antenna system. The microcell is effectively another base station on the carrier's network, and when a user involved in a call enters the building, the MSC would hand the call off to that microcell. If the carrier installs a microcell, they must also lease a dedicated link (typically a 1.544 Mbps private line) to connect that microcell back to their MSC. Typically systems installed by a cellular carrier will be designed to support only subscribers on that particular carrier's service.

A different model for a distributed antenna system comes about when a venue owner wants to improve all cellular services within a facility. In that case, the venue owner can install the distributed antenna system and charge the carriers a fee to connect to it. Those that do will have better service within that facility while those who do not can expect angry calls from their subscribers whose phones do not work. The other advantage for a company in deploying its own antenna system is that the company may also use it to distribute Wi-Fi and other radio signals within the facility.

12.9.3 Femtocells

A distributed antenna system is a relatively large-scale system designed for commercial installations. What about improving cellular service in a small office or a home? The solution now being proposed is femtocells. The prefix *femto* means one quadrillionth (i.e., 10^{-15}), and in the cellular world it refers to a very small cellular base station that connects back to the MSC over a broadband Internet connection. In this approach, a femtocell would be installed in the customer's home and connected to the customer's broadband Internet connection. When the user's cell phone comes within range of the femtocell, it automatically forwards a registration message over the Internet to their cell carrier's MSC that the user is now reachable through the home base station. If the user had a call in progress, that call would be transferred from the cellular network to a VoIP call over the Internet and delivered through the femtocell to their cellular handset. If the user placed a call from a cell phone while at home, the call would go through the femtocell and across the Internet as a VoIP call. See Figure 12-4.

The femtocell concept is very similar to consumer fixed-mobile convergence services like T-Mobile's HotSpot @Home, only the in-home connection uses cellular rather than Wi-Fi technology. The other advantage of the femtocell approach is that it will work with any cellular phone and does not require the dual mode Wi-Fi/cellular handset.

Femtocell deployments are only just beginning so it is difficult to gauge how big an impact they will have on cellular networks or on integrated Wi-Fi/cellular services. However, it is clear that they are another option for addressing the indoor coverage problem and extending the functionality of cellular technology. From the cellular carrier's standpoint the advantage of the femtocell is that it delivers the same functionality as the consumer fixed-mobile convergence services, but uses all cellular technology, rather than what they see as the less reliable Wi-Fi technology. The big question will be pricing. Will cellular carriers use the same pricing philosophy and stop the cellular billing when a call is handed off to the femtocell?

12.10 PBX-Controlled Convergence Solutions

While the cellular industry has been moving slowly to embrace Wi-Fi technology, PBX suppliers have been working feverishly to incorporate cellular services with their offerings. While it is impossible for a PBX system to truly hand off a call to a cellular network without the cooperation of the cellular carrier, they have developed a number of techniques that closely emulate the look-and-feel of an integrated network. While they cannot do much to reduce the cellular charges (in fact, they might actually increase them), if mobility and one number accessibility with integrated voice mail are the goals of the fixed-mobile convergence solution, the PBX suppliers might be in a better position to deliver that solution than the cellular carriers.

12.10.1 Integrated Wi-Fi/Cellular Handsets

Many of the PBX-controlled solutions depend on integrated Wi-Fi/cellular handsets, which are a relatively new phenomenon. The Wi-Fi Alliance's Web site lists close to 100 certified Wi-Fi/cellular handsets from Nokia, Motorola, Samsung, Philips, Sony-Ericsson, and others. However, most of the models listed are not available for sale in the US. The most widely heralded was Apple's iPhone, introduced in mid-2007, but at its initial introduction it was restricted to making voice calls only on the AT&T Wireless network. The Wi-Fi capability was solely for data access, not for VoIP calls.

12.10.2 Wi-Fi Functionality in Integrated WLAN/Cellular Handsets

Buyers should look closely at the Wi-Fi capabilities that are included in their Wi-Fi/cellular handsets. Many of the early models do not support the full range of features we look for in enterprise VoWLAN environments. In particular, many do not support 802.11a and g radios, WPA2-Certified encryption, or 802.11e/Wi-Fi Multimedia QoS. That means the integrated models will not work as well or as securely as Wi-Fi-only models.

12.10.3 Moving up to Management

Integrated Wi-Fi/cellular handsets are seen as a major element in moving WLAN voice into the management ranks. Most of the first wave of WLAN voice users have been non-management employees: nurses, retail clerks, and forklift operators. Those non-management employees are typically not given company-paid cell phones. Managers and salespeople who do have company-provided cell phones are not going to be happy with the prospect of carrying two separate handsets, one for the WLAN and one for cellular. They will probably leave the WLAN handset in a drawer and use the cellular device that works everywhere.

The key in a management-oriented wireless voice environment is to deliver seamless mobile voice capability, with an integrated look-and-feel, integrated voice mail, and an operating environment that is totally transparent to the underlying wireless network. An integrated Wi-Fi/cellular handset that requires the user to manually switch between the two and forwards missed calls to two separate voice mail systems clearly falls short of the mark.

12.11 PBX-Controlled Options: Manual Handoff

With regard to PBX-controlled options, the main dividing line will be whether the integration is transparent (i.e., automatic), or if handing off a call involves some action on the user's part. Each organization must define its own objectives for the service and then make a decision based on cost and functionality in light of their users' abilities and preferences. From a management standpoint, we are also concerned with ensuring that calls are carried on the most cost-effective service available. If a company provides a user with a dual mode Wi-Fi/cellular phone and he or she simply uses it to make and receive cellular network calls when in the office, the company will realize no cost savings.

12.11.1 Manual Handoff: Simultaneous Ring/Extension-to-Cellular

Simultaneous ring and extension-to-cellular are two closely related PBX features that aim at incorporating cellular service into the wired PBX environment. With this feature, when a call is placed to the user's desk phone, the PBX system simultaneously rings the call on the desk phone and on the cell phone; the user's cell phone number will be stored in the PBX, and the PBX will place a call to that number. When the call is answered on either device, the ringing is stopped on the other. See Figure 12-5.

With this feature, there are no automatic handoffs, nor is there a requirement for a Wi-Fi/cellular handset; in essence, this function will work with any cell phone on any

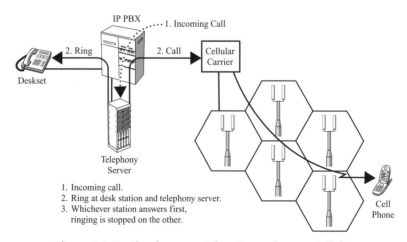

Figure 12-5: Simultaneous Ring/Extension to Cellular

cellular network as long as its number is stored in the PBX. One significant benefit of this arrangement is that users can provide one number to all of their contacts (i.e., the desk phone number), and be able to receive those calls at their desks or on their cell phones. Further, users can deactivate the voice mail feature on their cell phones and get all of their messages on the office voice mail system.

12.11.2 Manual Handoff

Some solutions also allow the user to transfer an existing call from one device to the other. If the user has answered the call on the desk phone, he or she can transfer it to the cellular phone with the push of a button and continue the conversation. It is up to debate whether or not this capability qualifies as an integrated solution, but it certainly provides integrated voice mail and one-number accessibility whether the user is in or out of the office.

That accessibility comes at a price.

- **Incoming Calls Answered on the Cell Phone:** All of the user's incoming cellular calls will be originating from the PBX. A call to the user's desk phone that was forwarded to his or her cell phone would involve:
  ○ A local call from the PBX.
  ○ An incoming call charge on the cell phone.
  ○ Usage on two PBX trunks for the duration of the call (i.e., hair pinning; see Figure 12-6).

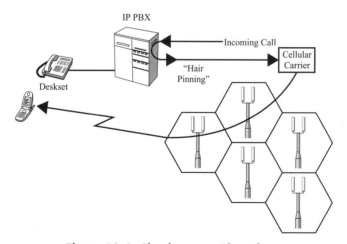

Figure 12-6: Simultaneous Ring Abuse

- **Cellular-to-Cellular Calls:** Normally, if one employee places a cell phone call to another employee's cell phone, that call is free (i.e., free in-net-to-in-net cellular calling). In this configuration, that call would be charged as an outgoing call from one user to the PBX, an outgoing local call from the PBX, and an incoming call to the other user's cell phone. Companies could instruct internal cellular users to call each other directly on the cellular network, but with the voice mail capability deactivated, there would be no way to leave a message.

12.12 PBX-Controlled Solutions: Automatic Handoff

The configuration for this type of solution would involve four primary elements.

1. **IP PBX:** The IP PBX would come with special software for the telephony server that would interface to the WLAN Voice Controller.

2. **Voice-Capable WLAN Infrastructure:** As calls within the facility would now be carried over the WLAN, a company will need a voice-capable WLAN infrastructure that incorporates all of the capabilities described in Chapter 11.

3. **WLAN Voice Controller:** This is the device that would attach to the IP PBX, manage the WLAN/cellular handsets, and interface with the IP PBX's telephony server. In the original Avaya/Motorola Seamless Convergence offering, Motorola had provided this element and it was called the Wireless Services Manager (WSM). Today DiVitas Networks is building a similar server called the "Mobile Convergence Appliance."

4. **Dual Mode WLAN/Cellular Handsets:** The final element will be handsets with the appropriate form factor and feature set for the user population. On the Wi-Fi side they must support the required WLAN voice features (i.e., 802.11a, b, g radio links, WPA2-based Security, 802.11e QoS). In particular, these handsets will require special software to interface with the WLAN Voice Controller to facilitate the handoff function.

12.12.1 Handset Registration

In an installation of this type, when one of the dual mode handsets comes within range of the WLAN, software in the handset would cause it to automatically register with the WLAN Voice Controller. The WLAN Voice Controller would in turn alert the IP

PBX's telephony server that the user is reachable through the wireless LAN. If the IP PBX received a call to that station, it would be delivered over the WLAN.

12.12.2 Availability Monitoring

The key trick will be knowing whether the station is still within range of the WLAN. As long as the station remained within the WLAN coverage area, it must regularly report its receive signal strength to the WLAN Voice Controller; those control messages are exchanged both during a call and when the station is idle.

12.12.3 Handoff

When the user moves out of the WLAN coverage area, the station would report declining WLAN signal power, and when a defined threshold was reached, the WLAN Voice Controller would report to the IP PBX's telephony server that the station is no longer reachable on the WLAN.

- **Incoming Call Handling:** Once the user is no longer reachable through the WLAN, if an incoming call were placed to the user's office number, the IP PBX would automatically forward the call to the user's cellular number. In effect, the cellular function works just like the extension-to-cellular feature.

- **In-Progress Calls:** If the user is involved in a call when he or she moves out of the WLAN coverage area, the IP PBX will automatically call the user on his or her cell phone number and transfer the call from the WLAN to the cellular connection.

Like the Simultaneous Ring/Extension-to-Cellular features we described earlier, this capability provides one number accessibility and integrated voice mail. If a call is forwarded to the cell phone and it is not answered, it goes to voice mail.

12.12.4 Policy Enforcement

One major advantage that an automated solution has over the manual variety is that the company no longer has to depend on the user to select the appropriate network. If the user is within range of the wireless LAN, any incoming or outgoing calls will automatically be routed over it; outgoing calls can be carried on any of the network services available through the PBX. As before, calls to the user's desk phone that are forwarded to the cell phone will hair pin through the PBX, busying two trunks, and will be charged as a local call from the PBX and an incoming cellular call. Each company will still have to determine how intra-company cell phone-to-cell phone calls should be handled.

12.12.5 "Dragged In" Call

The handoff is essentially a call transfer activated by the IP PBX in response to a command from the WLAN Voice Controller. While calls can always be transferred from the WLAN to the cellular network, the reverse is not true. If a user receives a cellular call placed directly to his or her cellular number (i.e., a regular cellular network call) and the user subsequently enters the WLAN coverage area, there is no way to transfer the call to the WLAN (i.e., the PBX has no way to know there is a cellular call in progress). That call is said to be "dragged into" the facility—it simply continues on the cellular network and usage charges apply.

12.12.6 DiVitas Networks

The seamless convergence idea has been resurrected by a start-up called DiVitas Networks. Their WLAN Voice Controller is called the *Mobile Convergence Appliance*. To provide dual mode handsets to work with their solution, DiVitas has developed software for Windows Mobile, Linux, or the Symbian operating systems. In essence, any dual mode device (i.e., handset, PDA, or laptop) could be equipped with software that allowed it to function with the Mobile Convergence Appliance.

12.13 WiMAX Broadband Wireless Access

Most of the interest in fixed-mobile convergence to date has centered on converging Wi-Fi and cellular, but there is another mobile wireless technology starting to take root called WiMAX. WiMAX, short for Worldwide Interoperability for Microwave Access, describes metropolitan area wireless networks built to the IEEE 802.16 standards. Where Wi-Fi has a defined transmission range up to 100 m, a WiMAX base station might support a cell radius of several miles. The technology is being promoted by a vendor group called the WiMAX Forum. Like the Wi-Fi Alliance, the WiMAX Forum looks to develop interoperability test suites to deliver multi-vendor solutions that will lead to lower cost through chip-level implementation. South Korea is pushing a similar technology called WiBro, and at some point the two will probably be merged.

WiMAX is a much more difficult technology to pin down. First, there are two versions of WiMAX, one for fixed location service and the other for mobile users. The technology can operate in licensed or unlicensed spectrum using a variety of channel sizes, and the standards describe a number of different services with different QoS capabilities. So while it is often described as a broadband data service, WiMAX has features to support both voice and data applications. The first US-based WiMAX deployments are getting underway in late 2007, but nationwide services are not likely to appear before 2010. In other countries the prospects for WiMAX are different given its competitive position relative to 3G, Wi-Fi, and wired broadband Internet services.

12.13.1 WiMAX/Wi-Fi Market Overview

The original version of the IEEE 802.16 standard, released in December 2001, addressed fixed-location broadband access systems operating in the 10–66 GHz frequency band. Given the nature of those higher frequency radio signals, a service based on that technology would require line-of-sight alignment of the antennas. In January 2003, the 802.16a standard was released, and it defined specifications for fixed location systems operating in the 2–11 GHz band; those lower frequencies would support non-line-of-sight (NLOS) operation.

12.13.2 Fixed Location WiMAX: IEEE 802.16-2004

The WiMAX standards now cover two different, incompatible versions. The focus of the original 802.16 committee was a metro area, fixed-location broadband wireless access technology that would compete with wired broadband Internet services like ADSL and cable modems. The fixed location standard eventually progressed to a version designated 802.16d, but the IEEE eventually dropped the letter suffixes and now calls it 802.16-2004. While it is called fixed-location, *non-mobile* would be a better description. With fixed-location WiMAX, a user could access the network from a stationary position anywhere in the coverage area; we refer to that capability as *nomadic* service.

12.13.3 Mobile WiMAX: IEEE 802.16-2005

Even before the fixed-location standard was finalized, the IEEE 802.16 committee began work on a mobile version of WiMAX. The Mobile WiMAX standard was originally designated 802.16e, but it was eventually renamed 802.16-2005. While it featured true mobile capability with base station-to-base station handoffs, it could also be used to provide fixed or nomadic broadband wireless access. So a mobile WiMAX offering could compete with wired ADSL/cable modem services, nomadic Wi-Fi Hot Spot services, and mobile 2.5G/3G cellular services. The WiMAX carriers have not disclosed the data rates or pricing plans they intend to offer, so the actual service specifications are essentially unknown.

12.14 WiMAX Deployment Plans

Carrier plans for WiMAX deployment vary widely based on the market. The plans seem to shape up as follows.

12.14.1 Pre-Standard Fixed Location Services

A number of broadband wireless carriers rolled out services prior to the ratification of the WiMAX standards, using equipment from companies like Alvarion and Navini Networks. In the US, the largest BWA carrier is Clearwire Communications, and they had about 250,000 subscribers as of mid-2007 (Note: At that time, the total number of broadband households in the US was over 50 million). Clearwire targeted smaller cities where ADSL and cable modem services were not widely available. The fixed-location service offered data rates between 512 Kbps and 1.5 Mbps at a monthly fee of $30 to $40 including modem rental. Their primary marketing agreement was with direct broadcast satellite carriers who could offer Clearwire's WiMAX-based Internet service in combination with their TV services to better compete with the cable companies.

12.14.2 Mobile WiMAX Service

While the mobile service was originally seen as a follow-on to fixed-location service, the US carriers seem to be bypassing the fixed-location version and going straight to Mobile WiMAX. In mid-2006, Clearwire and Sprint/Nextel both made major commitments to deploy widely-available mobile WiMAX services by the end of 2008. Both of those carriers had extensive frequency holdings in the licensed 2.5 GHz Broadband Radio Service (BRS) band that is the primary target for US WiMAX deployments. Recognizing the investments required and the gaps in each of their license holdings, in mid-2007 the two companies announced their intention to merge their deployments and develop a single WiMAX network they both would sell under the name Xohm. The service could compete with ADSL and cable modem suppliers in the fixed-location market (though indoor coverage may still be problematic), but also go after the higher-priced mobile 2.5G/3G data services.

12.14.3 *Fixed-Location Service in Lesser Developed Countries*

One very good potential market for fixed-location WiMAX is providing broadband Internet services in lesser-developed countries. The telecommunications infrastructure in many developing countries is based entirely on wireless technology. The telephone service is provided via cellular technology, and TV service is delivered by direct broadcast satellite. That means there is no telephone wire to support ADSL and no coaxial cable plant to support cable modems. WiMAX could provide a reliable, low cost mechanism to deliver broadband Internet service in those markets. That fixed location service might also play a role in extending broadband Internet access to rural areas in the US where ADSL and cable modem coverage is still sparse.

12.15 WiMAX Technology

Like Wi-Fi, the WiMAX standards address Layers 1 and 2 of the OSI reference model, however, the potential implementations of WiMAX are much different and more diverse. Where Wi-Fi is designed to operate in the unlicensed 2.4 and 5 GHz bands, WiMAX systems can operate in licensed or unlicensed spectrum anywhere between 2 and 66 GHz. Only the lower portion of that band (i.e., 2 to 6 GHz) will support non-line-of-sight (NLOS) operation, so that is the primary focus for WiMAX. Within that 2 to 6 GHz range, however, there are a number of potential radio bands:

- **US Licensed 2.3 GHz Wireless Communication Service (WCS):** There is 30 MHz of frequency between 2.305 and 2.360 GHz designated Wireless Communications Service that could be used for WiMAX. Unfortunately that spectrum is divided into two 15 MHz bands (2.305 to 2.320 GHz and 2.345 to 2.360 GHz), making it more difficult to utilize for a WiMAX type service.

- **US Licensed 2.5 GHz Broadband Radio Service (BRS):** In the US, the FCC has allocated 195 MHz of licensed radio spectrum between 2.495 and 2.690 GHz for Broadband Radio Service (Note: This spectrum was formerly called Multipoint Distribution Service [MDS]/Multichannel Multipoint Distribution Service [MMDS]). Sprint/Nextel and Clearwire are the two largest owners of BRS spectrum in the US.

- **International Licensed 3.5 GHz Band:** A swath of licensed spectrum roughly equal to BRS has been allocated in the 3.4 to 3.7 GHz range throughout most of the rest of the world, and that has the potential to support WiMAX systems outside of the US.

- **US Licensed 700 MHz Band:** With the migration to digital TV in 2009, the FCC is auctioning 84 MHz of spectrum in the 700 MHz band, the frequencies that had formerly carried UHF TV channels 52 to 69. The particular attraction of these lower frequencies is that they will travel about four times as far as a 2.5 GHz signal with the same amount of loss, and have better penetration capabilities. WiMAX is one of several services being considered for that band, though a WiMAX 700 MHz radio link standard would have to be developed.

12.15.1 Unlicensed Bands

There are also two unlicensed bands that might be used to support WiMAX services.

- **Unlicensed 5 GHz U-NII Band:** In the US, the 5 GHz Unlicensed-National Information Infrastructure (U-NII) band also has potential for WiMAX deployments. This is the same band used for 802.11a wireless LANs. The lower 100 MHz of that band (5.15 to 5.25 GHz) is defined for indoor operation only, but the other 455 MHz could be used for WiMAX. The most attractive portion would be the upper 100 MHz (5.725 to 5.825 GHz), which has the highest allowable transmission power. Most carriers shun the idea of operating a service in an unlicensed band, but if they are operating in a rural area, the chance of interference is probably quite low.

- **Unlicensed 3.6 GHz Band:** In the US, the FCC has recently opened an additional 50 MHz of unlicensed spectrum in the 3.65–3.70 GHz band for fixed-location wireless services. There are no operating rules for that band as yet, so it is still effectively unavailable.

12.15.2 WiMAX Radio Link Technologies

The 802.16 standards define three options for the radio link protocol, two of which use Orthogonal Frequency Division Multiplexing (OFDM). Only the OFDM versions are being developed.

Fixed-Location WiMAX:

- **OFDM:** Up to 256 OFDM sub-channels based on channel bandwidth.

- **SC-A:** Single carrier channel (not used).

Mobile WiMAX

- **Scalable OFDM Access (SOFDMA):** Up to 2,048 OFDM sub-channels based on channel bandwidth.

12.15.3 Development Plans

The 256 OFDM sub-channel option will be used for fixed-location WiMAX. The actual number of sub-channels is based on the channel bandwidth, which is adjustable from 1.25 to 20 MHz. Based on their members' product development plans, the WiMAX Forum is not going forward with certification plans for the SC-A option. SOFDMA is the radio technique that was developed for mobile WiMAX. The number of OFDM sub-channels varies from 128 to 2,048 depending on the bandwidth of the channel (see Table 12-4). The initial Mobile WiMAX deployments will use the 1,024 sub-channel format in a 10 MHz channel. The fixed and mobile WiMAX radio links are compared in Table 12-5.

Table 12-4: SOFDMA Scalability Parameters

	System Bandwidth (MHz)				
	1.25 MHz	**2.5 MHz**	**5 MHz**	**10 MHz**	**20 MHz**
Sampling Frequency (Fs, MHz)	1.429	2.85	5.714	11.429	22.857
Sample Time (1/Fs, nsec)	700	350	175	88	44
FFT Size (Sub-channels)	128	256	512	1,024	2,048
Subcarrier Frequency Spacing	11.1607 KHz				

Table 12-5: Summary of 802.16 Radio Links

	802.16-2004	**802.16-2005**
Spectrum	2–11 GHz	<6 GHz
Configuration	Fixed-Location/Non-Line-of-Sight	Mobile/Non-Line-of-Sight
Bit Rate	Up to 70 Mbps (20 MHz Channel)	Up to 25 Mbps (10 MHz Channel)
Modulation	256 Subcarrier OFDM	1,048 Subcarrier OFDM
Mobility	Fixed/Nomadic	Fixed/Nomadic/Mobile ≤75 mph
Channel Bandwidth	Selectable 1–25 to 20 MHz (20 MHz Initially)	Selectable 1–25 to 20 MHz (10 MHz Initially)
Typical Cell Radius	3–5 miles	1–3 miles

12.16 Other WiMAX Radio Link Features

12.16.1 Adaptive Modulation

The WiMAX radio link incorporates adaptive burst profiles, which adjust the transmit power, signal modulation, and Forward Error Correction (FEC) coding to accommodate a wide variety of radio conditions. In OFDMA, that can be done on a carrier-by-carrier basis.

12.16.2 Adjustable Channel Bandwidth

With WiMAX, the channel bandwidths are adjustable from 1.25 to 20 MHz. A 20 MHz channel is the same size as that allocated for 802.11a and g wireless LANs. The other sizes are important for carriers operating in licensed spectrum, as the license will specify the range of frequencies available.

12.16.3 FDD or TDD Configurations

To provide further flexibility, WiMAX defines both frequency division (separate inbound/outbound channels) and time division duplex configurations. There is also a mesh configuration defined, but no vendor seems to be building a product that makes use of it. The initial plans for Mobile WiMAX will use TDD in a 10 MHz channel.

12.16.4 Advanced Antenna Techniques

The radio link incorporates features to take advantage of advanced antenna systems to improve overall range and performance. A number of vendors are building WiMAX equipment that incorporates MIMO and beam-forming antenna systems to improve overall radio link performance.

12.17 MAC Protocol/Quality of Service (QoS)

Along with the radio link options, the WiMAX standards also describe a sophisticated media access control (MAC) protocol to share the radio channel among hundreds of users while providing a range of QoS capabilities for each user connection. Unlike the contention-based MAC protocol used in wireless LANs, WiMAX uses a request/grant access mechanism similar to DOCSIS cable modem systems. That mechanism eliminates inbound collisions and supports both consistent-delay voice services and variable-delay data services.

12.17.1 Integrated Encryption Capability

To ensure privacy, the WiMAX standard specifies that all transmissions can be encrypted. The standard allows for multiple encryption options, including 3DES and the Advanced Encryption Standard (AES). The initial deployments will likely use AES-based encryption.

12.17.2 Inbound Request/Grant Protocol

As with a cable modem system, outbound WiMAX transmissions are broadcast in addressed frames and each station picks off those frames addressed to it. Access to the inbound channel is controlled by the base station. Users wishing to transmit inbound must send requests on a contention-based access channel. Inbound traffic channel capacity is allocated based on transmission grants issued by the base station, thereby eliminating inbound collisions.

12.17.3 Service Profiles

Subscriber stations are identified by a 48-bit MAC address, and each data flow is identified by a 16-bit Connection ID. The request/grant access mechanism supports four primary types of service that could be provided.

- **Unsolicited Grant Service (UGS):** Dedicated, consistent delay (i.e., isochronous) service for real-time voice and video, where a station is allocated a dedicated portion of the inbound transmission capacity.

- **Real-Time Polling Service (rtPS):** Real-time service that operates like the 802.11e Hybrid Controlled Channel Access (HCCA), where the base station polls each real-time user on a regular schedule to provide consistent delay service.

- **Non-Real-Time Polling Service (nrtPS):** Variable-delay data service with capacity guarantees akin to frame relay's Committed Information Rate for high-priority commercial users.

- **Best Effort:** An IP-like best effort data service for residential Internet users.

12.17.4 Prospects for WiMAX?

The one thing you can say about WiMAX is that it looks great on paper. It operates in licensed and unlicensed spectrum, supports a range of channel sizes, can provide FDD or TDD operation, features integrated encryption, and has a full set of QoS options. The big question is whether the proponents will be able to translate that technology into a marketable service. While the introduction of WiMAX has been preceded by a marketing campaign that has run for half a decade, we are just now getting the first true WiMAX-based service in the US.

For the time being, we do not know what WiMAX service will cost, what data rates it will support, how well it will work indoors, or which of the QoS options defined in the standard will be supported. In short, WiMAX is still a concept rather than a service. Sprint/Nextel and Clearwire have committed to deliver the Xohm mobile WiMAX service to about 100 million potential customers in the US (i.e., about 1/3 of the US population) by year-end 2008.

WiMAX will be going up against well-established competition in both ADSL/cable modem and 3G areas. However, adding competition should spell more options and lower prices for the buyer.

12.18 Conclusion

An integrated WLAN/cellular service appears to be the next logical step in WLAN telephony. However, beyond the hype factor, this is not a simple matter from either a technical or a business standpoint. To begin with, the WLAN infrastructure must be capable of supporting enterprise-grade voice. That would mean a pervasive infrastructure with sufficient capacity and support for QoS, load balancing, security, and all of the features described in Chapter 11. The next requirement is integrated Wi-Fi/cellular handsets, and on the Wi-Fi side those would have to support the full range of features needed for enterprise service. Finally, a truly integrated solution would require a full buy-in from the cellular carrier, which would entail a signaling link that would make the user's private network appear to be an integrated component in the cellular carrier's infrastructure.

Even without the cellular carriers' cooperation, we can simulate integrated network features through the PBX. While those solutions can provide mobility, one number accessibility, and integrated voice mail, they can also result in far higher cellular charges. That is particularly true if we leave the selection of the network service to up to the user.

While the cellular carriers have been reluctant to make a commitment to Wi-Fi related solutions, that might be about to change. Even though it is only a consumer service, T-Mobile's HotSpot @Home offering was a milestone. AT&T and Verizon have made overtures that they are considering integrated solutions, and Sprint/Nextel has piloted an enterprise-oriented fixed-mobile convergence offering called Wireless Integration. So, while this book focuses primarily on the challenges involved in Wi-Fi voice, that might be just the first step in a mobile environment that incorporates Wi-Fi, cellular, WiMAX, and other wireless technologies in the future.

CHAPTER 13

Designing a Wireless LAN for Voice

Now that we have looked at the radio interfaces, protocols, and security issues involved in a wireless LAN, it is time to see what it takes to bring all of these elements together to actually build one. While setting up a small-scale WLAN might simply require the installation of a single access point (AP) and configuring a few NICs, a large-scale network will involve significant planning.

The design of a wireless LAN must address three primary issues: capacity, coverage, and ongoing operations. In this chapter, we will look at the first two, which are the major concerns in the design and implementation phase. In the final chapter, we will discuss the issues involved in maintaining and operating a voice-capable WLAN.

The initial concern in implementing a wireless LAN was providing coverage or ensuring that a user can get wireless access from any location within the facility. The problem is that the wireless LAN connection must also support the performance requirements of the application. A low transmission rate can be as bad as no service at all, and it is far more frustrating. To ensure capacity, we need to look at the number of users who will be active in each cell, determine their total requirements, and then arrange the cells to meet those requirements.

In this chapter, we will begin by reviewing some of the important characteristics of wireless LANs. Then we will look at the design variables we can control to ensure capacity and coverage. We will also list the major factors in defining requirements, focusing primarily on network capacity and how we translate that into the number of access points required. Then we will review the coverage issues and describe the steps involved in conducting an RF site survey. Finally, we will introduce some of the automated design tools that are available.

13.1 Designing Wireless LANs

The first thing to recognize about WLAN voice is that everything works better with a good radio signal. When the radio coverage is good (i.e., a high signal-to-noise ratio in all areas), devices operate at higher bit rates, which results in shorter transmission times, fewer retransmissions, and higher network throughput. In short, your basic goal should to be to provide the best signal quality throughout the coverage area.

There are a few fundamental facts to keep in mind when laying out a wireless LAN.

- **WLAN Voice Requires Dense, Pervasive Coverage:** The first thing to remember about WLAN voice is that capacity will be as important as coverage. Early data-oriented WLANs focused on providing some service throughout the coverage area. To support voice, we will require dense coverage throughout the service area. *Dense* means that all users have a high quality signal allowing them to transmit at the highest data rates and with the lowest incidence of retransmissions. For data users, the impact of adverse radio conditions will be slower response. For voice users, poor coverage means degraded voice quality, dropped calls, and no-service conditions.

- **Low Speed, Half Duplex, Shared Media Transmission Systems:** While wireless adds an important ingredient in terms of flexibility, WLANs have nowhere near the capacity of wired LANs. In a WLAN, all users in a given area share one relatively low-capacity half duplex channel on a contention basis. To compound that difficulty, radio signals decrease in power as a function of distance and physical obstacles. Metal, water, and reinforced concrete that impede radio transmissions are in abundant supply in office environments. The water content of human beings can also attenuate the signal. Further, radio energy reflected by file cabinets and other hard, flat metallic surfaces can create significant multipath. All of these factors affect reliability, and an unreliable network frustrates users and leads to complaints.

- **Fast and Slow Transmitters Share the Same Channel:** Unlike a wired shared media LAN where all users transmit at the same bit rate, in wireless LANs, users in areas with poorer radio coverage will get far lower transmission rates. In that shared channel environment, a user transmitting at 1 Mbps will impact the performance of other users as it will take longer for them to send a frame. Further, those marginal transmissions are more likely to fail resulting in retransmissions and declining network throughput.

- **Predicting the Location of Mobile Voice Users:** By definition, mobile users will be *mobile*. Just like the design of a cellular network, we have to predict where voice users will congregate and ensure we have adequate capacity to meet requirements in those areas. Cellular carriers typically initiate service in an area with fairly thin coverage, monitor usage patterns, and then increase capacity in areas where they see the highest volume of calling. Of course, cellular carriers can make some well-founded predictions about calling such as that there will be more calling in airports and downtown areas than in less densely populated suburban areas. Enterprise customers will have a difficult time predicting where users will attempt to access a service that has not been provided for them as yet. As a result, administrators must assume that they are not going to get it right on the first pass and will require tools to monitor usage, map growth, and a budget to increase network capacity in the areas that require it.

- **Getting it to Work Is Job One, Keeping it Working Is Also Job One:** The key performance metric in networking is: "What have you done for me lately?" User populations, application requirements, and traffic patterns change, and a WLAN solution must encompass plans for maintenance, ongoing performance monitoring, traffic measurement, and network upgrades.

13.1.1 Basic Network Design Questions

To begin a wireless LAN project, we must first identify and quantify a number of elements regarding the network service we are planning to provide. The first questions on that list will be as follows.

- How many users will we be providing service for, and can we categorize them with regard to tasks and expected calling volumes?

- Do we have users who will require both voice and data capabilities on the same device?

- In what areas will the WLAN voice service be available, and are there specific areas where we can anticipate heavy usage?

- What range of user voice devices will be supported (e.g., WLAN handsets, WLAN/cellular handsets, PDA, or laptop soft phone clients)?

- Will there be particular times of the day or particular conditions that will create the greatest requirement for network capacity?

- What level of network performance should be provided with regard to voice quality, dropped calls, and no-service conditions?

- Are there particular classes of calls or callers who should be given precedence over others?

- What is the potential business impact of degraded service, dropped calls, momentary outages, or a complete network failure?

- If such an event were to occur, what would be the required time frame to restore full service?

From that basic list of questions, many more will be generated. However, the starting point for any network design must be a fundamental definition of the service that will be provided, and an estimate of the volume of that service that will be required. Once we know the users, and the applications they will access, we can determine requirements for security, reliability, network management, and other features that we will have to support.

13.1.2 Cellular Layout of a Wireless LAN

A large-scale wireless LAN is actually a number of individual wireless LANs whose access points are interconnected through the wired LAN. Each wireless LAN will require a radio channel on which to operate. A small office installation may require only one channel and the only concern is that it will not interfere with wireless LANs in adjacent offices. A large-scale commercial installation will require a number of channels and they will be laid out over the coverage area in a pattern similar to a cellular telephone network. The channel will be selected in the access point, and all of the users in a particular area will share the channel that covers that area.

13.1.3 Association Process

When a wireless LAN device is powered up, it automatically searches all radio channels for access points in its immediate area. It selects the access point with the best signal-to-noise ratio, and attempts to associate. The weaker the signal from the access point, the lower the transmission rate that is supported. While an 802.11b device in close proximity to an access point will operate at 11 Mbps, devices farther away with poor path characteristics might operate at 1 or 2 Mbps. Many access points will allow administrators to specify the minimum data rates at which to allow users to associate.

13.1.3 Non-Interfering Channels: 3 in the 2.4 GHz Band and 23 in the 5 GHz Band

One of the critical variables in a WLAN is the number of channels that are available. The number of channels is based on the bandwidth of each channel and the range of radio spectrum allocated. Given the spectrum that is allocated in the 2.4 GHz band, there are only three non-interfering channels for 802.11b/g networks. The 5 GHz band allows for as many as 23 802.11a channels, though not all products support the full range of channels and some of those channels are restricted to either indoor or outdoor use.

13.1.4 Cellular Channel Layout: Co-Channel Interference

Besides providing coverage in all required areas, the other critical factor in the design is co-channel interference. Given the limited number of channels, particularly in the 2.4 GHz band, the same channels will have to be reused by access points in different parts of the facility. Those other transmitters on the same channel in different parts of the facility create co-channel interference; that interference will degrade the signal-to-noise ratio and cause client devices to operate at lower transmission rates. The basic rule in assigning channels is not to reuse channels in adjacent cells. This is far easier to accomplish with 802.11a's 23 channels than with 8021.11b/g's 3 channels. Network engineers can use directional antennas to shape the coverage area or to improve coverage in hard to reach areas. However, radio propagation is highly unpredictable and there may be additional interference from access points in other buildings during the winter months when there are no leaves on the trees to attenuate the signal.

13.1.5 Designing for Capacity and Coverage

When wireless LANs were first introduced, the focus of the design was to ensure that the signal reached all required areas. To provide the largest coverage area at the lowest cost, administrators would use a small number of access points set to the maximum transmit power. A single access point with a range of 107 m could theoretically cover an area of about 113,000 m^2. There could be hundreds of users in that space vying for access to that one shared channel. To increase network capacity, administrators must add more access points and simultaneously reduce the transmit power to limit co-channel interference. Of course, adding more access points also increases the cost of the network.

13.1.6 Ensuring Capacity as Well as Coverage

Providing bad service is almost as bad as providing no service at all. Proper design involves ensuring that there is signal coverage in all areas, but also ensuring that the network provides adequate performance for all users in a cell. This is particularly true in WLAN voice networks where lower transmission rates and more frequent retransmissions will increase delay and jitter. It should also be noted that wireless LAN design is an ongoing process. All of these issues must be revisited as usage grows and more access points/cells are needed to support the traffic.

13.1.7 Design Variables

There are five basic factors in the design that we can control to ensure there is adequate capacity as well as adequate radio coverage.

1. **Radio Link Interface:** The selection of the radio link, 802.11a, b, or g, defines the maximum transmission rate on the shared channel. An 802.11a signal will lose power at a faster rate than 802.11b/g.

2. **Cell Layout/Channel Assignment:** The number and placement of access points can allow network engineers to provide additional capacity.

3. **Power Levels:** With access points using the same channels in different parts of the facility, administrators must adjust the transmit power as new cells are added. Reducing transmit power also reduces the range, so more access points will be required to provide the same coverage.

4. **Limit Association Rates:** In order to keep low-speed users from impacting the performance of users with better signal strength, we can limit the range of rates at which users will be allowed to associate. For example, in an 802.11b installation, it is possible to only allow users to associate if their signal strength will support data rates ≥ 5.5 Mbps. Higher transmission rates require a stronger received signal, so restricting access to higher rates also decreases the coverage area of the access point. Note that restricting association rates without increasing the number of access points can create dead spots in the coverage area.

5. **Call Access Control/Load Balancing:** As all users in a cell will be sharing that one variable speed radio channel, it is important to limit the number of simultaneous calls that can be established and, preferably, to reroute excess calls to alternate cells that have available capacity.

Table 13-1: Indoor Performance Metrics for WLAN Client Adapter (2.2 dBi Omnidirectional Antenna)

Data Rate (Mbps)	802.11a (ft)	802.11b/g (ft)
54	60	90
18	150	120
11	—	160
6	210	300
1	—	410
Source: Fluke Network		

13.2 Basic Planning Steps in a VoWLAN Deployment

13.2.1 Define Current IT Infrastructure

The planning process should begin with an overall assessment of the existing wired and wireless infrastructure, and the expertise of the people responsible for the project. Any VoIP implementation requires people with expertise in both voice and data technologies, and with the introduction of Wi-Fi, radio expertise is added to that list. A successful deployment also calls for cross training so that the voice professionals recognize the technologies and challenges in the data area and vice versa. For a large organization, the VoWi-Fi implementation might be one of several major projects being undertaken during one planning cycle. If the organization is deploying its first IP PBX system, it might not be wise to attempt a VoWi-Fi project simultaneously.

13.2.2 Assess Current WLAN Infrastructure

Most organizations will have at least some type of WLAN already in place, but it will probably require a significant upgrade to support voice. A voice WLAN requires dense, pervasive coverage. Many enterprise WLANs today provide spot coverage in conference rooms and other shared areas, and the capacity is geared for low-volume casual data use.

If a company is currently using equipment from a particular vendor, it is time to review how well they have performed. Further, does that vendor's product line extend to the larger, centrally controlled systems that will be at the core of a WLAN voice network? Perhaps it is time to cut the ties to the incumbent vendor and start anew. On the other hand, if that network is serviceable for data applications, a separate voice network using a different switching system could be overlaid on top of it. That overlay approach will protect the investment in the existing infrastructure, but the network designer will have to weigh that benefit against the problems associated with operating two separate networks in the same facility. That dual network scenario can be particularly challenging if the company is supporting integrated voice/data devices, and the voice element is communicating on one network and the data device on another.

13.2.3 Define Applications

The first step in designing any network is to determine the desired outcome before setting out to do it. WLAN voice will be on the list, however, the administrator needs to define "voice" more precisely. Will the voice service be traditional business telephone service with calls that last an average of 3.5 to 4 minutes, or push-to-talk exchanges that last for a few seconds? How about conference calls that last for a few hours and could drain the battery in the WLAN handset? Are these routine calls or life-threatening emergencies as we might find in hospitals or first responder networks? Are these internal calls or conversations with customers where poor quality will reflect badly on the company? Do the two parties speak the same native language? That might sound like an odd question, but research has shown that better sound quality is required if the two parties are already struggling to understand one another.

If the network is also expected to support data applications, it is important to review that same series of questions on the data side. Are general office applications expected (e.g., email and Web access), or only a specific production application? How big an issue will it be if the network service is degraded or unavailable? In an office environment, the impact may be inconvenience, but in a production environment it may impact a company's ability to get the product out the door.

13.2.4 Define User Population

It is also critical to specify identifiable user populations and expected calling volumes for each. Particular care must be taken with user populations that will have both voice and data service. In defining users, it is also important to restrict access. If the WLAN supports visitor services, will visitors attempt to use Wi-Fi handsets or laptop softphones to make Skype calls over the network? If those calls are not allowed, the network engineer will have to incorporate a technique to identify and block them.

13.2.5 Define End User Devices

The network designer should also specify the types of end user devices that will be supported. Those devices could range from Wi-Fi handsets to dual mode Wi-Fi/cellular handsets, single or dual mode smart phones, and softphone clients running on PDAs or laptops. With data devices, it will be important to include details like operating systems and applications required.

13.2.6 Battery Life

Mobile devices work on batteries, and if the battery is dead, that user is off the network. As we noted earlier, Wi-Fi was not designed with battery conservation in mind, and it will be critical that the battery last for a full shift.

13.2.7 Specify Coverage Areas

The administrator will also need to define what areas will be provided with wireless LAN voice coverage. In an office environment, typically the network designer should look to provide coverage throughout the entire office. For specialized applications like materials handling, the designer might restrict coverage to the warehouse. If the warehouse network is designed for voice and the office network is not, the designer has to ensure that access points in the office network do not allow voice devices to associate. That can become problematic if the voice users are equipped with laptop softphones.

13.2.8 Quantify Users and Traffic Requirements

This is the area where network designers need to be the most creative. Mobile users can wander anywhere and the administrator has to make some prediction as to where they will most likely congregate and ensure there is adequate capacity in those areas. Common areas like building lobbies and cafeterias are obvious hot spots. Administrators might try to notice where people use cell phones. People love wireless, and their telephone usage may increase when WLAN voice is provided, as a user can now make a call when going between meetings rather than waiting to return to his or her desk.

The big thing to recognize is that the first guess will be incorrect some percentage of the time. A key element in the network will be monitoring systems that provide ongoing information regarding the volume of calling in each part of the network (i.e., each access point), the number of calls denied, and the overall quality of the calls that are carried out. That will be the raw material for planning growth and expansion. We will have more to say about network management requirements in the next chapter.

13.2.9 Determine the Available Budget

Everyone wants everything until they find out how much it costs. The network designer must determine the budget that has been allocated to the project, or prepare an estimated budget, before spending a lot of time designing a WLAN no one has the money to pay for. Remember to budget for LAN switch ports, LAN cabling, and power to support the access points!

13.2.10 Become Familiar with the Facility

It is a requirement to have a floor plan of the facility that includes distances, the type of construction materials used, and the location of existing LAN wiring closets. If the designer can get access to blueprints, that is ideal. You should still walk through the facility to verify the accuracy of the blueprints and to look for furnishings (e.g., metal storage cabinets) that are not reflected on blueprints but will cause signal degradation. WLAN planning tools can scan in a blueprint, specify capacity requirements in each area, and then make recommendations regarding access point placement; it is crucial to use the tool's ability to take furnishings into account as well.

13.2.11 Identify Security Requirements

It is important to identify the sensitivity of the information being stored and sent over the wireless network. Users will expect 100 percent privacy on telephone calls. That is particularly true in health care environments where there are strict privacy requirements defined by HIPAA. WLAN voice also introduces the potential for toll fraud or hackers getting unauthorized access to network facilities to place calls. The need to provide E911 location information on emergency calls is a growing requirement though none of the current regulations specifically addresses WLAN stations. The network designer must remember to define plans for visiting users, and ensure that they understand that they are not allowed to set up their own temporary WLAN inside the facility. Up until now, most security breaches have been the result of rogue access points. Designers must ensure the security plan covers all potential threats including professional criminals and industrial espionage, and not just casual eavesdropping.

13.2.12 Schedules

Designing a wireless LAN does involve a certain amount of trial and error. While everyone will want the wireless LAN installed yesterday, a good job takes time. If this is the first attempt at a wireless infrastructure, the network designer must be realistic about promising installation dates. It is possible to install a wireless LAN quickly, but that does not mean it will immediately meet the users' requirements for coverage and capacity.

13.2.13 Initiate the Technical Design and Product Selection

In the formal product selection process, the network engineer should focus on three major areas.

1. Infrastructure elements.

2. User devices.

3. Network management systems.

Many designs only consider the first two elements, and administrators assume that they will be able to figure out the network management problems when they arise. In planning a WLAN voice solution, it is absolutely imperative that the tools and techniques for administering voice devices, monitoring call volumes, and addressing quality and coverage problems be integral considerations in the equipment selection.

One of the biggest decisions will be whether to use a shared network or separate network approach for supporting voice devices. In either case, the architecture must include mechanisms to support QoS, security, and handoffs. Critical to those decisions will be an estimate of the network capacity required to support voice users. Hence it is important to have an estimate of the number of voice users, the expected calling volume from each, and an idea of where those calls are going to occur in the network.

13.3 RF Design and Mapping Tools

Before we look at the process of designing a WLAN, we should take a look at the tools that are available to assist in that design. Tools for developing a WLAN design are available from WLAN switch vendors and from third-party companies like Motorola/Wireless Valley and Ekahau. These tools can provide assistance in determining how many access points are needed, where they should be located, and what their power and channel settings should be. Automated tools can save network designers time, money, and headaches, and enable them to more easily deliver an enterprise quality solution.

When it comes to mapping the radio coverage pattern of a facility, there are two main approaches that can be followed.

1. **Predictive Design Tools:** Tools that predict coverage based on RF propagation formulas.

2. **Coverage Mapping Tools:** Tools that record coverage based on actual signal measurements.

13.3.1 Predictive Design Tools

A predictive tool allows network designers to input a floor plan of the area to be covered. Along with the physical dimensions of the facility, the tool should allow for a description of the building materials used, so that the model can predict material losses. The other critical factor will be to indicate whether it will be a 2.4 GHz or a 5 GHz network as the loss characteristics vary with frequency.

Depending on the capabilities of the program, it might also request inputs regarding numbers of users, bit rate per voice channel, and other characteristics that will affect the design. The tool then produces a map that indicates the recommended number and placement of access points and the signal levels and expected data rates throughout the area. By altering the input variables, the network designer can quickly get a range of possible configurations and an estimate of how functional the radio coverage plan will be.

When evaluating site planning tools, there are two major factors to consider, ease of input and quality of output.

13.3.2 Ease of Input

The administrator should get a detailed presentation of what is involved in the process of inputting information to create the design. Users have invested heavily in planning tools only to find that they lacked the time or expertise to use them effectively. It is essential to determine what level of customer support will be provided, and investigate the cost and availability of training classes. Prospective buyers should ask for references from exising customers and check to see what their experience was with the product.

13.3.3 Quality of Output

At the end of the day, indoor radio propagation is highly unpredictable. Even with the best design tool, network designers should not expect that the output will provide a design plan that will be put directly into service. Some degree of tuning will be required. In essence, network designers are looking at how much time and effort it is possible to trim from the process and how much better the network design will be as a result. Network designers should also consider the amount of work that will be involved in translating the output of the design tool into the input required for the WLAN switch to be used. In the end, it might be just as well to hire a competent WLAN design consultant and outsource the detailed RF design.

13.3.4 Coverage Mapping Tools

Where planning tools provide a design for a network yet to be implemented, a mapping tool is a tool to measure and document a WLAN network that is already installed. A mapping tool begins with a scaled facility diagram that is input to a laptop with a wireless LAN interface. A technician then carries that laptop into the WLAN coverage area, and indicates his or her location on the facility diagram. The tool then senses the access points it can detect at that location along with signal strength and signal-to-noise ratio. The technician then moves to another location in the coverage area and clicks that point on the facility diagram to indicate his or her new position, and the tool then records the access points it can see at that new position. The tool uses those measurements to predict what the signal strength and signal-to-noise ratio will be at every point between those two positions.

Once the survey is complete, the user can produce a map of the entire facility and a visual image of the actual signal coverage at every point. The tool can also produce a representation of the coverage being provided by each access point. That type of detailed record of the coverage can be a major boon in troubleshooting. If users are complaining about coverage problems in a particular area, an administrator can refer to the coverage map and determine what the coverage looked like at implementation. Some systems will provide a comparison of the original coverage and the current coverage. If there are substantial differences, they could have been caused by changes in the physical environment, antenna failures, or configuration errors in the access points.

13.4 Infrastructure Design: RF Site Survey

The traditional approach to WLAN design has begun with a site survey, the goal of which was to determine the optimum location of access points, the channels to use on each, and the optimal power setting. That design could be done manually using a scaled diagram of the facility or with a computerized tool. We will describe the manual process and indicate the areas that could potentially be automated.

When wireless LANs were built using stand-alone access points, this type of planning was critical. The introduction of centrally controlled WLAN switching systems has greatly simplified the design process, with the result that many analysts question whether such an extensive planning regimen is really justified. WLAN switch vendors like Aruba note that thin access points are relatively cheap, so they recommend that users should simply deploy them on a dense grid (e.g., 50 to 100 foot spacing) and allow their RF management software to auto-configure the network.

Other WLAN switch vendors like Cisco and Trapeze provide site survey tools that users can employ to plan out their networks. Others like Siemens-Chantry networks accept the output of third-party WLAN design tools like those from Motorola/Wireless Valley or Ekahau, and will sometimes partner with them on projects.

While using the term "site survey" gives the impression of producing a mathematically accurate plan for building a wireless LAN, in reality, a site survey is just providing a structure for what is essentially a trial-and-error process.

The following steps give a detailed account of how to conduct an RF site survey.

13.4.1 Begin with a Pictorial Description of the Facility

The network designer must secure a set of building blueprints or some other scaled floor plan of the facility in order to begin to understand the environment in which service will be provided. In a manual design, the administrator will be determining the initial location of access points using that floor plan as a guide. If you have access to a computerized design tool, the floor plan is the basic input. Rough, free hand drawings will not suffice, as the rendering must be to scale; the computerized tools will have a mechanism to identify the scale. The more detail that can be provided with the floor plan, the more accurate the outcome. Specifying the type of building materials used in different partitions will greatly enhance a computerized system's ability to estimate signal attenuation. If conducting a manual survey, it is useful to do some preliminary testing to determine signal loss in different environments before deciding on access point placement.

13.4.2 Do a Walk Through

Even with access to blueprints, it is still important to walk through the facility to verify their accuracy and to get the lay of the land. This is also a good time to note potential barriers that may affect radio propagation. Remember, radio signals are more severely attenuated by masses of dirt, concrete, metal, and water. Be on the lookout for metal racks and partitions, as these are items that would not be reflected on blueprints. It is also helpful to get the facilities maintenance personnel involved as they will probably have drilled through most of those walls and may have practical insights that are not reflected on the blueprints.

13.4.3 Identify Heavy Usage Areas and Capacity Requirements

In defining the coverage plan, the administrator should also indicate the areas where the greatest congregation of users is anticipated. Computerized design tools typically request capacity requirements (either in bps or number of simultaneous calls) in each part of the coverage area. When doing a manual survey, the administrator should indicate the areas where denser coverage is required, and plan to use smaller radius cells there.

13.4.4 Identify Preliminary Access Point Locations

This is the starting point of the actual WLAN design. By identifying coverage areas and ranges, the designer can select preliminary locations for access points to provide coverage. Plan for some overlap, typically 15 to 20%, because load balancing capabilities will require that there be an alternative access point available to handle excess calls. One important feature to investigate in evaluating load balancing schemes is which stations get diverted. The basic assumption in load balancing is that new call requests get passed off to alternate access points. However, it could be that some of the active callers are in marginal coverage areas and those connections should be handed off to provide capacity for new calls.

The designer will quickly recognize that the determining factors for the coverage plan will be impacted by the capabilities and design philosophies of the product being used. It is also important to consider the mounting options for access points. In office locations, the access points are typically suspended from ceiling supports. In warehouses or outdoor areas, an important first step is to look for existing mounting points like ceiling supports or light stanchions. It is far cheaper to attach an access point to an existing structure than to build a new tower.

The other factor to consider is coverage patterns. Omnidirectional antennas provide a circular coverage area (except where material obstructions impede signal propagation). Do not overlook the potential to use directional antennas or even radio repeaters to expand coverage areas. With regard to the problem of covering hard to reach areas, many WLAN switches now support mesh extensions. With a mesh access point, the access point does not have to be wire-connected to the LAN, but has the ability to relay messages from more remote mesh access points. A mesh extension requires careful consideration as messages relayed through multiple access points will experience greater delay, and the design of the mesh solution will have a great effect on the amount of capacity that is available in that mesh-covered area (e.g., will the client access and backbone connections use the same radio channel?).

13.4.5 Install Access Points in a Test Area

To test a plan, the designer should install a number of access points in a representative test area and evaluate the results. In an office building, it is generally recommended that the test be done on one floor of the building. Once the access points are in place and configured, the designer will need some kind of tool to test the signal power throughout the area. The simplest tool is to use the signal power indicator in a laptop, though it will be difficult to determine which access point is actually being seen. There are also free laptop tools like Net Stumbler and Kismet that provide more detailed information.

Ideally, a designer will have access to a wireless test set. Companies like AirMagnet, Network General, and Berkeley Varitronics sell a wide variety of wireless test sets and planning/surveying tools. At this point, the most useful tool would be an RF mapping tool that allows a user to map and visually display the RF coverage (e.g., AirMagnet Survey/Planner). With this in hand, the administrator can display the entire coverage pattern, but also look at the coverage area for each access point to determine overlay and potential co-channel interference.

In a multi-story building, it is also important to survey the floors above and below the test floor to measure the floor-to-floor signal leakage; the designer should go at least two floors above and below. It is important to also measure signal power in areas surrounding the building to assess security exposure. Finally, in campus arrangements it is important to determine if there is building-to-building interference. Note the presence of deciduous trees between buildings as the leaves will result in differing levels of signal attenuation at different times of the year. Once the administrator has arrived at a design for a single floor, he or she must test it with at least one additional floor.

Based on the results of the test installation and survey, the designer will probably have to make adjustments to the location and/or power settings in the access points. This is also the point to look at the potential for using external or directional antennas to address problem areas. Even if a computerized design tool was used to select the original placement locations, those tools use estimates of signal loss and typically cannot accurately account for factors like multipath. The most predictable RF environment is free space, so open offices and auditoriums will typically be easier areas to cover.

Factories and warehouses have lots of open space, but may have machinery and shelving; generally the shelves are made of metal. In a warehouse, a designer should consider the material stored on the shelves. If they are storing textiles they will be less of a problem than if they are storing metal parts. Either the shelf itself or the items on it can reflect and attenuate radio signals. A designer might consider doing loss tests in storage areas to determine signal loss through one, two, or more rows of shelves.

Even with a site survey, there is still a considerable amount of trial-and-error in fine tuning the configuration. Computerized tools can help in providing initial guidance, but predicting indoor signal propagation is still an imprecise science. The bottom line is that a designer will not know what the actual propagation picture is going to look like until they start putting the access points in the area and testing the results.

13.4.6 Expand the Configuration

A site survey performed on one floor as a representative of all floors assumes the physical arrangement of walls and partitions are the same on every floor and that there is no significant floor-to-floor interference. Hospitals are typically an exception to those rules. Wall arrangements vary widely from floor to floor, and walls on radiology and operating room floors generally have a metallic lining which will severely attenuate radio signals.

13.4.7 Develop the Detailed Implementation Plan

Once designer is satisfied with the planned access point locations, he or she should identify them on the facility diagrams along with the recommended mounting locations that the installers will need to complete the installation. That design plan should also include the WLAN channel to be used at each access point and the required transmit level.

13.4.8 Document Findings

Finally, the designer should create a detailed map of all equipment locations and settings and a log of the signal readings, signal-to-noise ratios, and supported data rates throughout the coverage area. It is important to take particular note of readings near the outer boundary of each access point coverage area. An RF mapping tool will create the best documentation, but manual testing can also be used. In any case, the designer will need a test set and not just the power indicator on a laptop. You will want to record the signal level and signal-to-noise ratio from the access point that should be covering that area, but he or she will also want to be able to test the cell overlap and potential co-channel interference. Again, that type of RF coverage record can be a great asset in ongoing troubleshooting efforts.

13.4.9 Do We Need a Site Survey?

Some analysts have suggested that a site survey is a waste of time and energy, as all we are doing is conducting an organized process of trial and error. By whatever name, some type of preliminary layout plan is going to be required to determine the initial placement of access points. However, when all is said and done, the final placement will involve a good deal of "walk around" design.

13.4.10 Pilot Test

Only the foolhardy would attempt to roll out a full-scale, production WLAN voice network on Day One. Given the range of unknowns with regard to radio coverage, call density, and troubleshooting requirements, a pilot rollout is highly recommended. The design of that rollout plan will be critical, because the designer will need to test the basic functionality of the different voice-enabled devices in the different coverage areas, along with the ability of the network support systems to maintain them.

If there are distinct groups of users who will be operating in different, well-defined areas, multiple group-based pilot tests might be considered. It is important to remember that the pilot test generally means supporting a limited number of devices, so this is not a test of the network's ability to support the anticipated calling volume. The ability of the network to support the basic user requirements is the obvious purpose of a pilot test, but in a WLAN voice deployment, one of the key elements to focus on is the effectiveness of the monitoring and support systems, and the ability of the help desk personnel to identify and resolve problems.

13.4.11 Rollout Planning

The pilot test is the opportunity to test out the components, identify and address deficiencies, and prepare to move into a production environment. Rollout planning is also a critical element in success. Little problems can become major nightmares when the numbers increase. Users love wireless, so the network manager will be under great pressure to get the network into production quickly; once the users start seeing the pilot models, everyone is going to want one. If infrastructure or management systems are not up to the task, problem handling can overwhelm resources. If there are glitches with a particular voice-enabled device, putting more of them in the users' hands will be a recipe for disaster.

A good rollout plan is the ultimate test of a design. Certain elements of network management systems will be common to all devices, so it is important to recognize what is being tested with each phase of the rollout. The user pressure to get in on the wireless network can be intense, so the rollout plan must also address user communications and manage user expectations, or a network manager may be pushed into a very unsatisfactory service introduction.

13.5 Access Point Issues: Range/Capacity Guidelines

Another basic issue in the design will be the trade-off between range and transmission rate. The variables in that equation are: frequency, transmit power, receiver sensitivity, and antenna design. The installation is the time to adjust the transmit power and choose the type of antenna to shape the coverage area. Receiver sensitivity is fixed, but it is possible to effectively reduce the sensitivity by defining minimum association rates (i.e., higher rates require a stronger received signal). Table 13-2 provides some general guidance regarding transmission rates and ranges for 2.4 GHz and 5 GHz networks.

13.5.1 Voice Traffic Engineering

Probably the most difficult factor to design for is the amount of network capacity that will have to be provided in each part of the coverage area. This will require an estimate of the number of users who are likely to show up in the same area and the volume of telephone traffic they will generate with their WLAN handsets. As we noted earlier, an 11 Mbps WLAN will support at least six to eight simultaneous calls, and a 54 Mbps network should support at least 20.

It is important to remember that the estimate for the number of calls assumes that all of the users are operating at the maximum bit rate for the network (i.e., 11 or 54 Mbps). WLAN transmission rates drop with distance, obstructions, and interference, and when the transmission rate for a user decreases, his or her voice transmissions will be keeping the network busy for a longer period of time. So once again, good RF coverage will be an essential element in a successful voice over WLAN deployment.

Table 13-2: Indoor Transmission Rate Versus Range (Cisco Aironet 1200 Access Point)

Data Rate (Mbps)	5 GHz	2.4 GHz	
	802.11a (6 dBi Patch Antenna) (ft)	802.11g (2.2 dBi Dipole Antenna) (ft)	802.11b (2.2 dBi Dipole Antenna) (ft)
54	45	90	—
48	50	95	—
36	65	100	—
24	85	140	—
18	110	180	—
12	130	210	—
11	—	160	160
9	150	250	—
6	165	300	—
5.5	—	220	220
2	—	270	270
1	—	410	410

Note: The 802.11a range is computed using a 6 dBi patch antenna while the 802.11b and g ranges are computed with a 2.2 dBi dipole antenna.

13.5.2 Station Utilization: Six CCS Busy Hour Load

One factor playing in our favor is the fact that most business phones are idle most of the time. Traditionally, we assume that a business line will generate six CCS (i.e., ten minutes or 0.167 Erlangs) of traffic during the busy hour of the day. That means that an access point that can support six simultaneous calls can support a far larger population of users each of whom is using the network only ten minutes out of an hour. Table 13-3 lists the number of potential users that could be supported in an area assuming each user is generating six CCS of traffic with a probability of blocking of one percent and five percent.

13.5.3 Access Point Issues: Placement and Powering

The typical installation plan calls for locating the access points high on a wall or on the ceiling. That provides for a better signal propagation transmission, better security, and increased reliability. Installing the access points higher up allows the signal to pass

Table 13-3: Maximum Users Per Access Point with Six CCS per User

Simultaneous Calls Per Access Point	Maximum Capacity in Erlangs P.01 (1% Busy Hour Blocking)	Maximum Users at Six CCS (0.167 Erlangs)/ User	Maximum Capacity in Erlangs P.05 (5% Busy Hour Blocking)	Maximum Users at Six CCS (0.167 Erlangs)/User
1	0.01010	<1	0.05263	<1
2	0.15259	1	0.38132	2
3	0.45549	3	0.89940	5
4	0.86942	5	1.5246	9
5	1.3608	8	2.2185	13
6	1.9090	11	2.9603	18
7	2.5009	15	3.7378	22
8	3.1276	19	4.5430	27
9	3.7825	23	5.3702	32
10	4.4612	27	6.2157	37
11	5.1599	31	7.0764	42
12	5.8760	35	7.9501	48
13	6.6072	40	8.8349	53
14	7.3517	44	9.7295	58
15	8.1080	49	10.633	64
16	8.8750	53	11.544	69
17	9.6516	58	12.461	75
18	10.437	63	13.385	80
19	11.230	67	14.315	86
20	12.031	72	15.249	91
21	12.838	77	16.189	97
22	13.651	82	17.132	103
23	14.470	87	18.080	108
24	15.295	92	19.031	114
25	16.125	97	19.985	120

above the cubicle walls and other potential path impairments. Security is increased because curious hands cannot reach them without a ladder. Finally, getting access points up in the air gets them out of the way of physical hazards. However, as there are no AC power receptacles in the ceiling, this usually means that we will have to provide power over the LAN interface using IEEE 802.3af.

13.5.4 IEEE 802.3af: Power over Ethernet (PoE)

IEEE 802.3af is a relatively new standard (approved June 12, 2003) that describes how to power computer equipment over the Ethernet interface rather than using a separate AC power source. The two initial applications are powering IP telephone sets and WLAN access points.

13.5.5 Basic Specifications

The power specification for 802.3af defines a continuous 350 mA power source that can range from 44 to 57 VDC. That would equate to an average power output of 15.4 W.

13.5.6 Configurations

The specification defines two potential configurations.

1. **End-Span:** In end-span power, the power source is provided by the LAN switch and is carried on the same conductors that carry the data signal (i.e., 1/2 and 3/6). The transmission signal is alternating current (AC) while the power feed is direct current (DC), which allows the two to be carried on the same conductors.

2. **Mid-Span:** In the mid-span option, a power hub is connected between the LAN switch and the station. In that case, the power is always provided on the two spare pairs (4/5 and 7/8)

When using Power over Ethernet (PoE), it is imperative to confirm that an uninterrupted power source is provided to the wiring closet where the LAN switch or the power hub will be installed.

13.5.7 IEEE 802.3ax PoE

There is a developing standard for PoE that will better than double the wattage that can be carried over the LAN cabling. Designated IEEE 802.3ax, the goal of this new standard is the ability to deliver about 60 W, using all four pairs in the cable. Technical problems may keep this standard from ever being ratified.

13.5.8 Heat Dissipation With PoE

One nasty surprise that came with the move to PoE was the problem of heat dissipation in wiring closets. A power supply is an AC-to-DC converter. As commercial power is AC and a PoE feed is DC, the PoE component in the LAN switch includes a power supply.

The problem is that computer power supplies are not very efficient. Typically, the efficiency of a computer power supply is in the order of 60 to 70 percent. That means that 30 to 40 percent of the power is being converted into heat. When using local AC power supplies with access points, those heat sources are scattered all over the office. However, with PoE all of that heat is concentrated in the wiring closet! If that heat is allowed to accumulate in the closet, it can severely damage the equipment installed there.

13.5.9 Air Conditioning Requirements

PoE is used primarily for IP/Ethernet handsets and WLAN access points, and there will typically be more handsets than access points. Part of the facilities planning for any system using PoE will be power and air conditioning plans for the wiring closets. The first step in that will be to determine how much heat will have to be dealt with. As heat generation is a factor of the power draw and the efficiency of the power supply, the basic thing to look at is the total wattage of the equipment in the closet. Some general guidelines to use are:

- **Up to 1,500 Watts:** Door grills can be used to vent the hot air into the building and the building's air conditioning will take care of the cooling requirements.

- **1,500 to 5,000 Watts:** Air conditioning will typically be required in the closets and a mechanism to remove the heat from the building.

- **Over 5,000 Watts:** A high-density solution like a rack-mounted cooling system will likely be required.

13.6 Conclusion

Wireless LAN design is a somewhat organized but largely seat-of-the-pants affair. Like any network project, the first point is to begin with a clear understanding of what to build including the number of users, applications to be accessed, capacity requirements, types of devices, and areas covered. Again, given the vagaries of radio propagation, providing coverage and adequate network capacity in all areas will involve some amount of trial and error. Filling in the dead spots means adding access points, external antennas, or other elements, and all of those increase the cost.

In the early days of WLANs, the design process was purely a trial and error process. As the industry has matured, we are seeing an expanded array of testing, monitoring, and mapping aids that can be added. Troubleshooting wireless networks can still be a frustrating experience, because there is no way to see the radio signal, and a troubleshooter can only surmise why there is a stronger or weaker signal at any point. However, while the vagaries of radio propagation will always confound us, we are getting the tools to give us a leg up on the task.

CHAPTER 14

Network Management in Wireless LANs

Once the wireless LAN is implemented and carrying traffic, the real job has just begun. When a network manager turns up a network, that initiates an ongoing process of maintaining, operating, troubleshooting, and growing the network infrastructure. All too often, networks are viewed as static entities whose configuration is pictured in concise box and line diagrams. The problem is that this puts the focus on the components, not the job. The job of the network is to provide service to users. The overall task of network management is the design and operation of all of the critical support systems needed to provide reliable and cost-effective service on an ongoing basis.

As with many areas in wireless LANs, network management capabilities are still developing, so network administrators will likely have to cobble together a management plan from a number of disparate parts. Some of those capabilities may be found in the control systems for a WLAN switch, while others may be added to existing management systems. It is likely that some additional tools will have to be purchased as well, so those should be considered in the overall project budget.

In this chapter we will begin with a definition of network management, and then look at the different functions that fall in that area including:

- Record keeping and administration.

- Performance monitoring.

- Capacity planning/network expansion.

- Troubleshooting and problem isolation.

- Security monitoring.

The most important issue to recognize about network management is that it is not a separate entity that can be added on to an existing network; network management systems are an integral part of the network. So in preparing a network design, the network engineer should not look simply at the physical elements involved in getting a bit from Point A to Point B, he or she must also focus on building an infrastructure for providing services.

14.1 Network Management

Network management includes all of the systems required to provide reliable and cost-effective service on an ongoing basis.

Network management is not one responsibility, but a whole series of tasks that are required to ensure the ongoing operation of a network. Network management tasks are typically divided into a number of areas, though the use of wireless technology does introduce new responsibilities.

14.1.1 FCAPS

The ISO Telecommunications Network Management Model and framework uses a somewhat different organization for network management tasks, which they refer to as FCAPS. FCAPS stands for:

- Fault management.

- Configuration management.

- Accounting.

- Performance management.

- Security.

In non-billing organizations, "Accounting" is sometimes replaced with "Administration". While "FCAPS" is a nice acronym, I still prefer my organization as it relates better to the way these functions are organized in most enterprise organizations.

14.1.2 Proactive and Reactive Management

There are two approaches to managing a network: Proactive and Reactive. Most organizations use a combination of the two. Proactive management is where the network manager defines performance objectives for the network, monitors whether those parameters are being met, and takes action without a specific request from a user.

Reactive management is where corrective action is initiated when a service complaint is registered at the help desk. Given the inherent difficulties involved in radio networks, proactive management is imperative in a wireless LAN.

14.1.3 Local and Remote

Most network management systems look to support users, devices, and networks in large central facilities. While those sites will certainly receive attention given the number of users impacted by an outage, it is important that we do not overlook smaller remote sites. Those smaller sites are also engaged in important, customer-facing functions, and generally do not have much (if anything) in the way of on-site technical support. Further, we generally cannot afford to dispatch a technician to those sites for each trouble call. If we are planning to deploy service to those locations, the support requirements should be addressed up front. It may be that the only cost-effective support plan involves contracting with a local service provider who will work in conjunction with central site network maintenance personnel.

14.2 Record Keeping and Administration

The first step in managing a network is knowing what you have, where it is, and how it is configured. Networks involve lots of details. That information will have to be updated with each new client device or infrastructure element that is added to the network. Most organizations will have some type of central record keeping system for all IT related equipment and network components. A well run organization should have an institutionalized process where the record keeping requirements for any new network or application will be identified and linked to that central repository.

A WLAN record keeping system should address three major areas:

1. Users.
2. Client Devices.
3. Infrastructure.

14.2.1 User Records

The administrator should have a record of every user who is authorized to access the wireless network. Typically, every company-provided laptop has Wi-Fi capability today, so that file should be a starting point; if softphone clients are supported on the network, those users will be a subset of laptop users. WLAN voice handsets will be issued by the IT department, and as they will have to be registered in the telephony server, they should be relatively easy to track. Hopefully the system for tracking wired voice handsets can be augmented to capture these as well. The tracking can become more challenging if client devices are shared by users at different times or on different shifts.

The adoption of SIP-based VoWLAN phones should not alter that arrangement to any great degree. As any handset used on the internal PBX system will have to be registered on the telephony server, the introduction of user-purchased SIP handsets should be essentially impossible (assuming an adequate authentication mechanism). If users elect to use Skype or other Internet-based phone services (and their use is endorsed by the organization), that traffic will be treated as "data" rather than "voice". However, the PC client may mark those packets for voice QoS, which could wreak havoc on capacity planning. Many WLAN switches now have the ability to police QoS settings, and the administrator should ensure that those systems can recognize those unauthorized voice users and remark their priority to "data".

The other challenging area will be visitor services. Today, visitor services are restricted to basic Internet data access, so the VoWLAN management systems will not be directly impacted. The use of softphones or SIP-based VoWLAN handsets could change the VoWLAN environment into the same freewheeling state that exists for visitor data services. One real concern is that a visitor might access illegal content such as child pornography sites while using visitor services; the voice equivalent would be to make obscene or threatening calls while using company facilities. Given that, it is imperative that the administrator has some mechanism to restrict access to legitimate visitors, that he or she limits their network access to the duration of their stay, and that he or she is able to identify which visitors are accessing what services in case problems should arise.

14.2.2 Client Device Information

There are a number of technical information elements an administrator must be able to track with regard to each client device. The administrator will need to know the type of device (e.g., handset, PDA, mobile computer, laptop), the MAC address of its WLAN NIC, and its telephone number, as well as the manufacturer, model, serial number, software release, purchase date, and maintenance history. The administrator also needs to know whom the device was issued to so that it can be deactivated if that employee leaves the company. If there are multiple users, the administrator must be able to deactivate an individual user rather than the device.

On the macro scale, the administrator should track the range of devices that are authorized for use on the network. There must also be a procedure to test new client devices that could eventually be added to the network. At the moment, there are relatively few Wi-Fi voice devices and the IT department will be purchasing all of them. However, with the growing popularity of user-purchased devices like Apple's iPhone that include both Wi-Fi and cellular capability, companies may be pressured into supporting those devices as well. Before being placed in that position, companies should have a policy regarding support for user-owned devices, and if they are supported, a plan to test and certify them and a published list of supported devices.

With the small range of devices available today, client device management should be fairly straightforward, but IT departments should be prepared to support a full range of client devices much as the cellular carriers do.

14.2.3 WLAN Infrastructure

Network documentation will be critical in managing the WLAN infrastructure. Administrator will need a complete inventory of all access points, controllers, and facility diagrams identifying their exact locations. The records for each access point should also include manufacturer, model, serial number, MAC address (i.e., BSSID), WLAN channel used, transmit power setting, purchase date, and maintenance history. The physical connection to the wired LAN, pair assignment, and special configuration details like external antennas should also be recorded. Needless to say, those records will have to be updated each time an access point is installed or removed.

Probably, the most critical element in the infrastructure records will be the RF coverage map. When the network is deployed, it is essential to get an accurate survey of the RF coverage in the facility. When doing a manual survey, the technician will probably have to limit the number of sample points and the information recorded when using an RF mapping tool that captures receive signal strength and signal/noise ratio from every access point that can be seen from each sample area. That tool provides a very detailed look at the RF coverage and the area being covered by each access point in the network.

That RF survey will be a major help in troubleshooting RF coverage problems down the line. If users begin experiencing problems in a particular area, the network technicians can determine how good the coverage in that area should be and what access points should be visible. They can also go to the area with a test set and determine if the RF coverage is significantly different from what was seen on the initial survey. Better tools can make that comparison automatically. Using such tools can provide essential clues regarding the cause of the problem such as changes in the access point settings, new material obstructions, or failed antennas.

14.3 Performance Monitoring

Performance problems in wireless LANs usually manifest themselves in three ways: poor call quality, dropped calls, and no service available. Poor call quality is normally a result of poor RF coverage, though excess traffic can also cause those problems if the call access control has not been managed effectively (i.e., by allowing too many calls to be placed through an access point). As we noted in Chapter 8, the standard measure of call quality is the Mean Opinion Score (MOS), a voice quality scaling system ranging from 1 to 5; the goal for business telephone service is a MOS of 4.0 or above. Where the original MOS score was generated by reading challenging phrases and having a panel of testers score them, the tradeoffs are now understood well enough that monitoring tools can provide a simulated MOS based on the voice encoding and packet loss encountered. The monitoring system can deduce the simulated MOS either from the WLAN traffic or by copying the Real-Time Control Protocol (RTCP) messages that will be embedded in the voice traffic stream.

14.3.1 Call Specific Information

The important thing to recognize about voice quality is that the more general the information, the less useful it becomes. The average MOS for all of the calls in the network could be well above 4.0, but there could still be hundreds or thousands of problem calls. An administrator must be able to gauge the user satisfaction for each individual call. Some users will report call problems, but many may not. So it is important to be able to track performance proactively rather than depending on following up on trouble calls. Resolving capacity or coverage problems may take weeks, so it is important to recognize when service levels in a particular area start to decline and begin taking action before the service becomes abysmal.

The first step in meaningful tracking is the ability to identify the quality of each individual call and be able to link that call to a particular client device and phone number. The client device will be identified by a MAC address, so to resolve that to a particular telephone number the network must either monitor the signaling messages that are used to set up that call, or else maintain a table that links the MAC address to the phone number. Among the elements to capture about that call are:

- Date and time of the call.

- Calling and called parties.

- Access point(s) used.

- Handoff history.

- Minimum/average/maximum MOS.

- Premature disconnect (i.e., no disconnect or BYE message seen).

Measuring call quality is challenging because if we merely look at the average MOS, we can get a very hazy view of the user's experience. A user might have experienced excellent call quality for 15 minutes but then had 15 seconds of random noise, resulting in the user hanging up. On the average, the MOS for that call was pretty good, but the user still had to hang up!

Call quality must be measured from the user experience, but no-service conditions are critical network problems. As we have noted, voice call activity can be volatile, so it is important to keep an ongoing look at capacity and quality by access point or area as well. Two fundamental metrics an administrator needs to track are all-channels-busy conditions and rejected calls by access point. An all-channels-busy condition means the access point has reached the call access control limit for the maximum simultaneous calls it is allowed to support; any additional call requests will be turned away (i.e., either given a busy signal or rerouted to a different access point). Another important metric is how many calls were rejected or redirected.

Regular all-channels-busy conditions are a clear indication that the some part of the network has reached its capacity. As adding new access points will take time, it is preferable to see those problems coming rather than trying to address them once they have already become critical. Many WLAN switching systems and external monitoring systems will allow an administrator to monitor ongoing call density by access point so he or she can hopefully detect and address those problem areas before they become critical. As with any ongoing monitoring system it is vital that someone is actually watching it. That means either a system must allow set thresholds that trigger alarms, or that the routine performance statistics are regularly reviewed.

Normally, access points will be monitored for rejected calls, but it is also important to monitor MOS by access point. If there are consistent quality problems in a particular area, it could be a symptom of an RF coverage problem that has not been recognized by analyzing quality on individual calls.

14.4 Capacity Planning

The outputs of the performance monitoring system will be the inputs for the capacity planning function. When there are quality or capacity problems in particular areas, that is generally an indication that a network upgrade is needed. Network upgrades generally mean additional expenditures. However, it might be the case that there are areas with unexpectedly low usage, and the fix might be a matter of rearranging existing assets rather than purchasing additional equipment.

Again, network modifications take time. The steps must be to first recognize and analyze the problem, then decide what steps need to be taken to address it, secure the purchase approval, order the necessary equipment (which might include equipment for the wired LAN as well), install the cable drops, connect the access points, and finally tune the new area. Once the upgrade has been done, it is crucial to also upgrade the facility records to reflect the changes. Those steps do not all happen over a three-day weekend!

The other major input for capacity planning will be IT business planning. Performance monitoring will alert the administrator to problems encountered in supporting the existing base of users. The other factor that will impact capacity will be new users. If the organization finds value in the VoWLAN capability, other user groups will want their people to have WLAN voice handsets. Fortunately, as a network manager operates the network he or she will begin to develop planning metrics based on existing usage patterns for different types of users.

That growth might come by way of additional users in the same coverage area, though new user groups may require that the coverage area be expanded to different parts of the building or campus. By looking at the types of functions represented by these new users and comparing them to the usage patterns of existing user groups, network manager can better estimate the additional capacity per user that will be required.

14.5 Troubleshooting/Problem Isolation

Troubleshooting is the one function most typically associated with network management. Certainly, it is the most intense function in that an administrator will be dealing with a user whose expectations have already not been met. Help desk personnel will be familiar with the wrath of unhappy users, but they will still require additional training and tools to address WLAN network problems.

The first difficulty is that radio transmission is still an iffy proposition. With 25 years worth of experience and an army of technicians at their disposal, cellular carriers still have problems maintaining service quality. Indoor signal propagation is even more precarious, and unlike a broken wire that can be seen, no one can see a bad radio signal.

The helpdesk personnel should first collect the information needed to isolate the problem. That would include the called and calling numbers. Hopefully, only one of those is a WLAN handset, or there will be the further question of which handset was actually creating the problem. It is important to know where the user was when he or she experienced the problem. If the user was in a stairwell or an elevator, the administrator may have to inform them of the inherent limits of radio transmission. Also, the administrator will need some mechanism to sectionalize the trouble; is it on the wireless LAN, the wired LAN, or the WAN?

Among the typical problems seen are:

- **No Service/Rejected Call:** This could be caused by a surge in usage, RF coverage problems, or a mis-configured client device (Has this phone ever worked?). Do not overlook obvious problems like dead batteries.

- **Poor Quality/Inconsistent Quality:** Generally RF coverage is the first culprit to come to mind when encountering quality problems, but quality problems could also be caused by a damaged client device.

- **Disconnects:** Like call quality problems, disconnects are typically a result of poor RF coverage, but it might also be that the user roamed into an area where the access points were all at capacity. Was the user moving when the disconnect occurred?

- **Wrong Numbers or Other Call Failures:** WLAN voice signaling protocols generally incorporate error detection and correction, so if the user is experiencing connection problems they are more likely related to configuration errors in the handset or the server rather than transmission problems.

Once the problem has been cleared, trouble ticket tracking can also provide significant insights. The ticket should reflect the particular area where the trouble was reported, and the access points that supported the call. If the same locations or access points keep appearing, that is probably a signal that the area should be looked at more closely.

14.6 Security Monitoring

The last network management concern is security monitoring. In Chapter 6, we looked at the recommended security options we should employ on WLAN voice systems, but security must be monitored. Most organizations have implemented security monitoring for their wired networks, and the WLAN should be managed under that same group. As with the Help Desk function, the security personnel will likely need additional training to deal with the range of wireless security threats and access to the WLAN monitoring systems.

The basic tool will be a wireless intrusion detection system (WIDS) that monitors for unauthorized WAN transmissions like rogue access points, malicious (i.e., hacker deployed) access points, and Ad Hoc networks. *Wireless* intrusion detection is actually a misnomer for these systems, as they only monitor for wireless LANs and not for other wireless security threats like those that might use cellular, WiMAX, or Bluetooth. The WIDS capability might come in the form of a stand-alone monitoring system like those from AirMagnet, AirDefense, or Airtight Networks, or it could be a function in the WLAN switch. In monitoring for unauthorized access points, there are a few things the system should do:

- Ignore, or mark as benign, those legitimate access points that are operating on nearby WLANs.

- Determine whether a questionable access point is connected to the wired network.

- Provide information to locate the unauthorized access point.

- Disable the access point, particularly if it appears to be attempting to perpetrate a man-in-the-middle attack. Generally the technique used to disable an unauthorized access point is a disassociation attack. The basic trick is to monitor clients associating with the access point and immediately send a disassociation message before they have the opportunity to send any traffic.

Besides monitoring for unauthorized access points, the monitoring systems should also be able to identify actual attacks against the wireless LAN. Among the recognized attack strategies are:

- **Radio Jamming:** Generating excessive noise in the 2.4 or 5 GHz radio bands will slow WLAN transmissions and can potentially shut down the network entirely. While there are 2.4 GHz radio jammers available on the Internet, network technicians should also check other equipment in the facility that generates energy in the 2.4 GHz band. Those frequencies are used by a variety of devices, particularly in the medical field. Also, be sure to check on leaky microwave ovens that generate pulsed signals around 2.45 GHz.

- **Association Floods:** A hacker could attempt to disable or overload access points by flooding them with association requests; this strategy is similar to a TCP *SYN Flood*. Currently, management frames and control frames are not encrypted, which leaves WLANs vulnerable to strategies that target management messages.

- **Disassociation Attack:** A hacker could launch a disassociation attack like the ones used to disable unauthorized access points. Again, unencrypted control messages leave the network vulnerable.

- **CTS Flood:** Another crafty way to disable a network is to target the NAV timers. Every WLAN station has a NAV time that can be set with an RTS or CTS message; no device will attempt to send while its NAV timer is set. Back-off timers are also frozen for the NAV duration. A hacker could simply bombard the network with CTS messages that trick the stations into thinking the network was perpetually busy.

Security researchers have also turned up a number of vulnerabilities in Wi-Fi drivers that could leave them vulnerable to hacking exploits like buffer overflows.

Along with monitoring the network, the security group must also be monitoring the literature to be aware of the new vulnerabilities and exploits as they appear. Among the better Web sites to monitor for wireless security issues are:

- SANS Institute: http://www.sans.org.

- Mitre Corp. (Common Vulnerabilities and Exposures): http://www.cve.mitre.org.

- Voice over Packet Security Forum: http://www.vopsecurity.org.

- Certstation Threat Management Advisory: http://www.certstation.com.

- Wireless Vulnerabilities and Exploits: http://www.wve.org.

- National Institute of Standards and Technology: http://www.nist.org.

14.7 Conclusion

Network management is a big area, and unfortunately there is no single product to buy that addresses all of the various concerns. Developing a network management system requires that a network manager think through the entire process of providing a service from ordering, to equipment configuration, training, monitoring, troubleshooting, and ongoing planning, and checking that there are systems in place to ensure that the service meets the users' expectations for quality, performance, and reliability.

The IEEE 802.11 Standards

WLAN Radio Link Protocols: 802.11a, b, g, and n

The transmission rates and frequency bands are summarized in Table A-1, which was also shown as Table 2-7 and Table 5-1.

802.11d

802.11d is a supplementary standard addressing regulatory considerations for radio links in other countries. The North American and ITU Digital Transmission Rates are given in Table A-2.

802.11e: Wi-Fi Multimedia/Wi-Fi Multimedia Scheduled Access

802.11e is an enhancement to the 802.11 MAC protocol to provide Quality of Service (QoS). There are two different options defined within 802.11e. The first is a contention-based mechanism called Enhanced Distributed Control Access (EDCA), which specifies four priority levels or Access Categories. The Wi-Fi Alliance specifies compatibility with that option as Wi-Fi Multimedia (WMM). The second option, called Hybrid Controlled Channel Access (HCCA), supports real-time services using a polled access. The Wi-Fi Alliance specifies compatibility with that option as Wi-Fi Multimedia-Scheduled Access (WMM-SA), though no vendors are known to be implementing it.

802.11h: EU 5 GHz Interference Requirements

802.11h is a standard describing management and control extensions for to the MAC and PHY layers of the 802.11a to ensure regulatory compliance in Europe. The standard defines Dynamic Channel Selection and Transmit Power Control mechanisms for 5 GHz operation in compliance with CEPT Recommendation ERC 99/23.

802.11i: Wi-Fi Protected Access/2

802.11i is an enhancement to the 802.11 MAC protocol to provide improved security. The original 802.11 standards included a simple encryption system called Wired

Table A-1: IEEE 802.11 Radio Link Interfaces

Standard	Maximum Bit Rate (Mbps)	Fallback Rates	Channel Bandwidth	Non-Interfering Channels	Transmission Band	Licensed
802.11b	11	5.5, 2, or 1 Mbps	22 MHz	3	2.4 GHz	No
802.11g	54	48, 36, 24, 18, 12, 11, 9, 6, 5.5, 2, or 1 Mbps	20–22 MHz	3	2.4 GHz	No
802.11a	54	48, 36, 24, 18, 12, 9, or 6 Mbps	20 MHz	23	5 GHz	No
802.11n (Draft)	≤289 (at 20 MHz) ≤600 (at 40 MHz)	Down to 6.5 Mbps (at 20 MHz)	20 or 40 MHz	2 in 2.4 GHz 11 in 5 GHz (40 MHz)	2.4 or 5 GHz	No

Table A-2: North American and ITU Digital Transmission Rates

	North America		ITU	
	Level	Bit Rate	Level	Bit Rate
Original Hierarchy	DS-0	64 Kbps	DS-0	64 Kbps
	DS-1	1.544 Mbps	E-1	2.048 Mbps
	DS-2	6.312 Mbps	E-2	8.448 Mbps
	DS-3	44.736 Mbps	E-3	34.368 Mbps
	DS-4	274.176 Mbps	E-4	139.264 Mbps
SONET/SDH Hierarchy	OC-1	51.84 Mbps	STM-0	51.84 Mbps
	OC-3	155.52 Mbps	STM-1	155.52 Mbps
	OC-12	622.08 Mbps	STM-4	622.08 Mbps
	OC-48	2.488 Gbps	STM-16	2.488 Gbps
	OC-192	9.953 Gbps	STM-64	9.953 Gbps
	OC-768	39.813 Gbps	STM-256	39.813 Gbps

Equivalent Privacy (WEP) that used a 40 or 104 bit RC4-based encryption. The 802.11i security standard incorporates the National Institute of Standards and Technology's Advanced Encryption Standard (AES) that provides a mechanism that is far superior to WEP. The Wi-Fi Alliance identifies compatibility with 802.11i as Wi-Fi Protected Access/2 (WPA/2) Certified.

Wi-Fi Protected Access (WPA)

While 802.11i was in development, the Wi-Fi Alliance recognized an immediate need for a security solution to address the deficiencies of WEP. To address this, they developed a solution called Wi-Fi Protected Access (WPA). WPA uses the same RC4-based encryption as WEP, but the key is changed on every packet to thwart brute-force attacks. That key changing solution is referred to as Temporal Key Integrity Protocol (TKIP). The major advantage of WPA is that it could be implemented quickly in most devices with a software download, while 802.11i typically required a hardware upgrade.

802.11k: Radio Resource Management

802.11k is a developing standard to define radio resource management that will specify how a client selects a channel. A key feature will be neighbor reporting, or the ability for a device to store information regarding nearby access points. Opportunistic key caching will allow clients to store encryption keys and speed handoffs. Ratification of this standard is expected in 2008.

802.11p: Vehicular Environment

802.11p is a developing standard for using Wi-Fi in moving vehicles. The standard is known as WAVE (Wireless Access for the Vehicular Environment) that would allow Wi-Fi equipped vehicles to communicate with roadside access points.

802.11r: Roaming

802.11r is a developing standard to support fast secure roaming between access points, particularly for voice users. The goal is to provide a secure handoff in 50 msec or less. It is proving difficult to develop this standard as security must be maintained during the handoff process. Ratification is expected sometime in 2009.

802.11s: Mesh Backhaul

802.11s is another developing standard for the backhaul portion of mesh-based wireless LANs. A mesh configuration provides the ability to extend Wi-Fi coverage throughout a campus or an entire metropolitan area. Pre-standard mesh systems are being marketed by a number of companies including Cisco, Nortel, Motorola/Mesh Networks, Tropos Networks, Firetide, and PacketHop.

802.11u

802.11u is another developing standard that will work with 802.11k's neighbor report. The overall goal is to support interoperability between WLANs and other wireless networks (e.g., cellular and WiMAX). It will also address requirements for load balancing traffic among multiple access points.

802.11v

802.11v is another developing standard to extend wireless network management to client devices. It will allow stations to perform management functions, such as monitoring, configuring, and updating, in either a centralized or distributed manner through a Layer 2 mechanism. One of the key capabilities will be client power control.

Other Related Standards

802.1x

802.1x is a framework for authentication protocols that allows device authentication in wired or wireless networks (Note: This is a ".1" standard, not a ".11" standard). The 802.1x framework defines the Extensible Authentication Protocol (EAP), which supports a number of authentication protocols including MD5, TLS, TTLS, PEAP, and Cisco's LEAP and EAP-FAST.

802.3af: Power over Ethernet

802.3af is a standard that defines how electrical power can be carried to a device over the Ethernet interface thereby eliminating the need for an AC power source at the device. It is used primarily to power IP telephones and 802.11 wireless LAN access points, and can deliver roughly 15 W of DC power over the LAN cabling.

LWAPP, CAPWAP, SLAPP

LWAPP, CAPWAP, and SLAPP are acronyms that represent three different protocols for managing access points in centrally controlled WLAN networks (i.e., a WLAN switch).

- LWAPP, the Light Weight Access Point Protocol, was developed by Cisco and may form the foundation for an eventual IEEE standard.

- CAPWAP, Control and Provisioning of Wireless Access Points, was originally an IETF response to Cisco's LWAPP that was proposed by Airespace. Ironically, Airespace was subsequently acquired by Cisco, and now CAPWAP has become a rallying point for Cisco's competitors in the WLAN switching area.

- SLAPP, Secure Light Access Point Protocol, is a more lightweight approach than LWAPP being proposed by Aruba and Trapeze, two other WLAN switch vendors.

802.15: Bluetooth

The IEEE 802.15 standards describe short-range (i.e., ≤10 m) wireless networks also known as Wireless Personal Area Networks (WPANs). This standard is generally referred to by the more popular name Bluetooth, and it is designed to address wireless networking of portable and mobile computing devices such as PCs, PDAs, cell phones, and consumer electronics. There are three different versions supporting data rates, of 1 Mbps to 3 Mbps (Bluetooth or IEEE 802.15.1), ≤55 Mbps (WiMedia or IEEE 802.15.3), and ≤250 Kbps (Zigbee or IEEE 802.15.4).

802.16: WiMAX Broadband Wireless Access

The IEEE 802.16 standards define a wireless "first-mile" connection, which could be used as an alternative to cable modem or DSL services for supporting high-speed Internet access. It overlaps with Wi-Fi Mesh networks as an alternative for providing metropolitan area wireless networks. There are two versions of WiMAX, a fixed location version specified in 802.16-2004, and a mobile version with handoff functionality specified in 802.16-2005. Like the Wi-Fi Alliance, the WiMAX Forum defines coordinated implementations of the IEEE 802.16 standards, and provides conformance testing to ensure multi-vendor interoperability.

Useful Tables

The tables that follow will be an important reference to the administrator or designer of WLAN voice. Table A-3 gives Ethernet cabling requirements. Table A-4, also given as

Table A-3: Ethernet UTP Cabling Requirements

EIA/TIA	ITU/IEC	10 Mbps (10-BaseT)	100 Mbps (100-BaseT)	1 Gbps (1000-BaseT)	10 Gbps (10G-BaseT)
Category 3	Class A	100 m, 2-pairs	N/A	N/A	N/A
Category 5	Class C	100 m, 2-pairs	100 m, 2-pairs	N/A	N/A
Category 5e	Class D	100 m, 2-pairs	100 m, 2-pairs	100 m, 4-pairs	N/A
Category 6	Class E	100 m, 2-pairs	100 m, 2-pairs	100 m, 4-pairs	55 m, 4-pairs
Category 6a	Class E_A	100 m, 2-pairs	100 m, 2-pairs	100 m, 4-pairs	100 m, 4-pairs

Table A-4: Voice Coding Options

Coding Technique	Bit Rate (Kbps)	Approximate Encoding Delay (msec)	Loss Tolerance (%)	Applications
G.711 Pulse Code Modulation (PCM)	64	0.13	7–10	Public Telephone Network, PBXs, and Most IP PBXs
G.726 Adaptive Differential PCM (ADPCM)	24, 32, or 40	0.4	5	T-1 Multiplexer Networks, and DECT Cordless Phones
G.729a	8	25	<2	WLAN and Wide Area Packet Voice Networks
G.723.1	5.3 or 6.4	67	<1	Limited Due to Coding Delay
G.722 Wideband (50 Hz to 7 KHz)	64	0.4	5	Radio Broadcasting and Conferencing Systems

Table 8-1, covers voice coding options. Table A-5, also given as Table 8-2, gives a typical representation of call quality levels. Table A-6, also given as Table 7-3, gives packet voice transmission requirements. Table A-7, also given as Table 11-2, gives theoretical maximum calls per WLAN. Table A-8 gives a conversion chart from dBm to mW.

Table A-5: Typical Representation of Call Quality Levels

User Opinion	R Factor	MOS (ITU Scale)
Scale Maximum	120	5.0
Attainable Maximum (G.711)	93	4.4
Business Quality	≥80	≥4.0
Marginal	70–80	3.6–4.0
Poor	50–70	2.6–3.6
Not Recommended	≤50	≤2.5

Table A-6: Packet Voice Transmission Requirements (Bits Per Second Per Voice Channel)

Codec	Voice Bit Rate (Kbps)	Voice Bandwidth (KHz)	Sample Time (msec)	Voice Payload (bytes)	Packets Per Second	Ethernet (Kbps)	PPP (Kbps)
G.711 (PCM)	64	3	20	160	50	87.2	82.4
G.711 (PCM)	64	3	30	240	33.3	79.4	76.2
G.711 (PCM)	64	3	40	320	25	75.6	73.2
G.722	64	7	20	160	50	87.2	82.4
G.722	64	7	30	240	33.3	79.4	76.2
G.722	64	7	40	320	25	75.6	73.2
G.726 (ADPCM)	32	3	20	80	50	55.2	50.4
G.726 (ADPCM)	32	3	30	120	33.3	47.4	44.2
G.726 (ADPCM)	32	3	40	160	25	43.6	41.2
G.729A	8	3	20	20	50	31.2	26.4
G.729A	8	3	30	30	33.3	23.4	20.2
G.729A	8	3	40	40	25	19.6	17.2

Note: Assumes 40 octets RTP/UDP/IP overhead per packet. Ethernet overhead adds 18 octets per packet. PPP overhead adds 6 octets per packet.

Table A-7: Theoretical Maximum Calls Per WLAN (20 msec Voice Sampling)

Codec	802.11b Network				802.11a or g Network			
	11 Mbps	5.5 Mbps	2 Mbps	1 Mbps	54 Mbps	36 Mbps	18 Mbps	6 Mbps
G.711	23	16	8	4	78	69	51	24
G.729A	30	24	14	8	92	86	73	45
G.723.1	44	36	21	13	138	129	110	66
Skype iLBC	28	22	13	7	89	83	69	40

Source: www.oreilly.com, Matt Gast, "How Many Voice Callers Fit on the Head of an Access Point," December 13, 2005. Available at http://www.oreillynet.com/pub/a/etel/2005/12/13/how-many-voice-callers-fit-on-the-head-of-an-access-point.html.

Table A-8: Conversion Chart (dBm to mW)

dBm	mW	dBm	mW
−1	0.79	15	31.62
0	1.0	16	39.81
1	1.26	17	50.12
2	1.58	18	63.10
3	1.99	19	79.43
4	2.51	20	100.0
5	3.16	21	125.89
6	3.98	22	158.49
7	5.01	23	199.53
8	6.31	24	251.19
9	7.94	25	316.23
10	10.0	26	398.11
11	12.59	27	501.19
12	15.85	28	630.96
13	19.95	29	794.33
14	25.12	30	1000 (1.0 W)

Source: http://www.soontai.com/cal_exunit.html.

Glossary of Acronyms

AAA: Authentication, Authorization, and Accounting

ACK: Acknowledgment

ADPCM: Adaptive Differential Pulse Code Modulation

ADSL: Asymmetrical Digital Subscriber Line

AES: Advanced Encryption Standard

AIFS: Arbitrated Inter-Frame Spacing (802.11e)

AM: Amplitude Modulation

ANSI: American National Standards Institute

AP: Access Point (802.11)

ARP: Address Resolution Protocol

ARQ: Automatic Retransmission Request

ATIM: Announcement Traffic Indication Message

ATM: Asynchronous Transfer Mode

BPSK: Binary Phase Shift Keying

BRI: Basic Rate Interface (ISDN)

BRS: Broadband Radio Service

BS/BTS: Base Station/Base Transceiver Station

BSS: Basic Service Set (802.11)

BSSID: Basic Service Set Identifier

CAVE: Cellular Authorization and Voice Encryption

CCK: Complementary Coded Keying (802.11b)

CCMP: Counter Code with CBC MAC

CCX: Cisco Compatible Extensions

CDMA: Code Division Multiple Access

CF: Contention Free

CIR: Committed Information Rate (Frame Relay)

CoS: Class of Service

CSMA/CA: Carrier Sense Multiple Access with Collision Avoidance (802.11)

CSMA/CD: Carrier Sense Multiple Access with Collision Detection (Ethernet)

CTS: Clear to Send

CW: Contention Window (802.11)

DCF: Distributed Control Function (802.11)

DES: Digital Encryption Standard

DHCP: Dynamic Host Configuration Protocol

DIFS: DCF Inter-Frame Spacing (802.11)

DOCSIS: Data Over Cable Service Interface Specification

DoS: Denial of Service

DSL: Digital Subscriber Line

DBPSK: Differential Binary Phase Shift Keying

DQPSK: Differential Quadrature Phase Shift Keying

DSSS: Direct Sequence Spread Spectrum

EAP: Extensible Authentication Protocol (802.1x)

EDCA: Enhanced Distributed Control Access (802.11e)

EIA/TIA: Electronics Industries Association/Telecommunications Industries Association

ESS: Extended Service Set

ETSI: European Telecommunications Standards Institute

FCAPS: Fault, Configuration, Accounting, Performance, and Security Management

FCC: Federal Communications Commission

FCS: Frame Check Sequence

FDD: Frequency Division Duplex

FDM: Frequency Division Multiplexing

FDX: Full Duplex

FEC: Forward Error Correction

FHSS: Frequency Hopping Spread Spectrum

FM: Frequency Modulation

FTP: File Transfer Protocol (TCP/IP)

FTTN: Fiber to the Neighborhood

FTTP: Fiber to the Premises

GMSK/GFSK: Gaussian Minimum Shift Keying/Gaussian Frequency Shift Keying

GSM: Global System for Mobile Telecommunications

HCCA: Hybrid Controlled Channel Access (802.11e)

HDX: Half Duplex

HTML: Hyper-Text Mark-up Language (TCP/IP)

HTTP: Hyper-Text Transfer Protocol (TCP/IP)

Hz: Hertz (Prefix Kilo = Thousands, Mega = Millions, Giga = Billions)

IBSS: Independent Basic Service Set

IDS: Intrusion Detection System

IEEE: Institute of Electrical and Electronics Engineers

IMS: IP Multimedia Subsystem

IP: Internet Protocol

IP Sec: IP Security

ISDN: Integrated Services Digital Network

ISM: Industrial, Scientific, and Medical

ITU: International Telecommunications Union

IV: Initialization Vector (RC4 Encryption)

LAN: Local Area Network

LDAP: Lightweight Directory Access Protocol

LEAP: Lightweight Extensible Authentication Protocol (Cisco)

MAC: Media Access Control (LANs)

MAN: Metropolitan Area Network

MDS: Multipoint Distribution Service

MIMO: Multiple Input-Multiple Output

MMDS: Multi-Channel Multipoint Distribution Service

MOS: Mean Opinion Score

MPLS: Multi-Protocol Label Switching

MSC: Mobile Switching Center

mW: MilliWatt

NAP: Network Access Point

NAK: Negative Acknowledgment

NAT: Network Address Translation

NAV: Network Allocation Vector (802.11)

NIC: Network Interface Card

NIST: National Institute of Standards and Technology

OFDM: Orthogonal Frequency Division Multiplexing

OSI: Open Systems Interconnection

PAN: Personal Area Network

PBX: Private Branch Exchange

PCF: Point Control Function (802.11)

PDA: Personal Digital Assistant

PEAP: Protected Extensible Authentication Protocol

PKI: Public Key Encryption

PLCP: Physical Layer Convergence Part (802.11)

PoE: Power over Ethernet (IEEE 802.3af)

PON: Passive Optical Network

POP: Point of Presence

PPTP: Point-to-Point Tunneling Protocol

PRI: Primary Rate Interface

PSK: Pre-Shared Key (or Phase Shift Keying)

PTT: Push-to-Talk (or Post, Telephone, and Telegraph)

PVC: Permanent Virtual Circuit

QAM: Quadrature Amplitude Modulation

QCELP: Qualcomm Code Excited Linear Prediction

QoS: Quality of Service

QPSK: Quadrature Phase Shift Keying

RC4: Ryvest Cipher 4

S/N Ratio: Signal-to-Noise Ratio

RADIUS: Remote Access Dial-In User Security

RF: Radio Frequency

RPE-LTP: Regular Pulse Excitation—Long Term Prediction

RSS: Received Signal Strength

RSVP: Resource Reservation Protocol

RTCP: Real-Time Control Protocol

RTP: Real-Time Transport Protocol

RTS: Request to Send

SDH: Synchronous Digital Hierarchy

SIFS: Short Inter-Frame Spacing

SIP: Session Initiation Protocol

SIMPLE: SIP for Instant Messaging and Presence Leveraging Extensions

SNMP: Simple Network Management Protocol (TCP/IP)

SONET: Synchronous Optical Network

SS7: Signaling System Number 7

SSID: Service Set Identifier

SSL: Secure Socket Layer

SVP: SpectraLink Voice Priority

TCP: Transmission Control Protocol

TDD: Time Division Duplex

TDMA: Time Division Multiple Access

TKIP: Temporal Key Integrity Protocol (802.11)

TLS: Transport Layer Security (802.1x)

ToS: Type of Service

TTLS: Tunneled Transport Layer Security (802.1x)

UDP: User Datagram Protocol

UMA: Unlicensed Mobile Access

UMTS: Universal Mobile Telecommunications Service

U-NII: Unlicensed National Information Infrastructure

UWB: Ultra-Wideband

VLAN: Virtual Local Area Network

VoIP: Voice over IP

VoWLAN: Voice over Wireless LAN

VPN: Virtual Private Network

VS-CELP: Vector Sum Code Excited Linear Prediction

VWLAN: Virtual Wireless LAN

WCDMA: Wideband CDMA

WEP: Wired Equivalent Privacy (802.11)

WIDS: Wireless Intrusion Detection System

Wi-Fi: Wireless Fidelity (802.11)

WISP: Wireless Internet Service Provider

WLAN: Wireless LAN

WPA: Wi-Fi Protected Access

WPAN: Wireless Personal Area Network (Bluetooth)

1xEV-DO: 1 Carrier, Evolution, Data Only or Data Optimized (CDMA Cellular)

1xEV-DV: 1 Carrier, Evolution, Data and Voice (CDMA Cellular)

1xRTT: 1 Carrier, Radio Transmission Technique (CDMA Cellular)

3GPP: Third Generation Partnership Project (WCDMA)

3GPP/2: Third Generation Partnership Project/2 (cdma2000)

Glossary of Terms

Access Point (AP): The base station in an 802.11 wireless LAN.

Acknowledgment (ACK): A protocol message sent to acknowledge correct receipt of a transmission.

Active Directory: Active Directory is an implementation of the IETF's Lightweight Directory Access Protocol (LDAP) directory service developed by Microsoft for Windows environments. In computer networks, directories maintain information on users, applications, files, and other resources available. Active Directory allows administrators to assign enterprise-wide policies, deploy programs to many computers, and apply critical updates to an entire organization.

Address Resolution Protocol (ARP): An IP-defined protocol that allows a station to discover the MAC address of another station when it knows the station's IP address. The process involves the device broadcasting an ARP request with the IP address it is trying to resolve onto the LAN causing the device with that address to respond.

Ad Hoc Network: A network configuration defined for 802.11 wireless LANs where two devices communicate directly with one another without using an access point. Also called an Independent Basic Service Set (IBSS) or a peer-to-peer connection.

Advanced Encryption Standard (AES): An encryption standard adopted by the National Institute of Standards and Technology. It uses an encryption algorithm called Rijndael, and it is used with the 802.11i security protocol.

Analog: The transmission of signals in the form of continuously varying waves. This is the natural form of the energy produced by the human voice.

Antenna: Any structure or device used to collect or radiate electromagnetic waves.

Antenna Gain: The increase in transmission power created by an antenna's ability to focus radio energy that is measured in dB relative to an isotropic source (dBi).

American National Standards Institute (ANSI): The official standards body for US telecommunications.

Amplitude Modulation (AM): A method of signal modulation that changes the amplitude (i.e., power) of a carrier based on the information to be sent.

Announcement Traffic Indication Message (ATIM): A management frame sent after a beacon message on an 802.11 WLAN where the access point indicates the sleeping stations for which it is holding buffered frames.

Arbitrated Inter-Frame Spacing (AIFS): An Inter-Frame Spacing is the interval a station must wait to transmit a frame on an 802.11 wireless LAN. AIFS was defined with the 802.11e Quality of Service standard and specifies that different waiting intervals are used to provide different levels of access priority.

Asymmetrical Digital Subscriber Line (ADSL): A high speed Internet access service that operates over copper telephone wire. The service can deliver bit rates ≤6 Mbps downstream and ≤1 Mbps upstream, depending on the range over which the service must operate.

Asynchronous: Any activity that is performed without reference to a synchronizing timing reference.

Authentication: The process of confirming that a user has a legitimate right to access computer-based information before that access is granted.

Automatic Retransmission Request (ARQ): A process in data transmission protocols where a receiver can detect transmission errors in messages and order the transmitter to retransmit those messages.

Bandwidth: The range of frequencies carried on a communication channel. Often misused in digital networks as a synonym for "bits per second."

Barker Code: An 11-bit spreading code that is used with the original 1 and 2 Mbps direct sequence spread spectrum radio link protocols for 802.11 wireless LANs.

Base Transceiver Station (BTS): The transmission tower and associated equipment in a cellular network base station or cell site.

Basic Service Set (BSS): A configuration defined in the IEEE 802.11 WLAN standards that utilizes a base station called an access point that supports a number of client devices. In home networks the access point is typically built into a DSL/cable modem router, while in an enterprise installation it is a stand-alone device that connects on an Ethernet interface to the wired LAN.

Basic Service Set Identifier (BSSID): The 48-bit MAC address of the wireless access point in an 802.11 wireless LAN.

B Channel: The basic transmission channel (i.e., *Bearer* Channel) in an ISDN; it carries a digital signal at 64 Kbps.

Beam Steering: A radio transmission technology that uses phased array antennas to concentrate a radio beam and aim it at a specific target, thereby increasing the strength and range of the signal.

Best Effort: A packet network service that provides no guarantee of delivery or quality of service. The public Internet is an example of a best effort network.

Binary Phase Shift Keying (BPSK): A signal modulation technique that changes the phase relationship of a carrier to send information. Binary means that two states are defined (e.g., 180° and 360°), so one bit of information can be sent per symbol.

Bluetooth: The popular name for the IEEE 802.15.1 standard for short range (≤10 m) wireless personal area networks (WPANS). The name comes from the Tenth Century Danish King Harald Blatland, called Bluetooth, who united the separate kingdoms of Denmark and Norway.

Bridge: A device for connecting LANs that operates at Layer 2 of the OSI model. A Layer 2 connection means that the bridge makes the decision to forward frames based on the MAC address found in the header of the LAN frame. Bridges were computer-based devices that have now been replaced with hardware-based LAN switches.

Broadband: The generic term used to describe a high-capacity transmission service. The term initially referred to any transmission channel with a bandwidth greater than voiceband (i.e., >3 KHz), however, there is no overall consensus over how wide a channel has to be to qualify as broadband.

Broadband Radio Service (BRS): A band of frequencies in the Personal Communications Services (PCS) spectrum (2.495–2.690 GHz) originally called Multipoint Distribution Service (MDS). Initially considered for wireless cable TV, it is now being used primarily for licensed WiMAX Broadband Wireless Access service.

Broadcast: A message or signal that is sent to all members of a network.

Buffer: A temporary data storage facility that is used to accommodate timing or traffic load differences between network elements.

Byte: In computer networks, an 8-bit unit of transmission possibly representing one character of information.

Cable Modem Service: A high speed Internet access service offered by cable television providers that operates over their coaxial cable or hybrid fiber/coax networks. Cable modems use a shared channel configuration where up to 200 devices may share one channel. The channel typically operates at 42 Mbps downstream and 12 Mbps upstream.

Carrier Sense Multiple Access with Collision Avoidance (CSMA/CA): The basic access protocol used with 802.11 wireless LANs. As WLAN stations cannot hear while they are sending on the shared radio channel, CSMA/CA uses a distributed control function that involves random pre-transmission waiting intervals to help avoid collisions.

Carrier Sense Multiple Access with Collision Detection (CSMA/CD): The access protocol used on Ethernet and 802.3 local area networks where stations can transmit on a shared channel any time they sense it is idle. If a collision is detected, the stations involved stop transmitting and attempt to retransmit after a random waiting interval.

Category *x* Cable (Where *x* = 3, 4, 5, 5e, 6, or 6a): The generic specification for the different grades of unshielded twisted pair (UTP) LAN cabling specified in the EIA/TIA 568 standard. See Table A-3.

Cell: In radio networks, the area covered by a base station. Limiting the transmit power of the base station allows radio channels to be reused in different parts of a larger coverage area.

Chips: Units of transmission in a direct sequence spread spectrum (DSSS) radio system. In DSSS radio, a user's bit stream is multiplied or spread, so each user bit is represented by a larger number of shorter duration bits called chips. As a result, the physical transmission rate on a DSSS channel is measured in chips per second (cps).

Cisco Compatible Extensions (CCX): A Cisco-developed set of extensions to the 802.11 wireless LAN standards. Cisco encourages other vendors to incorporate these enhancements in the WLAN designs, as part of the CCX program. Participating vendors can be found at: http://www.cisco.com/web/partners/pr46/pr147/partners_pgm_partners_0900aecd800a7907.html.

Co-channel Interference: In any cellular radio system, the interference from other users operating on the same channel in different parts of the coverage area.

Code Division Multiple Access (CDMA): A type of direct sequence spread spectrum radio technology used with cellular telephone systems that allows multiple users to share a single 1.25 MHz radio carrier.

Committed Information Rate (CIR): The average sustained transmission rate that a frame relay carrier guarantees to deliver on a particular virtual circuit. Transmissions in excess of the CIR are handled on a best effort basis.

Complementary Coded Keying (CCK): A type of direct sequence spread spectrum radio technology used with the 5.5 and 11 Mbps transmission rates in 802.11b wireless LANs.

Connectionless: A packet forwarding service that does not employ virtual circuits. In a connectionless service like IP, special packets called datagrams are forwarded individually router-to-router to their destination, but do not follow a predefined path. The services are typically best effort, and there is no guarantee that the packets will be delivered or that they will be delivered in the correct sequence.

Connection Oriented: A transmission system characterized by three phases: connection setup, information transfer, and connection tear down. All circuit switched services are connection oriented as are packet services like frame relay and ATM that use virtual circuits.

Contention Free (CF): In a wireless LAN, a period where stations are not vying for access to the shared channel. It is used in the PCF and HCCA options to support time-sensitive traffic. During contention-free periods, the access points polls those stations with time sensitive traffic on a scheduled basis.

Contentionless: A transmission system where each user is given the specific right to transmit for a portion of time on a shared transmission facility with the result that two devices cannot transmit simultaneously and cause a collision.

Contention System: A network where a number of users vie or contend for a shared transmission path and whose transmissions may collide with one another.

Contention Window (CW): In an 802.11 wireless LAN, the period of time during which stations vie for access to the shared radio channel.

Convolutional Coding: A form of forward error correction coding that works by taking the transmitted bit sequence, selecting a group of the most recently occurring bits, distributing them into two or more sets, and performing a mathematical operation (convolution) to generate one coded bit from each set as the output.

Counter Code with CBC MAC (CCMP): Defines the use of AES-based encryption in the 802.11i wireless LAN security standard.

Datagram: A special type of packet used in connectionless networks (e.g., IP networks). A datagram is a packet that finds its own path through a packet network rather than being forwarded over a virtual circuit.

Data Over Cable Service Interface Specification (DOCSIS): The standard for cable modems published by CableLabs. The standard addresses OSI Layers 1 and 2. The current versions are 1.0, which was the original document, 1.1, which added Quality of Service features for cable telephony to the Layer 2 protocol, and 2.0, which refined the QoS mechanisms and defined higher upstream and downstream transmission rates for Layer 1. Rev 3.0 is currently being finalized, and it will define downstream rates up to 1 Gbps by bonding multiple 6 MHz channels together.

DCF Inter-Frame Spacing (DIFS): In the Distributed Control Function of an 802.11 wireless LAN, the minimum amount of time a station must wait before it can send a frame on an idle channel.

D Channel: The signaling channel used in an ISDN subscriber interface. It carriers messages defined by ITU Recommendation Q.931.

Denial of Service (DoS): A type of hacker attack on a packet network or server that seeks to disrupt service by introducing a large number of spurious messages that serve to overwhelm the network's resources and degrade user performance.

Differential Binary Phase Shift Keying (DBPSK): A signal modulation technique that changes the phase relationship of a carrier to send information. Binary means there are two states defined (e.g., 180° and 360°) so one bit of information can be sent per symbol. The "Differential" part means that the encoding constellation is phase shifted (typically 45°) with each symbol to eliminate a 180° phase shift.

Differential Quadrature Phase Shift Keying (DQPSK): A signal modulation technique that changes the phase relationship of a carrier to send information. Quadrature means there are four states defined (e.g., 90°, 180°, 270°, and 360°) so two bits of information can be sent per symbol. The "Differential" part means that the encoding constellation is phase shifted (typically 45°) with each symbol to eliminate a 180° phase shift.

DiffServ (short for *Differentiated Services*): DiffServ is a class-based QoS mechanism for IP networks where packets can be assigned to a number of traffic classes; that classification is done by inserting a 6-bit value called a DiffServ Control Point (DSCP) in the Type of Service field in the IP header. DiffServ aware routers can then implement Per-Hop Behaviors (PHBs), which define the packet forwarding properties associated with each class of traffic. DiffServ is used to prioritize voice and video traffic in enterprise routing networks, and it is also used to guide the traffic prioritization in MPLS services.

Digital: A transmission or storage system that uses discrete values (e.g., 1s and 0s) to store or transmit information, rather than a system that relies on a continuous spectrum of values (i.e., an analog system).

Digital Encryption Standard (DES): An encryption standard used with virtual private networks. Different size encryption keys can be used, typically 56 bits or 168 bits (called 3DES or Triple DES).

Direct Sequence Spread Spectrum (DSSS): A type of radio transmission system used with CDMA-based cellular and 802.11b wireless LANs that uses a spreading code to multiply the user transmission rate. That spread signal is then carried on a wideband radio channel (i.e., 1.25 MHz in CDMA, and 22 MHz in 802.11b WLANs), which is less susceptible to noise, fading, or interference in one part of the radio band.

Distributed Control Function (DCF): The primary access control mechanism in 802.11 wireless LANs. Rather than having one station control access to the shared radio channel, all of the stations jointly control access using a system of back-off timers and waiting intervals.

Distribution System (DS): In 802.11 wireless LANs, the Distribution System is the wired LAN that connects all of the access points together. Typically, it is a switched Ethernet LAN.

Doppler: The upward or downward shift in the frequency of a signal as a transmitter is moves toward or away from the receiver.

Downstream: In a client-server connection, the transmission path from the server to the client.

Dynamic Host Configuration Protocol (DHCP): A TCP/IP protocol that is used to implement Network Address Translation (NAT). When NAT is used, devices are assigned a local address (i.e., 10.x.x.x, 192.168.x.x, or 176.16.x.x through 176.31.x.x) that they use within their local network. When the device needs to communicate over the public Internet, it sends a request to a DHCP server that lends it a "real" IP address to use for the duration of that session. The router or firewall translates between the "real" and local address on each packet.

Echo: In a voice communications channel, the reflection of a signal from the far end of the communications path. While all voice channels produce some echo, if the echo is delayed by a long transmission path or packet network congestion, it will be annoying to the speaker and may render the channel unusable.

Echo Cancellation: The function of removing unwanted return echo in voice transmission paths with long delays (e.g., >40 msec one way).

Electronics Industries Association/Telecommunications Industries Association (EIA/TIA): A US-based standards body that defines the standards for building cabling systems used for local area networks (i.e., EIA/TIA 568).

Encryption: The process of scrambling a message with a key so that it can only be read by another user who has the same key. The robustness of an encryption system is dependent on the design of the encryption algorithm and the length of the key.

Enhanced Distributed Control Access (EDCA): An enhancement to the 802.11 wireless LAN access protocol defined in 802.11e. EDCA defines four access classes or priorities to provide preferred access for voice and video over data transmissions.

Ethernet: The most widely deployed standard for local area networks that defines operating rates of 10 Mbps (the original), 100 Mbps, 1 Gbps, and 10 Gbps.

European Telecommunications Standards Institute (ETSI): A European standards organization charged with defining standards for EC countries. ETSI developed the GSM standard for cellular telephones and is responsible for coordinating the development of standards for European 3G wireless data services.

Extended Service Set (ESS): A configuration for larger-scale 802.11 wireless LANs where a number of access points are interconnected through a wired network (e.g., a LAN) and broadcast the same network name or SSID.

Extensible Authentication Protocol (EAP): An authentication protocol defined in IEEE 802.1x. EAP is essentially an authentication framework that can support a number of authentication protocols including MD5, TLS, TTLS, PEAP, and Cisco's LEAP and FAST.

Fault, Configuration, Accounting, Performance, and Security Management (FCAPS): The ISO Telecommunications Management Network model and framework for network management.

Federal Communications Commission (FCC): The primary interstate regulatory authority for telecommunications and radio systems for the US. The FCC was created by the Communications Act of 1934.

Femtocell: A very small cellular base station designed to improve service in indoor environments. Unlike a traditional cellular base station that connects to the cellular mobile switching center (MSC) on a dedicated link, femtocells connect to a broadband Internet connection and communicate with the MSC over the Internet.

File Transfer Protocol (FTP): A TCP/IP defined application layer protocol that describes how to provide batch communications.

Firewall: A server, possibly included in a router, that prevents unauthorized traffic or access requests from entering a network. Firewalls inspect each packet into or out of the network and decide whether to drop it or pass it based on a set of rules. In Internet VPNs, the firewall also encrypts traffic to ensure that it cannot be read by eavesdroppers. That encrypted VPN connection is called a *secure tunnel*.

Fixed-Mobile Convergence (FMC): The capability for networks to transparently hand off calls between cellular and private networks.

Forward Error Correction (FEC): A technique for correcting transmission errors that works on the principle of adding redundant bits in the transmitter. Using that redundant information, the receiver can detect and correct a certain percentage of the encountered errors.

Fragmentation: A feature of packet protocols that allows a long transmission to be broken into fragments that are transmitted individually. Fragmentation is used in IP networks so that messages from a network supporting a large maximum frame size can be delivered through a network supporting a smaller maximum frame size. Fragmentation is also used in 802.11 WLANs where long frames are more likely to suffer transmission errors.

Frame: The unit of transmission in a Layer 2 protocol. The term is also used to describe the unit of transmission produced by a multiplexer from one pass around its input channels (e.g., a SONET frame).

Frame Check Sequence (FCS): A 16- or 32-bit error checking algorithm that is computed using a cyclic redundancy check, and appended to a transmission frame so that the recipient can determine if any errors were introduced in the transmission system.

Frame Relay: A wide area packet switched data communication service used primarily for inter-site LAN/WAN networking. Frame relay provides permanent virtual circuits with delivery guarantees for a certain volume of traffic defined by a Committed Information Rate (CIR). Any traffic exceeding that commitment is handled on a best effort basis.

Frequency Division Duplex (FDD): A method of providing a full duplex transmission capability in a radio system that allocates separate radio channels for inbound and outbound transmissions. Cellular telephone networks use FDD.

Frequency Division Multiplexing (FDM): A method of sharing a communications channel by assigning different parts of the frequency spectrum (i.e., bandwidth) to different users.

Frequency Hopping Spread Spectrum (FHSS): A method of spread spectrum radio transmission that uses a wideband radio channel divided into a number of narrower band channels and the transmitter hops between them in a seemingly random pattern. The result is a radio system that suffers little from frequency selective fading and noise and is difficult to intercept or jam.

Frequency Modulation (FM): A method of radio modulation that works by changing the frequency of a radio carrier based on the information to be carried. It requires greater bandwidth than amplitude modulation, but provides vastly better signal quality once a minimum signal level (called the FM Threshold) is reached.

Full Duplex (FDX): The capability of a transmission system to send in both directions simultaneously.

Gateway: Any device that stands at the boundary of two networks allowing traffic to pass between them. The most typical use today is in IP telephone networks where it describes a device that connects IP telephony services to traditional circuit switched services. (Note: In the early TCP/IP literature, gateway was the term used to describe a router.)

Gaussian Minimum Shift Keying/Gaussian Frequency Shift Keying (GMSK/ GFSK): A method of radio modulation that encodes 1 and 0 bits by shifting the frequency of a radio carrier. (Note: This is also called Frequency Shift Keying.) A Gaussian filter is applied to limit the side lobes or radio energy that extends beyond the frequency limits of the channel.

Global System for Mobile Telecommunications (GSM): The worldwide standard for digital cell phones developed by ETSI and used in approximately 170 countries. The basic technology is TDMA-based and eight voice users are supported on dual 200 KHz radio carriers (one for inbound, one for outbound).

H.323: The ITU umbrella standard for streaming voice and video on packet-based networks and the signaling protocol to support it.

Half Duplex (HDX): The ability to transmit alternately in either direction on a transmission service.

Handoff/Handover: The process of transferring a mobile wireless connection from one base station to another.

Hertz (Hz): The number of cycles per second of a transmission signal. It is the primary measure of frequency, named for Heinrich Rudolf Hertz, the man credited with the discovery of radio waves (Prefix: Kilo = Thousands, Mega = Millions, Giga = Billions, and Tera = Trillions).

Hidden Node: A problem in 802.11 wireless LANs where stations cannot hear each others' transmissions and cause collisions because they think the radio channel is clear.

Hub: A device used to interconnect stations in a UTP-based shared media LAN. Hubs provide no switching or filtering functions, but merely broadcast the bits that are received on one port to every other port. Hence, all of the devices connected to a hub are sharing a single transmission channel (i.e., shared media).

Hybrid Control Function (HCF): The name for the MAC protocol defined in 802.11e to provide Quality of Service in wireless LANs.

Hybrid Controlled Channel Access (HCCA): A rarely deployed option in the 802.11e QoS standard with features similar to the original Point Control Function (PCF). In HCCA, the access point takes control of the transmission medium for periods of time during which it polls stations with time sensitive traffic. At other times the network operates in contention mode (i.e., EDCA mode).

Hyper-Text Mark-up Language (HTML): A TCP/IP defined presentation protocol that describes how to format Web pages to be displayed by a browser.

Hyper-Text Transfer Protocol (HTTP): A TCP/IP defined application layer protocol that describes how to download Web pages to a browser.

Inband: Any control information that is sent in the same transmission channel that carries the actual service. Most typically used to describe early forms of voice signaling.

Independent Basic Service Set: The official name for the Ad Hoc or peer-to-peer operating mode in 802.11 wireless LANs.

Industrial, Scientific, and Medical (ISM): A band of unlicensed radio frequencies allocated by the FCC and used for a wide variety of applications including wireless LANs, cordless phones, baby monitors, and Bluetooth devices. There are actually two ISM bands: 902 MHz to 928 MHz and 2.4 GHz to 2.4835 GHz. The 802.11b and g wireless LANs use the 2.4 GHz band.

Initialization Vector (IV): The initial portion of the encryption key in RC4 type encryption systems.

Institute of Electrical and Electronics Engineers (IEEE): A professional organization of Electrical and Electronics Engineers responsible for the development of standards for local area networks, power distribution, and other areas.

Integrated Services Digital Network (ISDN): An ITU-developed plan from the 1980s that looked to convert the entire public telephone network to an end-to-end digital environment delivering a wide array of voice and data services.

Inter-Access Point Protocol (IAPP): A protocol for managing handoffs between access points in a wireless LAN. It is defined in IEEE 802.11f.

International Telecommunications Union (ITU): The primary international standards body for telecommunications that is organized under the United Nations.

Internet: A large-scale, public packet switching network that is operated by a group of carriers called Internet Service Providers (ISPs). The term is used more generically to describe any collection of networks interconnected by routers (e.g., a private Internet).

Internet Control Message Protocol: A control protocol that operates in conjunction with IP and is used to provide testing and other network alarm/maintenance functions. The most widely recognized ICMP functions are Ping and Trace Route.

Internet Protocol (IP): An OSI Layer 3 protocol used in the Internet and other router-based networks. The basic function of IP is to provide a non-guaranteed (i.e., best effort) packet forwarding mechanism to connect together a variety of networks including wired and wireless LANs, frame relay, point-to-point private lines, and ATM services.

Intranet: An enterprise application that uses Web servers and Web browsers to disseminate internal information. Today the term intranet is used in a variety of ways including a general moniker to describe an organization's entire internal network.

Intrusion Detection System (IDS): In network security, an IDS is a server or an application that monitors network traffic looking for suspicious patterns of behavior known to be typical of hacking attempts. The monitoring can be done based on stored patterns called signatures or on anomalies, which are activity patterns the device has not encountered previously.

IP Address: Currently, the 32-bit address that is used in router networks and the Internet. It is normally written in dotted decimal notation (e.g., 143.251.87.128) where each digit is a decimal representation of 8 data bits (i.e., from 00000000 = decimal 0 to 11111111 = decimal 255). The address is divided into a Network ID, which identifies the network on which the device is found, and a Host ID, which identifies the specific device. The 32-bit address structure is used in IP Version 4, but as that address space is being exhausted, the plan is to upgrade to IP Version 6 which will feature a 128-bit address.

IP Multimedia Subsystem (IMS): An architectural framework for delivering Internet Protocol (IP) multimedia to mobile users. The major interest around IMS comes from its ability to implement fixed-mobile convergence or the capability to transparently hand off calls from cellular to private networks while delivering a consistent look-and-feel to the user.

IP PBX: A LAN switch equipped with a telephony server so that it can provide voice telephone service to IP/Ethernet compatible telephones.

IP Phone: A telephone handset that digitizes analog voice signals and transmits them in a series of IP packets typically over an Ethernet interface.

IP Security (IPSec): A suite of security protocols for IP networks that provide the ability to securely transport information over a public Internet service by encrypting the traffic in a secure tunnel.

Isochronous: A transport service that provides consistent delay (i.e., no jitter) as traffic is carried through the network. All circuit switched services are inherently isochronous.

Jitter: The variation in packet-to-packet transit delay that can degrade the quality of real-time voice and video transmission sent over a packet network.

Jitter Buffer: IP voice networks, a buffer located at the end of the IP transport (e.g., in an IP telephone, IP soft phone, or an IP telephony gateway) that is used to eliminate jitter and reestablish the isochronous nature of the voice transmission. Packets that arrive at the buffer are stored and then played out according to the timestamps embedded by the real-time transport protocol (RTP). Jitter buffers may use a fixed capacity or they can adjust their capacity based on the network performance.

LAN Switch: A device for interconnecting LANs that operates at Layer 2 of the OSI model. Layer 2 connection means that the LAN switch makes the decision to forward the frame based on the MAC address found in the header of the LAN frame. This same function was formerly provided by bridges.

Lightweight Directory Access Protocol: A directory protocol described in IETF RFC 4510. In computer networks, directories maintain information on users, applications, files, and other resources available. LDAP is an open industry standard for updating and accessing information in the directory.

Lightweight Extensible Authentication Protocol (LEAP): An authentication protocol developed by Cisco that can be used with 802.1x authentication systems. A number of vulnerabilities were discovered with LEAP and hence it is no longer recommended.

Local Area Network (LAN): A high-speed local data communications network used to connect computing devices in a building or campus environment.

Management Information Database (MIB): Collections of data elements used in TCP/IP's Simple Network Management Protocol that define the management information maintained for each particular type of networking equipment.

Mean Opinion Score (MOS): A subjective scale for measuring voice quality particularly in VoIP networks. The range of values is 1 to 5 where 5 is the best. It is generally held that a score above 4.0 is suitable for enterprise voice applications; a traditional PCM-encoded, circuit switched voice connection is rated about 4.4.

Media Access Control (MAC): The lower portion of the Layer 2 protocol in local area networks that is responsible for formatting and addressing transmission frames and introducing them onto the transmission channel.

Media Access Control (MAC) Address: This is the address used in LANs and other Layer 2 networks. Specifically, it is a unique 48-bit number that is burned on the device's Network Interface Card (NIC). The first three octets are known as the Organizationally Unique Identifier (OUI), and the other three octets are assigned by the organization in question. MAC addresses are usually shown in hexadecimal format, with each octet separated by a dash or colon (e.g., 00-08-74-4C-7F-1D). The IEEE expects the MAC-48 address space to be exhausted no sooner than the year 2100.

Metropolitan Area Network (MAN): A network or network technology whose range of coverage is an entire metropolitan area, as opposed to a Local Area Network (LAN) whose range is a single customer facility. In the radio arena, WiMAX is designed as a MAN technology while Wi-Fi is designed as a LAN technology.

MilliWatt (mW): A measure of transmission power equivalent to 1/1000 of a Watt.

Mobile Switching Center (MSC): The central office facility in a cellular telephone network that manages the base station network and connects to the wired telephone network.

Multipath: The distortion in a received radio signal resulting from the reception of multiple images or echoes of the transmitted signal; as those signals are at the same frequency but with different phase relationships, they can effectively cancel each other out. Those echoes are created by the radio signal reflecting or refracting off hand, flat surfaces in the environment.

Multiple Input-Multiple Output (MIMO): An antenna technology that uses multiple transmit (output) antennas and multiple receive (input) antennas. In a MIMO transmission system, the data stream is first divided into multiple lower speed steams called transmit chains. The transmit chains are then sent through separate radio transmitters whose antennas are separated by a short distance. Spacing the transmit antennas apart (i.e., spatial multiplexing) allows all of the transmitters to operate in the same frequency band. At the input end, the receivers will be able to distinguish the individual transmit chains as each will exhibit different multipath characteristics (i.e., a different radio fingerprint). The receive power of all of the images received on all of the input antennas can then be added together. MIMO systems can send multiple signals in the same frequency band and the receive signal strength for each is greatly increased; that results in a significant increase (i.e., 3 to 5 times) in the transmission rate, range, and reliability of the system.

Multi-Protocol Label Switching (MPLS): A major addition to the IP protocol that will allow router networks to direct traffic onto specific routes called Label Switched Paths (LSPs). The LSPs act like virtual circuits and can be used to manage traffic flows, or as a means for a carrier to provide VPN services to users. Those MPLS-VPN services offer multiple classes of service for voice, video, and data traffic with guarantees for worst-case performance with regard to delay, jitter, and packet loss.

National Institute of Standards and Technology (NIST): A non-regulated Federal agency with the U.S. Dept. of Commerce chartered to promote innovation and develop standards for the government networks. NIST-endorsed standards like the Advanced Encryption Standard (AES) are also used by commercial organizations.

Negative Acknowledgment (NAK): A protocol message returned by a recipient to indicate that a message was received with errors. The error detection is typically done using a cyclic redundancy checking (CRC) algorithm, and on receipt of the NAK message, the transmitting station may resend the message (i.e., an ARQ process).

Network Access Control: The function of controlling admission to a computer network based on security policies. Beyond authenticating a user's identity, the NAC system implements end point integrity checks where a server confirms the user's device is correctly configured and that its security software is up-to-date before allowing it to connect to the network. That confirmation lessens the likelihood a client station will unwittingly introduce a virus into the network.

Network Address Translation: An IP address handling technique that was developed to deal with the shortage of 32-bit IP Version 4 addresses. When NAT is used, devices are assigned a local address (e.g., 10.x.x.x, 192.168.x.x, or 176.16.x.x through 176.31.x.x) that they use within their local network. When a device needs to communicate over the public Internet, it sends a request to a DHCP server that lends it an address to use for the duration of that session. The local/network IP addresses are translated as packets that pass through a router or firewall.

Network Allocation Vector (NAV): A special timer defined in the 802.11 wireless LAN protocol that allows a station to reserve transmission time on the channel. Every station must have an NAV, and when it is sent with a control message, the station will assume the channel is busy for that length of time. It is used by the access point to reserve the channel for time sensitive voice traffic (i.e., HCCA) or in the RTS/CTS protocol when different stations cannot hear each others' transmissions (i.e., the hidden node problem).

Network Interface Card (NIC): A hardware card installed in a computing device that allows it to be attached to a local area network. The NIC generates the LAN protocol and provides the MAC address for the station. The NIC used in wireless LANs is sometimes called a Radio NIC.

Octet: In computer networks, an 8 bit unit of information possibly representing one character of information.

Open Systems Interconnection (OSI) Reference Model: A framework for developing protocols that would provide for multi-vendor interoperability. The framework defines seven layers or sets of functions that would have to addressed to provide full compatibility. While no real products use the scheme (data networks are typically built using the four-layer TCP/IP model), the OSI model provides a standard reference point for describing compatibility issues.

Orthogonal Frequency Division Multiplexing (OFDM): A form of spread spectrum radio transmission that divides a wideband radio channel into a number of subcarriers onto which information is encoded. It is used with 802.11a and g wireless LANs and with the WiMAX Broadband Wireless Access standard (IEEE 802.16). The advantages of OFDM are: bandwidth efficiency, reduced sensitivity to frequency-selective noise and fading, and greater tolerance to multipath distortion.

Out-of-Band: Any control message that is sent using capacity other than that which is used to carry the actual user information. In voice networks that would typically be a signaling message sent without using the transmission capacity allocated to carry the voice. An ISDN D Channel is an example of an out-of-band signaling technique.

Packet: An addressed unit of transmission generated by a Layer 3 protocol.

Packet Switching: A network technology that operates on the principle of dividing user traffic into segments called packets that are individually addressed and forwarded through a network of shared transmission links and switches.

Personal Digital Assistant (PDA): A handheld computer typically running the Palm, WindowsCE, or Windows Mobile operating system.

Physical Layer Convergence Part (PLCP): The initial Layer 1 transmission header sent before an 802.11 wireless LAN frame.

Point Control Function (PCF): An optional and rarely implemented mechanism in the 802.11 wireless LAN protocol that can support time sensitive voice traffic. The basic idea is that the access point takes control of the transmission medium for periods of time during which it polls stations with time sensitive traffic. At other times the network operates in its traditional contention mode (i.e., DCF mode). The 802.11e QoS standard features a similar operating mode called Hybrid Controlled Channel Access or WMM Scheduled Access, but no vendors appear to be developing it.

Point of Presence (PoP): A carrier's operating facility, usually a long distance carrier or an Internet Service Provider (ISP).

Post Office Protocol (POP): A TCP/IP defined application layer protocol for retrieving email messages from a server.

Power over Ethernet (PoE): A standard from the IEEE (802.3af) that defines how electrical power can be carried to a device over the Ethernet interface thereby eliminating the need for an AC power source at the device. It is used primarily to power IP telephones, 802.11 wireless LAN access points, and security cameras.

Pre-Shared Key (PSK): In 802.11 wireless LANs, an option for implementing encryption keys where the key is manually entered into the client device and into the access point.

Primary Rate Interface (PRI): An ISDN subscriber interface for high capacity devices (e.g., PBXs or remote access servers) that operates at 1.544 Mbps and provides 23 B

Channels plus a 64 Kbps signaling channel (i.e., D Channel) or simply 24 B Channels. The international version operates at 2.048 Mbps and provides 30B + D.

Pringles Antenna: A simple radio waveguide, constructed from the container used for Pringles potato chips, that improves reception of 2.4 GHz radio signals. It is often used for eavesdropping on wireless LANs (Note: Coffee cans work better).

Private Branch Exchange (PBX): A specialized telephone switch used to reduce the number of central office connections and provide calling features in larger (i.e., ≥50 employees) commercial installations.

Protected Extensible Authentication Protocol (PEAP): One of the authentication protocols supported with the Extensible Authentication Protocol. PEAP was designed to overcome the difficulties associated with managing certificates in client devices.

Protocol: A codified set of rules defining how some part of information exchange between devices should be conducted.

Protocol Stack: A set of software elements used to generate a full set of network protocols (e.g., TCP/IP).

Public Key Encryption (PKI): An encryption mechanism that provides for the secure distribution of encryption keys for each communication session. It is used with virtual private networks (VPNs) and with secure socket layer (SSL) communications.

Push-To-Talk (PTT): A type of radio service where transmission requires the user to push a button to transmit and then release it to receive (i.e., the walkie-talkie capability). This capability was first included in cellular networks by Nextel (now a division of Sprint).

Quadrature Amplitude Modulation (QAM): A method of signal modulation that encodes 1 and 0 bits by changing the amplitude of a carrier wave. Quadrature means that two carriers, 90° out of phase, are used. QAM systems are prefixed by a number that indicates the number of signal states that can be defined (e.g., 8-QAM carrying 3 bits per symbol, 16-QAM carrying 4 bits per symbol, 32-QAM carrying 5 bits per symbol, and 64-QAM carrying 6 bits per symbol).

Quadrature Phase Shift Keying (QPSK): A signal modulation technique that changes the phase relationship of a carrier to send information. Quadrature means there are four states defined (e.g., 90°, 180°, 270°, and 360°), so 2 bits of information can be sent per symbol.

Qualcomm Code Excited Linear Prediction (QCELP): A low bit rate voice coding system developed by Qualcomm and used with CDMA-based cellular telephone networks. Current versions of QCELP can encode voice at 8 Kbps or 13 Kbps.

Quality of Service (QoS): Any capability in a packet switching network to give priority to certain classes of traffic over others with the goal of improving the performance of those selected classes with regard to delay, packet loss, and possible jitter. Different network technologies will use different mechanisms to achieve these ends. For example, routers will typically maintain different queues for each traffic category, while wireless LAN networks assign shorter pre-transmission waiting intervals to higher priority classes.

Queue: A queue is literally a line; in computer networks, it is a line of messages or operations waiting to be processed.

Radio Frequency: Electromagnetic energy in the band between 20 KHz and 300 GHz.

Real-Time Control Protocol (RTCP): A support protocol used in conjunction with RTP to provide information on the performance of the RTP connection. Each station in the RTP connection will send an RTCP sender and/or receiver report every five seconds that includes statistics like number of packets lost, round trip delay, and jitter performance. RTCP uses a different port address than RTP, so a management system can intercept those reports and use them to drive a performance monitoring system.

Real-Time Transport Protocol (RTP): This is the TCP/IP protocol used to transport real-time voice and video. The main function of RTP is to identify the payload encoding (e.g., G.711 or G.729A voice) and append sequence numbers and timestamps so that the receiver can process the packets in the correct order and reestablish the timing continuity.

Received Signal Strength (RSS): The power in decibels (dB) of a signal when measured at the receiver.

Regular Pulse Excitation—Long Term Prediction (RPE-LTP): The low bit rate voice coding system used with GSM-based digital cellular networks. The GSM standards define a full-rate option that operates at 13 Kbps and a half-rate option that operates at 6.5 Kbps.

Request to Send/Clear to Send (RTS/CTS): A protocol option in 802.11 wireless LANs where transmissions must be preceded by an exchange of control messages (one RTS, and one CTS) which reserve the transmission channel so that one frame can be sent. The RTS/CTS feature can be activated in the access point when excessive collisions are detected and the problem stems from hidden nodes. It is activated automatically when 802.11b and g users are active on the same network (i.e., 802.11g Protection Mode).

Remote Access Dial-In User Security (RADIUS): A class of protocols that provide authentication for network users.

R-Factor: An ITU-developed voice testing technique specified G.107 that scores voice quality on a 120-point scale where a score of 80 or above is considered acceptable for enterprise telephone service.

Rogue Access Point: In a wireless LAN, an access point that is installed by a user or other unauthorized party without the knowledge of the networking department. The problem with a rogue access point is that it will typically create a security exposure that the networking department will not be aware of.

Router: A specialized packet switch for forwarding IP datagrams. This is the switching device used in the Internet and in end user LAN/WAN networks.

Ryvest Cipher 4 (RC4): An encryption algorithm developed by Ron Ryvest and used with the Wired Equivalent Privacy (WEP) WLAN encryption system.

Secure Socket Layer (SSL): A security feature in browsers that allows them to establish an encrypted connection to a server. SSL is used to send sensitive information (e.g., credit card numbers, PIN Codes, etc.) over insecure networks like the public Internet.

Secure Tunnel: A technique for encrypting information sent through a public Internet or other insecure networks. Secure tunnels are used for remote access, LAN/WAN communications, and have been used for providing security in early wireless LANs. IPSec is the most widely used secure tunnel protocol, and it can be generated in PCs, laptops, firewalls, and some communicating PDAs.

Service Set Identifier (SSID): An identifier code or name for an 802.11 wireless LAN network.

Session Initiation Protocol (SIP): A signaling protocol for voice over IP and other multimedia applications that is defined in IETF RFCs 3261–3264.

Short Inter-Frame Spacing (SIFS): In the Distributed Control Function of the 802.11 MAC protocol, the amount of time a station must wait before it can send an ACK message on an idle channel.

Signaling: In telephone networks, the process of setting up and tearing down connections and activating features (e.g., call forwarding, transfer, etc.).

Signaling System Number 7 (SS7): The common channel signaling system used to connect calls and provide features in the public telephone network. The term actually relates to the protocol used on that network, but it has been generalized to describe the entire signaling network.

Signal-to-Noise Ratio: The basic measure of noise performance on a communications channel. The reading is expressed in decibels (dB). As it is a ratio of good signal to bad noise, higher S/N values indicate a channel with better noise performance.

Simple Mail Transfer Protocol (SMTP): A TCP/IP defined application layer protocol that describes how to send email messages to a server.

Simple Network Management Protocol (SNMP): A TCP/IP developed protocol for forwarding network management messages to/from a control station.

Softphone: A program that runs in a desktop PC, laptop, or PDA that allows it to function as an IP telephone.

SpectraLink Voice Priority (SVP): A set of quality of service enhancements developed by SpectraLink (now part of Polycom) to provide better voice performance over wireless LANs. Access point vendors who incorporate these features in their designs are designated VIEW-Certified.

Subnet: A convention in IP addressing where a user can divide the User ID address space to define specific groups of addresses; in practice, each subnet is typically a separate virtual LAN. Traffic passing between subnets must go through a router, access control, encryption, and any other security features that may be implemented.

Subnet Mask: A special address filter used in IP stations that allows the device to distinguish between devices on its own network, devices on different subnets, and devices on other networks.

Switched Virtual Circuit (SVC): A temporary, user-initiated connection in a packet switching network, or the packet switching equivalent of a dial-up connection.

Synchronous: Any operation that is performed in conjunction with a timing reference.

Synchronous Optical Network (SONET): A set of standards defined by Bellcore (now called Telcordia) to provide multi-vendor compatibility for high capacity fiber optic terminals and networks.

Synchronous Transmission Signal (STS-x): The transmission levels defined in the SONET standards. The basic level is STS-1, which operates at 51.84 Mbps, and all other levels are direct multiples. The STS designation is used when the signal is in an electrical rather than an optical format.

T-1 Carrier System: A 24-channel digital carrier system first introduced in the early 1960s that operated at 1.544 Mbps on two pairs of copper wire with repeaters spaced at intervals of 3,000 to 6,000 feet. The term T-1 is also used in place of DS-1 as a generic name for a 1.544 Mbps transmission facility. T-1/DS-1 channels are used for a variety of applications including point-to-point private lines, Internet access connections, and voice network access.

T-3: A term used to describe a transmission channel with a capacity of 44.736 Mbps; the correct term is DS-3. T-3/DS-3 channels are used for a variety of applications including point-to-point private lines and Internet access connections.

Telephony Server: A special server used in IP telephony environments that allows a LAN Switch to provide connection management and telephony features for IP handsets.

Telnet: A TCP/IP defined application layer protocol that describes support for character-oriented (e.g., VT-100 type) interactive terminals.

Temporal Key Integrity Protocol (TKIP): A security feature for 802.11 wireless LANs defined in the Wi-Fi Protected Access (WPA) security plan where the encryption key is changed for each packet.

Third Generation Partnership Project (3GPP): The vendor organization assembled by ETSI to support 3G wireless data services based on the Wideband CDMA standard.

Third Generation Partnership Project/2 (3GPP/2): The vendor organization assembled by ANSI to support 3G wireless data services based on the cdma2000 standard.

Throughput: The total volume of good data delivered over a data communications system in a given unit of time (e.g., per second or per minute) after taking into account all protocol overhead, delays, and retransmissions.

Tie-Line: A dedicated channel connected between two PBX systems that is used to carry voice calls.

Time Division Duplex (TDD): A method of providing a full duplex transmission capability on a shared radio channel by allowing each end to transmit alternately. A descriptive name for this operation is "ping pong."

Time Division Multiple Access (TDMA): A method of sharing a radio channel where different users take turns transmitting on the same channel.

Trace Route: An ICMP function that operates like a Ping, but records the addresses of all of the routers the message passes through. It is used to troubleshoot routing problems in IP networks.

Transmission Control Protocol (TCP): An OSI Layer 4 protocol used with IP to provide reliable, connection-oriented services. Where IP provides best effort packet forwarding, TCP ensures that all of the parts of the message are delivered, in the correct order, and corrects errors and omissions.

Transport Layer Security (TLS): An authentication protocol used with 802.1x Extensible Authentication Protocol that requires digital certificates in all client devices.

Tunneled Transport Layer Security (TTLS): An authentication protocol used with 802.1x Extensible Authentication Protocol that does not require digital certificates in client devices.

Unicast: A service that delivers a packet to a single location.

Unlicensed Mobile Access (UMA): A technique for transferring calls between cellular and private networks developed by Kineto Wireless and adopted by the 3G Partnership Project as a short-term option for providing fixed-mobile convergence.

Unlicensed National Information Infrastructure (U-NII): The US name for the 555 MHz of unlicensed radio frequency allocated by the FCC in the 5 GHz band. The 5 GHz band is a relatively recent addition to the unlicensed spectrum, and as such the full band is not available in all countries.

Upstream: In a client-server connection, the transmission path from the client to the server.

User Datagram Protocol (UDP): A protocol in the TCP/IP set that can be used as an alternative to TCP. Where TCP ensures that the entire series of packets that constitute a message are delivered in order and error-free, UDP provides simple unreliable delivery. UDP is typically used to carry voice traffic, where retransmissions to recover lost or errored packets would be impractical.

Virtual Circuit: A logical connection in a packet switching network that gives the appearance of a real circuit (i.e., a circuit switched connection) while in fact traffic from different users is being dynamically interspersed on a shared facility. Virtual circuits typically provide for security and ensure that all messages are delivered in order.

Virtual Local Area Network (VLAN): A feature of LAN switches that allows a user to divide different groups of users into software-defined partitions that act like separate LANs. The feature can also be used for security as each virtual LAN is associated with an IP subnet, and communications between subnets must be sent through a router.

Virtual Private Network (VPN): The term VPN is used in three different ways: (1) An Internet-based remote access configuration that uses basic Internet service and ensures security through the use of encrypted secure tunnels. (2) An MPLS-based Internet service for inter-site communications that uses virtual circuits to ensure security and provides multiple classes of service levels with performance guarantees for delay, jitter, and packet loss. (3) In public network voice services, VPN is a bulk rate pricing plan for intra-company long distance that provides dialing features similar to a private tie-line network.

Virtual Wireless LAN (VWLAN): A feature of wireless LANs where an access point can be made to broadcast different network names or SSIDs and operate different virtual networks with different features over the same wireless LAN channel. This feature is often used to allow different groups of users with different security capabilities/requirements to share the same wireless LAN infrastructure.

Voiceband: An analog transmission channel capable of supporting one voice transmission signal; specifically a voiceband channel carries frequencies in the range between roughly 300 and 3,100 Hz.

Voice over IP (VoIP): The idea of carrying a digital voice transmission on a local or wide area IP network.

Walsh Code: A spreading code for direct sequence spread spectrum radio transmissions that is used with CDMA-based cellular telephone networks.

Wavelength: The distance spanned by one cycle of a signal. This is essentially a different way of measuring frequency.

Wideband CDMA (W-CDMA): The GSM plan for 3rd Generation cellular voice and data services that uses direct sequence spread spectrum radio technology on a 5 MHz channel.

Wi-Fi: See *Wireless Fidelity*.

Wi-Fi Alliance: The vendor group organized to support the adoption of 802.11 wireless LAN technology, define coordinated implementations of the IEEE 802.11 standards, and provide conformance testing to ensure multi-vendor interoperability. The Wi-Fi Alliance was originally called the Wireless Ethernet Compatibility Alliance (WECA).

Wi-Fi Certified: The certification provided by the Wi-Fi Alliance to indicate that a device has undergone the interoperability regimen specified to ensure compliance with a particular IEEE 802.11 standard.

Wi-Fi Multimedia (WMM): A certification developed by the Wi-Fi Alliance to indicate full compliance with the Enhanced Distributed Control Access (EDCA) option in the 802.11e QoS standard.

Wi-Fi Multimedia-Scheduled Access (WMM-SA): A certification developed by the Wi-Fi Alliance to indicate full compliance with the Hybrid Controlled Control Access (HCCA) option in the 802.11e QoS standard.

Wi-Fi Protected Access (WPA): A certification developed by the Wi-Fi Alliance to indicate compliance with specific security features available in 802.11i for wireless LANs. WPA provides improved security through a software upgrade and used the same encryption algorithm as WEP (i.e., RC-4), but changed the key for each packet, thereby thwarting brute force attacks. That key-changing feature is called the Temporal Key Integrity Protocol.

Wi-Fi Protected Access—Version 2 (WPA2): A certification developed by the Wi-Fi Alliance to indicate full compliance with the new security features available in 802.11i wireless LANs, including the support for encryption based on the Advanced Encryption Standard (AES).

WiMAX: The name adopted by the vendor organization that supports broadband wireless access as defined in IEEE 802.16 standards. There are two versions of WiMAX, a fixed location version specified in 802.16-2004, and a mobile version with handoff functionality specified in 802.16-2005. Like the Wi-Fi Alliance, the WiMAX Forum defines coordinated implementations of the IEEE 802.16 standards, and provides conformance testing to ensure multi-vendor interoperability.

Wired Equivalent Privacy (WEP): A rudimentary encryption system developed with the original 802.11 wireless LAN specifications. WEP uses a 40- or 104-bit RC-4 encryption and did not provide a mechanism for distributing new encryption keys. Given its vulnerabilities, it is not recommended for enterprise installations. WEP is now being replaced by two different implementations of the 802.11i security protocol designated WPA and WPA2.

Wireless Fidelity (Wi-Fi): The easily identifiable marketing name for 802.11-compatible products certified by the Wi-Fi Alliance.

Wireless Internet Service Provider (WISP): An Internet Service Provider that uses 802.11 wireless LAN technology to provide Internet access from defined locations called Hot Spots.

Wireless Intrusion Detection System: A security system for enterprise networks that uses a number of sensors linked to a controller to monitor and locate unauthorized installations of wireless LAN technology.

Wireless LAN Switch: A new class of products for building large-scale, enterprise-grade wireless LAN networks. These systems typically utilize limited function thin access points whose operations are managed by a central controller. The systems generally feature network configuration tools, dynamic RF management capabilities, rogue access point detection, and the ability of devices to roam seamlessly between access points.

1xEV-DO: The acronym stands for 1 Carrier, Evolution, Data Only or Data Optimized, and it is one of the enhanced versions of the wireless data services supported on CDMA-based cellular telephone networks. The service will require a separate 1.25 MHz CDMA carrier, and will support maximum data rates up to 2.4 Mbps downstream and 153 Kbps upstream. In practice, downstream data rates of 300 Kbps to 500 Kbps are typical.

1xEV-DO Rev. A: The enhanced version of 1xEV-DO that will support maximum data rates up to 3.1 Mbps downstream and 1.8 Mbps upstream. In practice, downstream data rates of 500 Kbps to 700 Kbps are expected.

1xEV-DV: The acronym stands for 1 Carrier, Evolution, Data and Voice, and it is one of the enhanced versions of the wireless data services supported on CDMA-based cellular telephone networks. The service can share 1.25 MHz CDMA carriers with voice services, and will support downstream data at rates ≤2.4 Mbps and ≤153 Kbps upstream. No networks based on this technology have been deployed.

1xRTT: The acronym stands for 1 Carrier, Radio Transmission Technique, and it is the first 2.5G/3G the wireless data services supported on CDMA-based cellular telephone networks. The service can share 1.25 MHz CDMA carriers with voice services, and will support symmetrical data at rates ≤144 Kbps. Actual downstream data rates are in the range of 50 Kbps to 70 Kbps.

Index